Space and Defense Policy

This edited volume introduces the reader to the role of space in military and defense strategy, and outlines some of the major foreign and domestic actors in the space arena, as well as constraints of law and treaties on activities in space. It also addresses science and technology as they relate to space policy.

The book addresses three main questions:

- How does the realm of space fit into strategic thinking about national security?
- How does policy regarding space develop and what considerations, both in the United States and abroad, figure prominently in calculations about space policy?
- How do different states/nations/actors regard the role of space in their national security calculations and how do these policies impact each other?

This book fills a niche in the space policy field, providing insights into space and strategy from international experts from the military, academic, and scientific communities. A unique feature of the book is the chapter on science and technology, which utilizes the latest information available concerning space utilization and exploration.

This book will be of much interest to students of space policy, defense studies, strategic studies, and US politics and national security.

Damon Coletta is Professor of Political Science at the United States Air Force Academy, Colorado. **Frances T. Pilch** is Professor of Political Science at the United States Air Force Academy, Colorado. She is the Division Chief for International Relations and National Security Studies.

Space power and politics
Series editors: Everett C. Dolman and John Sheldon
School of Advanced Air and Space Studies, USAF Air, Maxwell, USA

Space and Defense Policy

Edited by Damon Coletta and
Frances T. Pilch

Routledge
Taylor & Francis Group

LONDON AND NEW YORK

EISENHOWER CENTER
FOR SPACE AND DEFENSE STUDIES

First published 2009
by Routledge
2 Park Square, Milton Park, Abingdon, Oxon OX14 4RN

Simultaneously published in the USA and Canada
by Routledge
270 Madison Ave, New York, NY 10016

Routledge is an imprint of the Taylor & Francis Group, an informa business

Typeset in Baskerville by Wearset Ltd, Boldon, Tyne and Wear
Printed and bound in Great Britain by TJI Digital, Padstow, Cornwall

British Library Cataloguing in Publication Data
A catalogue record for this book is available from the British Library

Library of Congress Cataloging in Publication Data
Space and defense policy / edited by Damon Coletta and
Frances T. Pilch.
p. cm. – (Space power and politics)
1. Astronautics, Military. 2. Astronautics and state. 3. United States–
Military policy. 4. Security, International. I. Coletta, Damon V.
II. Pilch, Frances T.
UG1520.S65 2008
358'.8–dc22   2008032770

ISBN10: 0-415-77732-1 (hbk)
ISBN10: 0-203-88306-3 (ebk)

ISBN13: 978-0-415-77732-2 (hbk)
ISBN13: 978-0-203-88306-8 (ebk)

This book is respectfully dedicated to
Senator Wayne Allard and the Honorable Peter B. Teets,
Fathers of the Eisenhower Center for Space and Defense

Contents

Illustrations

Figures

Tables

Boxes

Contributors

David Christopher Arnold is Commander of the 22nd Space Operations Squadron, Schriever Air Force Base, Colorado. Lieutenant Colonel Arnold earned a PhD in History at Auburn University in Auburn, Alabama, in 2002. Texas A&M University Press published his dissertation on the evolution of satellite command and control as *Spying from Space: Constructing America's Satellite Command and Control Systems.*

Preston Arnold, USAF (Retired), is a former instructor in the Department of Political Science at the United States Air Force Academy. While at the Academy, he was the lead instructor in the course on Space Policy.

Damon Coletta is Professor of Political Science at the US Air Force Academy. He was first trained as an electrical engineer, after which he went on to earn a Masters in Public Policy from the Kennedy School of Government (1993), specializing in Science and Technology Policy, and his PhD in Political Science from Duke University. He co-edited the eighth edition of *American Defense Policy* (Johns Hopkins University Press, 2005) and is currently completing a book on technology-sharing in international coalitions.

Michael Gleason is a former spacecraft mission controller for Air Force Space Command. Lieutenant Colonel Mick Gleason has more than 5,100 spacecraft sorties to his credit. He is an Assistant Professor of Political Science at the Air Force Academy and a PhD candidate at George Washington University.

Tom Graham is a former Special Assistant to the President for Arms Control, Nonproliferation and Disarmament. Ambassador Graham has participated in a senior capacity in every major arms control and nonproliferation negotiation in which the United States took part from 1970 to 1997. He is the author of *Disarmament Sketches* (2002) and *Common Sense on Weapons of Mass Destruction* (2004), and co-author of *Cornerstones of Security* (2003). Currently, he is Chairman of the Cypress Fund for Peace and Security and Chairman of Thorium Power, Ltd.

Steve G. Green is a Professor of Management at the United States Air Force Academy. His 21-year Air Force career included system program office tours performing acquisition, cost analysis, and business management on three major space systems. He was awarded a Doctor of Business Administration (DBA) from the United States International University and an MS from the University of Southern California.

Roger G. Harrison is the Director of the Eisenhower Center for Space and Defense Studies at the United States Air Force Academy. Ambassador Harrison served as the US Ambassador to Jordan 1990–1993, and as the Director of Near East South Asia Center for Strategic Studies at the National Defense University in 2001. He holds the Wesley W. Posvar Chair in Political Science at the United States Air Force Academy. He was awarded his PhD in Political Science from the Claremont School of Graduate Studies.

Peter L. Hays is a retired Air Force Lieutenant Colonel who, since September 2004, has been a senior policy analyst with Science Applications International Corporation, supporting the plans and programs division of the National Security Space Office. He edited the 2005 and 2006 National Security Space Plans and serves as Chief of Staff for the year-long National Defense University Spacepower Theory Study.

Kurt A. Heppard is a Professor of Management at the United States Air Force Academy. While on active duty in the Air Force, he was an acquisition officer for the MILSTAR satellite program and served as an Education with Industry Fellow. He received his MBA from the Anderson Graduate School of Management at the University of California, Los Angeles and completed his doctoral degree at the Leeds School of Business at the University of Colorado, Boulder.

Darren Huskisson is the Chief, Cyber and Space Law at United States Strategic Command. In this capacity, Major Huskisson provides advice on military space and network activities to the command. He has served as a Judge Advocate in various positions in the Air Force since 1996 and has been an Assistant Professor of Law at the United States Air Force Academy. He holds a JD from the University of Nebraska and an LLM from McGill University.

Jonty Kasku-Jackson, a Colorado-licensed attorney with an Intelligence background, is currently the Course Director for the National Security Space Institute's Advanced National Space Security Studies course. Prior to this assignment she supported AFSPC/A3 where she participated in a variety of interagency studies on space. She attended Oregon State University. After serving five years as an Air Force Intelligence officer, she left the Air Force to attend the New England School of Law in Boston, MA where she received a JD.

Timothy Lawrence, Lieutenant Colonel USAF, is the Space Systems Research Center Director and the Navigation Guidance and Control Division Chief in the Department of Astronautics at the United States Air Force Academy. He was awarded his PhD in satellite engineering by the University of Surrey, UK.

Bruce Linster is a Professor of Economics at the United States Air Force Academy. He has published in the fields of game theory, experimental economics, evolutionary economics, and public choice. A retired USAF lieutenant colonel and former C-130 pilot, he earned his MA in Applied Economics and PhD in Economics at the University of Michigan.

John M. Logsdon is Director of the Space Policy Institute at George Washington University's Elliott School of International Affairs, where he is also Research Professor and Professor Emeritus of Political Science and International Affairs. His research interests focus on the policy and historical aspects of US and international space activities. He is the author of *The Decision to Go to the Moon: Project Apollo and the National Interest* and is general editor of the eight-volume series, *Exploring the Unknown: Selected Documents in the History of the US Civil Space Program*. He is a member of the NASA Advisory Council. He was awarded a PhD in Political Science from New York University in 1970.

James Clay Moltz is an Associate Professor in the National Security Affairs Department at the Naval Postgraduate School. His most recent book is *The Politics of Space Security: Strategic Restraint and the Pursuit of National Interests* (Stanford, 2008). Dr. Moltz is also chairman of the Space Futures Working Group at the NASA Ames Research Center. From 1993 to 2007 he served in various positions at the Center for Nonproliferation Studies (CNS) of the Monterey Institute of International Studies, including CNS deputy director and editor of *The Nonproliferation Review*.

Douglas J. Murray is one of the founders of the Eisenhower Center for Space and Defense Policy at the United States Air Force Academy. Brigadier General Murray served as the Permanent Professor and Chair of the Department of Political Science at the Academy until his retirement in 2007. He was awarded his PhD in International Relations from the University of Texas at Austin.

Xavier Pasco is a Senior Research Fellow at the Fondation pour la Recherche Stratégique (FRS) based in Paris, where he is in charge of the Department of Technology, Space and Security. Prior to 1997 he was a researcher at CREST (*Center for Research and Evaluation of the Relationships between Strategies and Technology*) associated with the Ecole Polytechnique. He received his PhD in Political Science from the University of Paris at the Sorbonne.

Eligar Sadeh is an Assistant Director for the Eisenhower Center for Space and Defense Studies at the United States Air Force Academy. He served as Assistant Professor of Space Studies in the School of Aerospace Sciences at the University of North Dakota. He is currently the editor of *Astropolitics: The International Journal of Space Politics and Policy* and editor of *Space and Defense: Journal of the Center for Space and Defense Studies.* He is the author of *Space Politics and Policy: An Evolutionary Perspective.*

William W. Saylor is a Visiting Professor in the Department of Astronautics, United States Air Force Academy. He is a Distinguished Graduate from the United States Military Academy at West Point, with a Nuclear Engineering area of concentration. He holds a Masters in Nuclear Engineering and Magnetic Fusion Technology from the Massachusetts Institute of Technology. He has been a Senior Scientist at SAIC, Inc. supporting numerous DoD and NRO space activities.

Ken Siegenthaler is a Professor in the Department of Astronautics at the United States Air Force Academy. His primary interests are lasers, remote sensing, and small satellites. He is a faculty mentor on the Air Force Academy FalconSAT Program. He has previously served in the positions of Director and Chief Scientist of an Air Force Research Laboratory. He was awarded a PhD and Masters Degree in laser physics from the Air Force Institute of Technology.

Robert L. Tremaine is currently Associate Dean for Outreach and Performance Support at the Defense Acquisition University's West Region Campus. During his active-duty career, he managed seven major air, missile, and space and development programs. He is the author of numerous articles, including "Delivering acquisition training to the space professional community," in the *Defense AT&L* magazine, March–April 2006. He earned an MS from the Air Force Institute of Technology (AFIT) and completed a joint military fellowship with the Harvard Business School.

David A. Turner is currently the Director of the Aerospace Corporation's Center for Space Policy and Strategy, which conducts policy and strategy analyses for a number of customers in the national security and civil space community. He holds a BSc in Aerospace Engineering from Syracuse University and a Masters degree in Science, Technology and Public Policy from George Washington University.

Brenda Vallance was a career Air Force officer who served in the Intelligence career field. After serving as a Professor of Political Science at the United States Air Force Academy, she became Dean, School of Behavioral and Social Sciences at St. Edward's University in Austin, TX. Brenda has numerous articles and presentations on the Russian military and Russian civic organizations, and was the co-editor of *American Defense Policy,* seventh edition. She received her PhD from the University of California, Los Angeles.

James Vedda is a senior policy analyst at the Aerospace Corporation's Center for Space Policy and Strategy in Arlington, Virginia. Previously, he was assigned to the Office of the Secretary of Defense, working on space policy and homeland defense issues. His published writing has appeared in journals such as *Space Policy, Space News, Space Times, Ad Astra, Space Energy and Transportation, Space Business News,* and *Quest.* James holds a Masters degree in Science, Technology, and Public Policy from George Washington University and a PhD in Political Science from the University of Florida.

Elizabeth Waldrop is a Space Law Instructor and Legal Subject Matter Expert at the National Security Space Institute (NSSI), Headquarters Air Force Space Command (AFSPC), Colorado, where she provides space education and training programs for space staff members, joint planners, and operators. Lieutenant Colonel Waldrop also advises the AFSPC Commander and his staff on space law. She attended Duke University and was awarded her LLM in Air and Space Law from McGill University.

Thomas G. Ward, Jr. currently works for Science Applications International Corporation (SAIC) on a part-time basis as a Senior Engineer, specializing in missile defense. From 1966 until retiring in 2001, he worked for the Central Intelligence Agency, serving as the principal intelligence officer of the Strategic Defense Initiative Organization (later the Ballistic Missile Defense Organization) and as the CIA's Officer-in-Residence at the United States Air Force Academy. He holds a Bachelor of Engineering Science in Chemical Engineering from Johns Hopkins University and a Doctor of Philosophy, also in Chemical Engineering, from Princeton University.

James J. Wirtz is a Professor in the Department of National Security Affairs, Naval Postgraduate School, Monterey, California. He is editor of the Palgrave Macmillan series, *Initiatives in Strategic Studies: Issues and Policies,* section chair of the Intelligence Studies Section of the International Studies Association and President of the International Security and Arms Control Section of the American Political Science Association.

Foreword

More than 50 years ago as the United States Air Force Academy was developing its curriculum, the faculty of the Department of Political Science realized that an integral part of the cadet course of study must include the history, evolution, organization, structure, and process of American defense policy. Not only was a course developed, but work was begun on a textbook, since at that time there was a paucity of good materials to support the course. The result was the first issue of *American Defense Policy*. More than an anthology, the book organized a collection of seminal pieces and commissioned new ones around a unifying and integrating framework. The success of the effort is evinced in that *American Defense Policy* is now in its eighth edition, has an international reputation and has played a significant role in promoting, if not creating, the field of American defense policy studies.

More than a dozen years later, as the international security environment was changing, becoming more complex and populated by many more participants, another faculty in the Department of Political Science realized that their students required a better understanding of other nations' defense policies. The realization led to a course in comparative defense policy and a supportive text *Comparative Defense Policy*. That publication, which led to three editions of *Defense Policy of Nations*, became the basis for a new field of defense learning. Each edition developed an audience of defense policy students and practitioners worldwide. As this foreword is being written, another Academy Political Science faculty is planning a new edition that will compare aspects of global defense policies, as they have had to address the challenges of the twenty-first century in a post 9/11 world.

Space and Defense Policy is the next phase and series in the evolution of the study of America's defense policy, building upon the foundations of the other two series. This book marks the initial effort to define what we hope will be an emerging field of space and defense studies. It is also the initial publication of the Eisenhower Center for Space and Defense Studies, a center committed to conducting inquiry into new and emerging aspects and realms of defense policy learning while sustaining past initiatives dating back to that first work published 50 years ago. The Eisenhower Center for Space and Defense Studies, named after the first American president to recognize the policy challenges and opportunities of space, is hosted by the Department of Political Science at the United States Air Force Academy. It continues the innovative work of the Center for the Study of Defense Policy, but is informed by a new recognition of the vital role of space in national defense. The Center arises from a compelling need, frequently voiced, to define space as a strategic environment and to develop an appropriate theory, strategy, policy, and paradigm for understanding space. Its founding principle is that if space is to serve mankind, we must think of it as a new medium, not as merely an extension of land, sea, and air. We need to think outside the gravitational pull of the Earth. The mission of the Eisenhower Center is threefold: to build an intellectual foundation for space policy, to define space policy studies curriculum for civilian and military institutions of higher learning, and to help develop a generation of young men and women with a vocation for space. The Center is to be a common learning resource for academics, the policy community, and the private sector in training, research, planning, and operations. It does this by developing a community of space professionals, critical thinkers, innovators, and visionaries through annual forums, workshops, research, and internships programs, an online journal, fellowships, and a summer space seminar for college students from across the nation.

The *Space and Defense Policy* textbook is a significant part of the Center's programs. Look on it not merely as an introduction to space policy studies, but as an invitation to become part of the initiative. This book is intended for many audiences. If you are student of space, this text will provide you with a framework for analysis built upon policy considerations. If you are a college student trying to decide on a career, this book will open up the possibilities of a vocation for space whether technical or non-technical, scientific or commercial, government or private. If you are a researcher, this book offers an agenda of critical issues to be addressed. If you are a policy-maker, this book is a primer. And if you are a teacher, let this book inspire your students to inquire where few have inquired before.

Brigadier General Douglas J. Murray, USAF (ret.)
Professor Emeritus/Eisenhower Center for Space and Defense Studies
Former Permanent Professor and Head
Department of Political Science, United States Air Force Academy

Acknowledgments

Space and Defense Policy emerged as the first major publishing venture of the Eisenhower Center for Space and Defense, founded by Ambassador Roger Harrison and Brigadier General Douglas J. Murray, USAF (ret.), and located in conjunction with the Department of Political Science at the United States Air Force Academy in Colorado. The editors are deeply grateful to those who participated in initial discussions concerning the organizational principles of this volume. Among those are Preston Arnold, Michael Gleason, and Deron Jackson. Many other colleagues in the Department of Political Science were enlisted to help with the completion of this project, and we owe a debt of gratitude to the entire department for its support.

After authors were selected for the desired topics, the Eisenhower Center for Space and Defense sponsored a forum at which each author presented his or her chapter, after which the chapters were discussed and critiqued by the distinguished attendees. We appreciate the contributions of all those who attended this forum, and also those who subsequently reviewed one or more chapters. In particular, the support and attention of former Under Secretary of the Air Force, Peter B. Teets, are noted. Roger Harrison, Director of the Eisenhower Center for Space and Defense, has been an invaluable resource and cheerleader along with Douglas J. Murray. Deron Jackson and Eligar Sadeh have contributed in countless ways to the success of this book and of the Center itself. Just as invaluable has been the work of Tracy Hicks, who keeps the Center running with efficiency and good humor. We thank Everett Dolman for pointing us to our publisher, Routledge Press, and our editors and assistants at Routledge for a job well done.

This acknowledgment would not be complete without noting the support of the Dean of the Faculty of the United States Air Force Academy, Brigadier General Dana H. Born, and the Chair of the Department of Political Science, Colonel Cheryl Kearney.

We would like to thank our families for their encouragement and patience as this volume was completed – in particular Jonan and Grayce Coletta and Lance, Junko, and Rich Pilch.

Permissions

The following organizations gave permission to use previously published material.

Hays, Peter, Brenda J. Vallance, and Alan R. Van Tassel, eds., with a foreword by Brent Scowcroft. *American Defense Policy*, Seventh Edition, fig. 4 p. 13. Copyright 1996 The Johns Hopkins University Press. Reprinted with permission of the Johns Hopkins University Press.

Humble, Ronald, Gary Henry, and Wiley Larson (1995) *Space Propulsion Analysis and Design*. NY: McGraw-Hill, table 8.1 and figure 8.2. Copyright by the McGraw-Hill Companies, Inc.

Futron Corporation (Bethesda, MD). Satellite Manufacturing Report (January 2005). Available at www.futron.com/resource_center/friends_of_futron/report_archives/satellite_manufacturing_report.htm. All rights reserved by Futron Corporation.

Introduction

Thinking about space and defense

Damon Coletta

When Alfred Thayer Mahan published *A History of Sea Power* in 1890, he did more than link a variety of historical success stories to a single, coherent geographic theme. Though that part was important, the really enduring aspect of Mahan's work was its impeccable timing. *A History of Sea Power* captured and invigorated the Zeitgeist of a rising star in world politics. Mahan was able to sort out centuries-old trends as well as dramatic reversals that marked the flow of geopolitical influence. He understood that in an age of rapid technological developments, a nation's fortune would not be divorced from its physical and cultural geographic inheritance.

This book on *Space and Defense* carries these principles, if not Mahan's specific conclusions, to the present day. Space, from low Earth orbit to geosynchronous station to interplanetary distance, brings a heady mix of risk and opportunity, not unlike that which the sea held for powerful nations in centuries past. If the United States is to maintain a leading position in the international system it requires a framework for understanding how this new medium, alongside land, sea, and air, will shape opportunities with other peoples for military competition and, at the same time, economic and scientific cooperation.

As this intellectual enterprise gains momentum, it is worth reviewing past strategic revolutions in America when success in geopolitics seemed to hinge on being the first to dominate a new medium or field an advanced weapons system. In each case the interplay between academic thinking and reality offers sobering lessons for the United States as it seeks security – or hegemony – by mastering physical geography.

First, an old sea story

In the decades leading up to Mahan's *History of Sea Power*, foreign political developments ushered the United States to the wings of the world stage. During the 1860s, Great Britain resisted intervening on behalf of the Confederate States of America, passing on a final opportunity to divide and weaken the United States. A French effort to control a canal across the Isthmus of Panama collapsed in 1889 before a tidal wave of mud, malaria, and yellow fever. Spain, the last of the European powers traditionally engaged in containing US expansion, loosened its grip on Cuba in the 1890s as it slipped into its own constitutional crisis.

In the same epoch, transcontinental railroad and stagecoach lines sewed up the final land frontiers between eastern and western cities. US industrial production surged in the last decades of the nineteenth century, on its way to passing that of Great Britain years before World War I. Between 1880 and 1900 the American economy absorbed nearly nine million immigrants, ensuring that America's new labor force had historical and cultural ties to Asia and Europe. As production grew, trade became a defining political issue. American capital, labor, and agriculture took a keen interest in import competition, raw materials, and expanding markets overseas.[1]

Mahan's work entered and helped redefine this milieu, highlighting physical, cultural, and political factors that would shape the future for the United States, just at the moment when this youthful champion from the New World would burst forth across the seas and oceans. Sea power, indeed, propelled US expansion first into the Caribbean. Gunboat diplomacy, dollar diplomacy, and the subsequent Good Neighbor policy in wider Latin America were predicated on American control of the sea.

US capacity to tip the scales in World War I, a clash of unprecedented destruction among the Old World powers, and the lasting ambition articulated by President Wilson to make the world safe for democracy simply could not exist without force projection across the oceans. Growing US interest on the far shores of the Pacific and the Atlantic led to even greater challenges: a two-front war against the Japanese empire and Nazi Germany. Out of geographic necessity, victories on both counts featured massive US forces attacking from the sea.[2]

Nuclear weapons and advances in rocketry at the end of World War II altered the strategies, but not the fundamental pattern of competition between the two remaining superpowers. A vital aspect of containment against the Soviet Union – a land power – played out over the oceans. Japan became a great barrier in the Pacific. An array of ocean-monitoring devices deployed in the Greenland–Iceland–United Kingdom gap awaited any breakout attempt by the Soviet Navy into the North Atlantic. Turkey guarded access to the Mediterranean. During the 45-year stand off, various US fleets attempted to thwart communist expansion in Korea, Taiwan, Lebanon, and Vietnam or to head off Soviet influence in the Persian Gulf, the Gulf of Aden, and of course the Caribbean, where the United States intervened to prevent the establishment of Soviet air, missile, or naval bases in Cuba, Nicaragua, and Grenada.

Sizing up the 100-year legacy of Mahan demonstrates that he got it right in a profound sense: by 1890 it was high time that American leaders from government, industry, and academia turned their attention toward the sea. Yet, before citing too many parallels with the present situation and declaring our own time ripe for a strategic thrust toward space, caution is in order. In an age of cooperation among the military services for land, sea, and air, single pivots for history, like sea or space power, will justly draw heavy criticism as being overly simplistic.

An important follow-on question asks how much of the nation's resources should flow to space operations. In the extreme, if every business, every school, and every public budget maximized space capability regardless of the costs, the purported means to greatness would swamp cherished ends, much as occurred with state structures honed to make war in the twentieth century before they were consumed by fascism.[3]

Although America's journey through the twentieth century may be told from a sea power perspective, it is also difficult to deny that multiple causes propelled the United States' rise to globalism.[4] Overemphasis on a government-coordinated effort to conquer space could squeeze out or distort spending on other priorities such as welfare, health care, transportation, or education. Within the military, it might privilege the Air Force over the Army even if "boots on the ground" and cultural training could be just as important as information dominance and precision strikes for future US success abroad.[5] Within the American economy, too much central control for maximum investment in space could impinge on private entrepreneurship in an epoch when spin-on from the private sector is just as crucial for national innovation as technological spin-off from highly capitalized government programs. Finally, within the international system, an American obsession to master space could provoke resistance from other states, particularly if, in line with Mahan's previous world view regarding the oceans, space became a medium for exerting international leadership based on coercive military and economic power, divorced from the legitimacy of US goals.

Another Mahan moment

Compared to a century ago, of course, the Zeitgeist for America has changed considerably. Although rumors of the superpower's demise after failed intervention in Vietnam were premature, the United States is no longer an exuberant liberal champion, dashing across the world stage in the manner of Teddy Roosevelt's Great White Fleet or Woodrow Wilson's American Expeditionary Force.[6] America's capacity for force projection in the information age far exceeds what it was in the early twentieth century, but the strategic demand and requirements for nurturing domestic political will have also expanded.[7]

Certainly, the trouble stabilizing Iraq after dethroning Saddam Hussein punctuated the shift. Leading writers on America's role in the world, such as Robert Kagan and Francis Fukuyama, reinforced the neoconservative line before the war of early 2003, arguing that military might, could and should be used to tidy up areas of the world that the march of history toward liberal polities and free markets left in its wake. Tellingly, a few years later these intellectuals joined several others in warning about the dangers of hard power divorced from soft power and international legitimacy.[8] Nor was the lesson lost on practitioners. The Bush administration toned down its rhetoric for the second term. Triumphal photo-ops on aircraft carriers gave way to more solicitous trips by the Secretary of State and the President himself to important capitals such as Paris, New Delhi, and Beijing. In both the language and pacing of diplomacy, the United States after 2003 recovered the prudence and patience for multilateral approaches that it had demonstrated in the Middle East and Europe after the Cold War.

Compared to Mahan's time, the United States today is a mature, maybe even aging, power. Beyond asserting its rights on the world stage or flexing its muscle to defend a fragile democracy, the United States now has the international system to consider. While it cannot appoint itself as world policeman over intergovernmental institutions and global markets, it nevertheless bears a disproportionate interest in, and responsibility for, international order. While it cannot rush headlong to take on all comers without spreading itself thin, the United States nevertheless cannot avoid preparing for the next crisis, which might involve state and non-state actors from any quarter. What Mahan can teach us about our time is to be on the lookout for propitious moments in history when the geopolitical landscape and emerging technology combine to create extraordinary opportunity along with increasing risk in world politics.

This book on space and defense policy seeks in part to address growing sentiment that the United States and the world are embarking upon another Mahan moment. In 1890 the great geopolitical question was whether and how the United States would take its place among the European great powers. Today, with forces deployed around the globe, no major personnel increases in the offing, and some $400 billion already spent over the last five years in attempts to stabilize Afghanistan and Iraq, the question is whether and how the United States will maintain its ascendancy in the twenty-first century. Can the country leverage its lead in space to facilitate US influence without provoking the sort of resistance by others that would dramatically extend costs of conflict for all concerned?

At the same time, technological developments promise cheaper access to space and broaden the range of feasible activities in that extreme environment. A sort of quantum leap occurred in Mahan's era, with fleets upgrading from wood and sail to steel and steam. Today, with a growing amount of hardware available for deployment in orbit, and advancing software for signal processing, military services and commercial organizations think more in terms of performing constellations, not just individual platforms. Constellations for navigation, communication, and environmental monitoring combine the precision of a high-performance satellite with wider area coverage and less down-time. Lowering the marginal cost and risk of low Earth orbit missions has already translated into progress for the

International Space Station. Admittedly, that project has been plagued by delays and cost overruns, but in principle it opens the way for other types of maintenance and assembly operations performed in space, which in turn means more massive, more powerful, and more mission-flexible objects may someday occupy permanent stations above the Earth.

Expanding capabilities for space technology are not for the leader alone. Declining launch costs on the margin and improving commercial payload performance mark the entry of new actors into space. Russian launch facilities from the Cold War era are still open and in many cases successfully compete for international contracts.[9] These contracts sometimes involve potential rivals to the United States. Foremost on the list of space competitors may be China since it has demonstrated a keen interest in long-range missile technology, computer designs on American payloads, and the systems management of Apollo-like manned missions. Among the great powers expanding their role in space are allied countries, particularly France, which attempts to leverage prior technical accomplishments with capital from partner nations. Recent projects sponsored by the European Union, for example, included Earth-observation satellites, planetary probes, and the Galileo navigation constellation that will complement and partially substitute for the US-run Global Positioning System.

Meanwhile, the commercial path to space has been on an upward, if not altogether smooth, trajectory. The projected boom in space-based telecommunications did not in fact materialize in the first years of the twenty-first century, but expectations for long-term growth in the tens of billions of dollars are still intact.[10] The number of suppliers for commercial launch services has grown, if slowly, while the number of commercial payloads on orbit is climbing faster. The year 2005 saw financial commitment to build the first "space-port" in the state of New Mexico and the first winner in the million-dollar X-prize series was crowned, in this case for a privately designed and funded re-usable launch vehicle. Commercial space opportunities are growing more numerous, but also higher in quality. The US-led war in Afghanistan in 2002 featured a strategic buy-out of private company satellite imagery so that this information would not be available for enemy planners.[11]

Developments in commercial space, in turn, point toward new possibilities for non-state strategic actors. As the conventional war against the Taliban wound down in late 2002, some analysts speculated about how the next great terrorist attack might be planned to include weapons of mass destruction. Missiles, crop-dusters, or barges filled with radioactive or other toxic substances might acquire targets scouted by a one-meter-resolution commercial imaging system and navigate on a commercially available positioning signal. Sophisticated terrorist networks also pose a threat to critical infrastructure of space-based communication and navigation which is located on the ground. The rise of commercial entities along with additional military installations sharpened the question of whether all important sites could be protected from sophisticated terrorist networks that might exploit tactical innovations from the 9/11 attacks.

In any case, expanding commercial interest in space, advances in space technology relating to military transformation, and international concern about America's role in the world following 9/11 and the Iraq War all make it an auspicious time – a Mahan moment – for considering space in the same breath as geopolitical futures. As the editors set about designing this volume on space and defense, they paused to consider previous cases when geography and technology combined to revolutionize military power and international relations.

Air power, nuclear weapons, and information operations

The overarching lesson arising from other technological revolutions in strategic studies warns against either extreme optimism or pessimism. For each revolution, much of the subject matter seems technical in nature, but the outcomes of greatest concern are fundamentally political. Following undeniable technological successes for air power, nuclear weapons, or information operations, it seems that other actors on the world scene and specialized interests below the state level took a hand in determining the ultimate political significance of each technical achievement.

As was the case with Mahan's focus on sea power, many factors determine the success of national strategies based around new capabilities – a fact illustrated by the interdependence of the various capability-based revolutions on one another. The historical development of each so-called revolution in global power relations never played out exactly as enthusiasts anticipated. Certain obstacles and opportunities simply could not be foreseen years beforehand. Yet, in each case, air power intellectuals, nuclear weapons specialists, and information operations futurists performed Mahan's service: they highlighted important questions early on; they created new vocabularies; influenced research programs; and launched debates to find answers before it was too late. In short, there was a benefit to analyzing the emerging role of each new technology in geopolitical context.

Air power and new ways of war

The invention of powered flight at the dawn of the twentieth century inaugurated an enduring controversy about the role of air-breathing platforms in the contest among nations. Compared to the tonnage of goods transported by ships and trains, air cargo would always amount to just a small portion of national commerce. In terms of speed and access, however, airplanes held potential for bringing devastating military effects to bear on opponents in crisis or war. The air power revolution, more than the nuclear or information revolutions, was orchestrated by relatively young leaders in the air service, and for the American case after 1947, the movement was sustained largely within an independent air force.[12]

Historical examples from the battleship bombing trials of the interwar years to the "Shock and Awe" campaign of Operation Iraqi Freedom make it difficult to ignore changes in national power projection wrought by advancing technologies, organizations, and doctrines of air power. At the same time, the cases illustrate how the intensity of air power development within a single service led to sand traps in the theoretical discourse. Gifted champions of air power who sought to maximize strategic effects of the new technologies often pushed for changes that made superiors in their own service uncomfortable and threatened traditional roles and missions for the Army and Navy.[13] The air power debate that consumed the career of Billy Mitchell remained sufficiently heated and disorderly to endanger Air Force generals during the Gulf War and Kosovo campaigns, in part because each of the services jealously guards its bureaucratic turf, but also because each service defends a legitimate policy position.

Thinking about air power reinforces the usual lessons of multi-causality for geopolitical outcomes: the importance of remaining alert for interactive effects, side effects, or blow-back at the strategic level, and the value of posing questions early in the process of acquiring national capability through a single line of technology or a single geographic dimension. More specific to the intellectual history of air power, the intervention of academia – first the circle of military colleges and federal research corporations then leading civilian institutions – worked to civilize the debate. The sterile, rational idioms of case comparison, statistical analysis, and modeling combined with the joint spirit of recent reforms to tamp incendiary rhetoric among the services. This is all to the good, for the difficulties of post-war stabilization in places such as Baghdad, Fallujah, and southeastern Afghanistan have presented new challenges to air power. Relatively few analysts possessed the resources or inspiration to study air power's role in mountain-based guerrilla warfare or urban insurgency before the US response to 9/11, and these few tended to agree that much problem-solving lay ahead.[14]

With respect to issues of space and defense policy, the story of air power teaches that eventual solutions for "space power" are unlikely to come from a single service or entirely from the military. Rather than an iconic father of space power, we are likely to see several constitutional congresses attempting to resolve diverse organizational interests and intellectual perspectives. Air Force Space Command may embody the United States Defense Department's executive agent for space, but over time, space power will go farther and faster if stakeholders from other services and other sectors of society receive an early invitation to participate in concept development. The method of social scientific policy analysis can be abused to obfuscate important concerns and alienate practitioners. "Too academic" is a common death sentence for proposals inside the

Pentagon.[15] Yet, the case of air power demonstrates that when done with appropriate respect for service cultures and command experience, rules of analysis – hallmarks of academic inquiry – serve as a lingua franca across bureaucratic or societal fault lines, improving policy deliberation about national defense through space power.

Nuclear weapons and muscle bound coercion

In the United States, there was some debate – echoing the one over air power – over whether nuclear arms were just another weapon to facilitate existing military operations or something qualitatively different. During the first full decade with nuclear bombs, President Eisenhower actually weighed in on both sides of the argument.[16] Even so, the acts of civilian leaders spoke loudest, and from Harry Truman through Ronald Reagan and the end of the Cold War, those actions adhered to the line that, unlike other wars, "a nuclear war can never be won and must never be fought."[17] In short, nuclear weapons revolutionized international conflict and American grand strategy.[18]

All-out attack with thousands of nuclear warheads represented an inescapable threat, but the possessor of nuclear strength was muscle-bound. This was not just because the two powers could punish each other with a devastating second strike if the other moved first. As Thomas Schelling predicted in 1966, under circumstances of mutual assured destruction, superpower competition and attempts at coercion continued.[19] The nuclear states were muscle-bound because, try as they might, it was very difficult to actually follow through with the winning strategy. At least as academics modeled the Cold War, the key to victory lay with highly refined command and control which could be used to maneuver the other side into a type of checkmate. The Soviets would concede, refusing to press further, when they realized that they could not escalate enough to win a crisis involving conventional forces or even limited nuclear forces, without bringing about general war.

Plenty of practitioners and academics went on to express serious doubts about escalation dominance as a practical national strategy. They saw too many barriers to fine control for both sides, and if the Soviet premier and the American president insisted on using nuclear brinkmanship to protect national interests, it seemed that the complications would at some point bring mutual destruction by miscalculation. At first blush, this blossoming of research and debate on nuclear issues seemed to model what was missing for so long from the service-based development of air power. Yet, the wider acceptance and deeper influence of academic rigor might have contributed to certain blind spots that should give pause to strategists who will wrestle with matters of space and defense.

Overall, broad discussion on nuclear weapons in the West successfully integrated knowledgeable communities to assist government leaders and sometimes challenge them with novel ideas. However, each level of analysis pointed toward policy paralysis. Tension between the superpowers would continue, but it was simply too dangerous for one to press a serious advantage against the other. Stability became the watchword for any responsible official. Yet, despite hundreds of journal articles and books, thousands of position papers, and all of the heavy theorizing, actual events did not cooperate.

The confident prediction for stasis, from the Eisenhower administration's unsuccessful rollback policy in the 1950s until late in the days when the Soviet Union was already failing, serves as a useful warning for future planners on space and defense. It would not be surprising if space power theorists soon built a scientific explanation for how predominance in space will determine geopolitical position in coming decades. Nevertheless, just as the nuclear balance did not in the end prevent decisive changes in the Cold War contest, stability in the measurable distribution of space assets should not imply stasis at the geopolitical level. Scope of influence for American or rival ideology, quality or quantity of members in military alliances and economic partnerships – in other words, the substance of international politics – can change even if space power says it should not.

Maintaining a healthy respect for the primacy of politics means recognizing the contingent nature of uni-dimensional theories about geopolitical outcomes, whether such theories are nuclear or space-based. Just underneath the surface of apparently fixed state structures, political leaders constantly flow in and out of power. Coalitions within and across state boundaries may change. National objectives and preferred means for grand strategy also change over time. In the future, we may see asymmetric responses to the global reach and precision strike capabilities of space-directed weapons. Present discussions among allies over the European Union's Galileo satellite navigation system may set a trend in which long-term commercial interests force adjustments in military planning for space. A space and defense cadre will have to remain vigilant against the possibility that new circumstances on Earth alter "proven" connections between US space technology and subsequent foreign policy options for other states in the international system.

The ongoing revolution in information operations

Especially after the dramatic victory of US-led forces in the 1991 Gulf War, and subsequent enthusiasm over another revolution in military affairs, it is reasonable to ask why space and defense should not be conceptualized as a mere subtopic under defense transformation. Why are space operations not simply categorized as contributing factors under America's information edge, with the latter taking pride of place alongside America's nuclear edge?[20]

During the great nuclear debates, it is true that core arguments flowed from the technological novelty of placing small packages with enormous destructive potential on fast bombers and missiles. Yet, very quickly in analyses of force structure and deterrence, geographic context emerged as an important consideration. The fundamental rule governing arrangement of the strategic triad into land-based missiles, nuclear submarines, and nuclear bombers testified to the ways in which geography constraints shaped the realization of a second-strike deterrent and stoked fear about what a first-strike force would look like.

In a similar vein, defense analysts have, for some time, explored the implications of quantum improvements in information technology for transformation of national defense. Important debates centered on the opportunities presented by high-capacity, lightweight memory chips, ultra-fast computer processing, or global, high-bandwidth communications. Yet, the realization of these systems as instruments for international cooperation or competition will also be shaped by geographical context, in this case the physical properties of space. The unique perspective, high above targets on the Earth's surface, and the unique operating constraints at various orbital distances are likely to influence how critical information technologies are developed for deterrence or assurance and how they are perceived by other states. With the geography connection in mind, then, the authors and editors of this book endeavor to organize a groundbreaking conversation about defense that will lead actors in the world, and the United States in particular, toward responsible and successful application of space power.

A framework for space and defense policy

In the context of space and defense, success is defined in grand strategic terms, that is, the perspective of a state actor employing power to achieve peace, prosperity, and justice for itself while supporting a legitimate order for the international system.[21] Given this point of departure, it makes sense to organize an original volume around themes taken from political analysis.

The book begins with the idea that space power is poised to influence policies affecting the national defense of many states. International interaction may be broken down into two basic categories: competitive maneuver, for advantage with respect to other states, and establishment of norms for mutually beneficial cooperation. Chapter 1, "Space and grand strategy", elaborates the connection between these two areas. Chapter 2 supplements the discussion by explaining the scientific vocabulary used for global debates on space and strategy. Chapter 3 offers insights into the multidimensional economic importance of the space sector and presents a picture of some of the interests of commercial actors in space. Chapter 4 presents the institutions and legal codes for regulating space activities worldwide, while Chapter 5 introduces the main international organizations that deal with space as a transnational issue.

Chapter 6 offers a method of analysis for how states formulate space policy, leading on to several chapters that detail special roles of sectors within the state. Chapter 7 introduces the importance of space in US military policy past and present. Recognizing the central role space assets play in collection of strategic and tactical intelligence, Chapter 8 uses a historical perspective on intelligence systems. It lays out enduring trade-offs for military and civilian agencies in this field while avoiding black areas delimited by classified information. Chapter 9 is a discussion of the space acquisition process in the United States for complex and expensive space systems, underscoring current policy dilemmas involving investment, technical risk, and accountability that will hound decision-makers in space and defense for years to come. Chapter 10 turns to the role of the civil sector in US space policy, emphasizing the historical importance of NASA and the critical nature of future choices in this arena for US defense.

Chapters 11 and 12 focus on foreign space perspectives, discussing respectively the relatively developed space programs of Russia and China, and the uniquely European programs that seek to reconcile national and collective interests within the transatlantic relationship. Finally, Chapter 13 analyzes the impact of these and other foreign space perspectives on US policy. An outstanding tutorial on space, "A brief introduction to astronautics," written in conjunction with this textbook, can be accessed through the Eisenhower Space and Defense Policy website, located at www.usafa.af.mil/df/dfps/csds/index.cfm.

Roger Harrison, director of the Eisenhower Center for Space and Defense Studies, concludes our study with an exploration of the nature of US space policy and the policy-making process. He examines to what extent they can be characterized as coherent and rational, and whether one can already discern movement toward a grand strategy that balances national desire for competitive advantage against the benefits of international cooperation in space.

Notes

1 Peter Gourevitch, *Politics in Hard Times: Comparative Responses to Economic Crises* (Ithaca, NY: Cornell University Press, 1986).
2 Robert Art, *A Grand Strategy for America* (Ithaca, NY: Cornell University Press, 2005).
3 Andreas Dorpalen, *The World of General Haushofer* (New York: Farrar & Rinehart, Inc, 1942).
4 Stephen Ambrose and Douglas Brinkley, *Rise to Globalism: American Foreign Policy since 1938* (New York: Penguin Books, 1997).
5 Thomas Barnett, *The Pentagon's New Map: War and Peace in the Twenty-First Century* (New York: Putnam, 2004).
6 Paul Kennedy, *The Rise and Fall of the Great Powers: Economic Change and Military Conflict from 1500 to 2000* (New York: Random House, 1987).
7 Stephen Brooks and William Wohlforth, "American Primacy in Perspective," *Foreign Affairs*, Vol. 81, No. 4 (2002): 20–33.

8 Robert Kagan, "A Tougher War for U.S. is One of Legitimacy," *New York Times* (January 24, 2004): B7; Francis Fukuyama, "After Neoconservatism," *New York Times Magazine* (February 19, 2006), available at: www.uvm.edu/~wgibson/Fukuyama.pdf; Francis Fukuyama, *After the Neocons: America at the Crossroads* (London and New York: Profile and Vintage Books, 1987).

9 David Radzanowski and Marcia Smith, "93062: Space Launch Vehicles: Government Requirements and Commercial Competition," CRS brief (Washington, DC: Congressional Research Service). Available at: www.fas.org/spp/civil/crs/93-062.htm.

10 See Bruce Linster, "Space and the Economy," in this volume.

11 Jim Krane, "Satellite Photos of Iraq for Sale," *Space.com* (March 24, 2003), available at: www.space.com/businesstechnology/technology/satellite_pictures_030324.html.

12 Robert Pape, *Bombing to Win: Air Power and Coercion in War* (Ithaca, NY: Cornell University Press, 1996).

13 Mark Clodfelter, *The Limits of Air Power: The Bombing of North Vietnam* (New York: Free Press, 1989), chapter 1.

14 Troy Thomas, "Aerospace Power in Urban Fights," *Aerospace Power Journal* (Spring 2002), available at: www.airpower.maxwell.af.mil/airchronicles/apj/apj02/spr02/thomas.html.

15 As a social scientist and a former consultant at the Center for Naval Analyses, Thomas Barnett, provides memorable insights on debate culture at the Pentagon. Thomas Barnett, *The Pentagon's New Map.*

16 Robert R. Bowie and Richard H. Immerman, *Waging Peace: How Eisenhower Shaped an Enduring Cold War Strategy* (Oxford: Oxford University Press, 1998).

17 Ronald Reagan, "1984 Address to a Joint Session of Congress on the State of the Union," available at: www.presidentreagan.info/speeches/reagan_sotu_1984.cfm.

18 For a contrary minority view, see John Mueller, *Retreat from Doomsday: The Obsolescence of Major War* (New York: Basic Books, 1989).

19 Thomas Schelling, *Arms and Influence* (New Haven, CT: Yale University Press, 1966).

20 Andrew Krepinevich, "Cavalry to Computer: The Pattern of Military Revolutions," *National Interest*, Vol. 37 (1994): 30–42; Joseph Nye, Jr. and William Owens, "America's Information Edge," *Foreign Affairs*, Vol. 75, No. 2 (1996): 20–36; George Friedman and Meredith Friedman, *The Future of War: Power, Technology, and American World Dominance in the 21st Century* (New York: St. Martin's Press, 1998).

21 Recent contributions from the perspective of grand strategy include: Robert Art, *A Grand Strategy for America* (Ithaca, NY: Cornell University Press, 2004); John Lewis Gaddis, *Surprise, Security, and the American Experience* (Cambridge, MA: Harvard University Press, 2004); Walter Russell Mead, *Power, Terror, Peace, and War: America's Grand Strategy in a World at Risk* (New York: Knopf, 2004); and Robert Lieber, *The American Era: Power and Strategy for the 21st Century* (Cambridge, UK: Cambridge University Press, 2005). John Ikenberry (2001) and Amitai Etzioni (2004) are particularly eloquent in arguing for the strategic value of legitimate order: John Ikenberry, *After Victory: Institutions, Strategic Restraint, and the Rebuilding of Order after Major Wars* (Princeton: Princeton University Press, 2001); and Amitai Etzioni, *From Empire to Community: A New Approach to International Relations* (New York: Palgrave Macmillan, 2004).

1 Space and grand strategy

James J. Wirtz

Strategy is the process whereby available resources are controlled and utilized in a militarily effective and politically compelling way to achieve national objectives, usually in the face of determined opposition. If the citizens of a nation are fortunate, their strategic thinking also will be informed by a coherent grand strategy which encompasses more than just plans to orchestrate the use of force. According to Edward Mead Earle, grand strategy "so directs and integrates the policies and armaments of the nation that the resort to war is rendered unnecessary or is undertaken with the maximum chances of victory."[1] Grand strategy harnesses or reflects political, economic, military technical, social, and demographic trends to advance national interests. It is a political statement about a nation's memories and expectations because it reflects a vision of a people's past, their purpose, and their future. Grand strategies always advance a plan to enhance the prospects of national survival, and most offer a culturally appealing vision of how to preserve or enhance national grandeur. Grand strategy uses the resources of the moment to shape the future so that it better suits national interests.

Grand strategies can take a variety of forms. Some are symmetrical. They seek to outperform competitors by beating them at their own game. Others are asymmetrical. Instead of matching an opponent's strengths, they seek to maximize military effects by exploiting an opponent's weaknesses. Asymmetric strategies seek maximum political impact with minimal military effort. Some grand strategies call for the forward deployment of military forces, while others call for forces to be kept in a central reserve as part of a strategy of "offshore balancing," an option available to states that enjoy a geopolitically insular position.[2] Alliances, international institutions, or the efforts and interests of other states can dominate some grand strategies, while others embrace stark autarchic concepts to guide their approach to the future. Some grand strategies posit a "year of maximum danger," or are motivated by the idea that "time is not on our side," ideas that cultivate preventive motivations for war.[3] In contrast, others are based on the ideas of containment and deterrence and that an opponent's position is unsustainable in the long run. Those who adopt containment as the guiding principle of grand strategy believe that if current trends continue, their opponents will inevitably be defeated by their own internal decay.[4]

A century ago, statesmen and military thinkers were preoccupied with the way a country's "war potential" could be harnessed to achieve national objectives. They included several important measures of national strength in their estimates of war potential: the size of a country's military and population, and the speed and extent to which it might be mobilized in wartime; demographic trends; national output of steel and cement; the extent and density of railroad and road networks; and whether or not a state possessed a modern chemical industry or manufacturing sector. Today a host of new capabilities, some of which were the stuff of science fiction just a few decades ago, are considered to be key determinants of a nation's future economic, political, and security prospects. The information revolution, specifically whether or not a country possesses the human and technical capital to exploit the opportunities made possible by advances in computers and communications, is seen as a key determinant of future national prosperity. For countries that can direct these emerging technologies toward military applications, the information revolution is producing a revolution in military affairs (RMA). A new type of warfare is emerging, based on the availability of a global precision-strike and surveillance complexes. The George W. Bush administration's emphasis on defense "transformation" reflects the judgment that future military success will not be found in "legacy" systems such as aircraft carriers, piloted fighter aircraft, or main battle tanks, but in new applications of technology, organizations, and doctrines to the battlefield. New technologies also are having a profound impact on societies, as people embrace an ideology of independence and direct action made possible by widespread access to the Internet.

An important factor in measuring a nation's contemporary and future war potential is access to and mastery of space. The presence of Earth-orbiting satellites that are used for reconnaissance and communication purposes have clearly contributed to both the information revolution and the RMA. Access to space, space systems, and ongoing space operations have for decades enriched life on planet Earth, fostering scientific progress, economic growth, and security. But the emerging military, economic, scientific and political "geography" of space raises a host of questions for strategists. Does space represent such a unique geography that traditional strategic concepts have little or no applicability when it comes to understanding military operations in space? Will space operations shape terrestrial military or political competition, or will developments on Earth and in space fail to have much influence on the other? Is space an enabler of economic growth and military prowess, or is it something more? Most importantly, what would constitute a US grand strategy for space?

This chapter will address these issues by first exploring the different geographies addressed by theorists interested in land, air, and maritime warfare and the different types of grand strategies that emerged from these different physical settings. It will then explore space as a geography of strategic interest. The chapter will conclude by offering some observations about the nature of a grand strategy that fully recognizes the importance of space for contemporary and future US national security.

Strategic geography

Although the dialectical nature of military or political struggle is a constant no matter where combat occurs, the physical environment shapes the nature of combat.[5] The fact that combat takes place in different geographic settings is largely responsible for differences in warfare at sea, on land, and in the air. Different strategic geographies lead to the adoption of different weapons and doctrines, and produce military institutions with different values and traditions.[6] As Daniel Moran notes, the impact of strategic geography produces highly observable effects on military organizations:

> The armies of the world resemble each other far more than they do the navies and air forces that fight under the same flags. A modern army, for that matter, resembles an ancient one more than it does a modern navy, since its most basic problems have not changed at all from one millennium to the next. The same is true for the other major military branches. Before any armed force can come to grips with its opponent, it must first master the immediate challenges of its physical environment. Ships must float. Aircraft must (however improbably) remain suspended in the air. Armies must propel themselves as best they can across an unyielding landscape rich with obstacles large and small. War*fare*, the *making* of war, is first of all about making the most of one's chances within the constraints imposed by Mother Nature.[7]

Strategic geography has an enduring impact on political and military competition and the institutions that exist to wage war.

In addition to shaping organizations and the conduct of warfare, different strategic geographies have produced different types of strategic theorists and theorizing. Although the best-known theorist of war, the Prussian philosopher Carl Von Clausewitz, used observations from early nineteenth-century land warfare to develop an explanation of war itself, suggesting that any variety of grand strategy can be derived from any strategic geography, this chapter highlights different types of strategic theory generated by focusing on different physical settings.[8]

Some theorists explain victory on the modern battlefield. Stephen Biddle's work on the effectiveness of combined arms operations offers compelling insights into the sources of victory in modern land combat.[9] Biddle offers a theory of victory for land combat, a theory that can be used as the basis of grand strategy. Some take a broader perspective by explaining how a successful campaign undertaken in a single strategic geography – here Giulio Douhet's work comes to mind – can lead to victory in war by nullifying an opponent's military strategy.[10] Douhet's grand strategy offers a path to victory by literally circumventing the bulk of an opponent's military capability by directly attacking their political commitment to war. Others take an even larger perspective by highlighting how the mastery of a strategic geography can not only produce victory in war, but national prosperity and an enduring ability to frustrate future rivals. Alfred Thayer Mahan remains the darling of admirals everywhere because he offered a compelling vision of how a blue-water navy and mastery of the world's oceans not only would produce victory in war, but would also enhance national grandeur and wealth, especially when compared to non-seafaring competitors.[11]

Theory of battlefield success: land warfare

Terrain is the key feature of land warfare. Terrain channels the movement of forces along relatively predictable paths allowing natural obstacles like swamps, high ground, or forests to be incorporated into defenses. Because of the impact of terrain, the movement of units, deployment from marching to fighting positions, and supply of ground forces become dominating issues for ground commanders.[12] By exploiting terrain, however, commanders can increase their combat capabilities by creating force multipliers. The proper use of cover, for example, gives forces on the defense what many consider to be a 3:1 advantage over forces on the offensive that must leave their positions and expose themselves to enemy fire to complete their mission. Surprise and maneuver can serve as force multipliers for units on the offense because they allow them to strike enemy units at unexpected times and places.

Because land forces operate over vast expanses of territory and because communications can become chaotic even on the modern battlefield, responsibility for crucial decisions in combat is often thrust upon the most junior noncommissioned officers.[13] To coordinate what are in fact thousands of independent decisions made in battle spaces isolated by terrain and the chaos of war, armies rely on doctrine to guide soldiers in combat. Doctrine is a highly refined and debated set of principles for the conduct of military operations that are communicated widely across armies to guide soldiers in the planning and conduct of military operations. It may at first appear counterintuitive that armies would disseminate information that logically should remain secret, but having everyone "read off the same sheet of music" in the chaos of battle has proven to be a key contributor to victory.

If terrain and doctrine play critical roles in land warfare, the use of scores of different types of weapons in land battle distinguishes it from air war and, to a lesser extent, war at sea. Armor, infantry, air power and air defense, to say nothing of logistics, intelligence, communications, and reconnaissance support, all have to be coordinated in the conduct of modern, combined-arms operations. To be used to their full effectiveness, these operations have to exploit available force multipliers and deny force multipliers to the opponent. Land warfare is complex. Commanders have to recognize the effects of terrain, the importance of doctrine, and integrate and use all available weapons into a synergistic campaign to prevail. Offense and defense also are useful terms when it comes to describing land operations, but they have less importance when it comes to explaining air and maritime operations.

Stephen Biddle's "modern system of force employment," is a type of grand strategy because it offers a succinct explanation of victory in land warfare that takes into account the complexity of ground combat. Combined-arms operations are the core premise behind Stephen Biddle's explanation of victory and defeat in war. Biddle suggests that the militaries which master combined arms operations can quickly and decisively defeat less professional and proficient opponents, and that this result is relatively insensitive to the numerical and technological balance on the battlefield. Combined arms operations are extremely difficult to organize because they require the coordinated use of artillery, armor, infantry, air, reconnaissance, and command and control assets. When mastered, however, combined arms operations create a lethal synergy that simply annihilates less adept yet similarly armed combatants. In Biddle's view, US success against Iraq in the 1991 Gulf War, for example, was not produced by superior American technology, but by the fact that American units were much better at combined arms operations than their Iraqi opponents. By contrast, al Qaeda fighters in Afghanistan gave a better account of themselves because they made good use of cover to conceal their movements, much as German storm troopers learned to advance under fire by hiding from the "rain of steel" that literally forced troops to go underground during World War I.[14]

As grand strategy, Biddle's ideas suggest that combined arms operations are a living art and that only well-educated, well-organized, professional officers who think about combat in an unromantic, non-ideological manner can expect to prevail on today's battlefield. Victory is mostly determined by the quality of a military's human, not material, capital. Cultivating and maintaining human capital is difficult to achieve, and its success can only be measured after battle has been joined. It is easier to believe in materialistic reassurances of victory, that shiny new weapons will overwhelm all potential opponents. Nevertheless, Biddle suggests that battle still rewards those who think and practice how best to employ all of the human and material capabilities at their disposal. In a sense, Biddle offers a grand strategy based on the achievement of combat proficiency: a professional, well-educated military, practiced in the art of combined arms operations, is the basis of victory in land war.

Victory in war: air warfare

The stalemate that emerged during World War I drove theorists to search for a technical, operational, or strategic way around the "defensive dominance" and carnage of trench warfare.[15] While some looked to armor (Blitzkrieg) and grand maneuver (the so-called strategy of the indirect approach)[16] or better defenses,[17] others hoped that the primitive aircraft of the day might offer a solution to grinding and indecisive trench warfare. In 1921, the Italian strategist Guilio Douhet wrote his classic treatise on air warfare *Il Dominio dell'Aria* (*Command of the Air*) that offered a coherent theory of air operations and a vision of single-service victory by using air power to its fullest effect. Although nearly 40 years would pass before the advent of nuclear weapons and long-range delivery systems turned Douhet's vision into reality, his strategy is directed at a "fluid" physical setting and promises victory in war through strategic innovation and dexterity, not just battlefield success. In other words, it offers a broader strategic vision than Biddle's work and addresses a geography that lacks natural obstacles to channel combat.

Because the atmosphere over the Earth, like the seas that cover most of the world's surface, lacks the terrain features that dominate land warfare, the notions of offense and defense are not especially useful when it comes to describing war in the air or war at sea. There is not a lot of talk about terrain-based force multipliers when it comes to air warfare. When terrain is no longer an obstacle, the military balance per se remains as the key factor that shapes the battlefield. This is the starting point for Douhet: a favorable military balance can create a position of command of the air, a situation whereby one side in a conflict can use the skies for its own purposes while denying the opponent similar access to the air. The goal of air operations is to gain command of the air, primarily through an attritional campaign to destroy an opponent's air defenses and air forces, shooting them down in air-to-air combat or destroying them on the ground in a sustained bombing campaign. As in maritime strategy, one might posit various degrees of command of the air, e.g. a situation where one can obtain local air superiority for limited amounts of time or a situation of command-in-dispute when both parties are battling for supremacy. But there are no defense operations, just competing offensive operations designed to sweep the opponent from the skies.

A second component of Douhet's strategic thinking is related to doctrine and highlights the differences between air and land operations. Douhet noted that effective air operations require the placement of all air assets under a unified and centralized command to maximize their offensive striking power. Air doctrine is thus not intended to deal with a plethora of independent decision-makers, but is instead intended to centralize operations to degrade the opponent's air operations as quickly as possible. This sort of idea sounds benign enough, but in reality it carries with it enormous bureaucratic implications. Inter-service rivalry in the United States since the interwar period has often revolved around the control of air assets and US strategy in both World War II and the Vietnam War was shaped by the machinations of air power enthusiasts.[18] This doctrinal tendency toward centralization also stands in stark contrast to land operations, which require that literally thousands of people work on a relatively ad hoc basis to synchronize their operations in the face of local challenges.

Unleashed from the limits of movement on land, air forces that control the skies can literally use them as a highway to undertake what Thomas Schelling has called the "diplomacy of violence."[19] Air forces can go directly to striking counter-value targets, literally leapfrogging land forces that remain relatively unscathed to destroy urban industrial centers or other targets valued by opposing leaders and societies. This was the key strategic innovation offered by Douhet. Once they were controlled, the highways of the air could be used to strike directly at an enemy's society, eventually causing the enemy population to break ranks with their leadership in a quest to avoid further death and destruction.[20] Although historians and strategists have called into question the efficacy of this type of coercive strategy,[21] and US theorists generally believe that air power is best used in a strategic counter-military role,[22] Douhet's notion that air power could be used for strategic effect by directly attacking the opponent's war potential (i.e. war industries) and will to fight has intrigued strategists for decades.

If it emerged upon the scene today, Douhet's *Command of the Air* would be considered to be an asymmetric strategy because it offers a method to circumvent the bulk of an opponent's military forces to undertake military operations with the specific purpose of upsetting the enemy's political calculus favoring war. The notion that a new fluid medium of movement could literally circumvent an opponent's defense establishment, thereby creating a weapon that could quickly knock the opponent out of war with a minimum of friendly casualties, is an objective that continues to generate a strong appeal today. The rise of new fluid geographies of space and cyberspace probably creates opportunities for asymmetric strategies that would mirror Douhet's strategic thinking.

Peace, prosperity and command of the sea

A third type of grand strategy is probably best characterized by the work of the American naval strategist, Alfred Thayer Mahan.[23] Unlike some contemporary theorists who depict the world's oceans as a barrier,[24] Mahan saw the sea as a highway that could be exploited for economic gain and strategic mobility, shared with friends and allies, and denied to a state's enemies. He also viewed the balance of capabilities as the determining factor in naval warfare: concepts like command of the sea, command in dispute, and *guerre de course* (commerce raiding) were used to understand naval strategy long before Douhet borrowed them to explain war in the air.

Events on land, however, have tended to influence war at sea. Naval battles tended to take place near significant land features as navies attempted to seize and exercise command of the sea to project power ashore to affect the battle on land. Additionally, blue-water naval battles rarely occurred given the difficulty of finding the opponent's fleet in the enormous expanse of the world's oceans. The expanse of the sea itself creates an environment in which it is possible to hide significant fleet and ship movements, e.g. the United States uses nuclear-powered submarines to hide its nuclear forces under the Atlantic and Pacific oceans. But with these caveats in mind, naval theorists have never found notions of offense or defense to be particularly useful in tactical or strategic thinking about war at sea.[25]

While Mahan, like Douhet, argued that command of the sea (air) was key to naval warfare and that the goal of the great power navies was to wrest command of the sea from competing fleets, the most important feature of his grand strategy was the way his strategic thinking linked sea power to national greatness. According to Mahan, the history of great power relations is marked by a cyclical rise and decline of great powers which generally pitted continental powers against maritime powers. Not every country that borders the world's oceans, however, is a maritime power. A nation's people and government have to take to the sea; they have to integrate their economy into maritime commerce and communications; and they have to protect their access to the world's oceans with a significant naval capability. In this great power competition, the maritime power has an advantage because it can tap resources on a global basis in its battle against its continental rival, and through its command of the sea it can prevent the land power from access to outside resources. The maritime power also has strategic mobility in the sense that it can attack the continental power at a variety of places along its periphery, an advantage that compensates for the land power's interior lines of communication.

In peacetime, this global reach favors the maritime power because it allows it to trade with partners around the world and to gain access to raw materials across the entire planet. The maritime power can help its allies prosper by allowing them to participate in global trade, building a powerful and prosperous coalition to deter hostile action. Mahan thus describes how the mastery of a single strategic geography not only can help win a major conflict, but can also create the basis of national prosperity and enduring security. For Mahan, exploiting, and if need be controlling, the world's ocean highways is the key to national greatness and the key to victory in the cyclical rise and decline of great powers.

A grand strategy for space?

The survey of land, air, and maritime environments would suggest that there are several different types of grand strategy that might be applied to the "geography of space." Biddle's theory of land warfare suggests that a parallel theory of victory in military space operations might be required, while Douhet's thoughts on air power might be applied to demonstrate how dominant space operations can trump opponent's capabilities in other combat geographies. Alternatively, a grand strategy that links future American economic prosperity, technological supremacy, and success in war might be the best way to energize US scientists and planners to exploit the full potential of space operations. In other words, does space represent a tertiary theater of operations in support of terrestrial warfare, or does it offer a key not just to future battlefield victory, but also to national prosperity and security?

Any grand strategy for space must characterize the physical setting for space operations. Does space constitute a barrier that will constrain or channel economic or military activity, or does it act as a highway, facilitating movement and communications? From a theoretical and practical question, this is a key assumption that will underlie grand strategy, but a variety of conflicting estimates exist about how the "geography" of space might shape operations. Today, space acts more as a highway, allowing global access for surveillance and communication systems that provide an order of magnitude improvement in coverage compared to land, air, or maritime alternatives. Some observers have argued, however, that orbital mechanics constrain the conduct of space operations, much as terrain constrains the conduct of land operations. According to Everett Dolman, "defensive positions" might be created in space by physics alone, or at least by the fact that all spacecraft either operate or pass through "Earth space" (which ranges from the lowest viable orbit to the area just beyond geostationary orbit).[26] Dolman suggests that Earth space can and should be patrolled and policed using available technology. Colin Gray offers an equally plausible contradictory estimate. He notes that the infinite depth of space might create an advantage for the defense, allowing assets to hide virtually anywhere, while orbital mechanics makes trajectories obvious, and hiding, nearly impossible.[27] Other scientists suggest that any sort of combat in space would create a lethal halo of debris around Earth that would destroy what is in fact a fragile environment.[28]

A variety of other considerations also could shape the nature of grand strategy. Access to space also can be facilitated by terrestrial location (e.g. access to near-equatorial launch latitudes), giving some states an offensive advantage or at least an ability to launch large payloads at less cost than their competitors. This is no small consideration, given the enormous costs involved in gaining access to space with today's technology. The fact that space is an extremely inhospitable environment suggests that for the foreseeable future the environment will pose a greater or at least a more continuous threat to humans in space than enemy action. It also is unlikely that doctrine will take a central role in space operations, at least not until communications become a factor in operations undertaken beyond the physical setting of the Earth–Moon system.

Sophisticated military operations are currently undertaken in space today. The United States, for example, relies on space platforms for a host of reconnaissance, targeting, early warning and communications capabilities, and missions. But because no combat operations are undertaken in space today, it is difficult to estimate which aspects of the physical setting of space will shape future institutions, doctrine, and operation.[29] It is thus probably premature to follow Biddle's lead in crafting a theory of space battle operations for the simple reason that no one has yet fielded a space warfighting force. Similarly, it would probably be premature to follow in Douhet's footsteps by embracing a strategic theory of asymmetric space operations, although the idea that space operations can be used to circumvent an opponent's terrestrial forces will probably exert an increasing appeal as space access technologies slowly improve. From a practical perspective, Gray's observation about the relatively limited role played by space operations in the first Gulf War will probably apply to space operations for at least the next couple of decades:

> In 1991, space power demonstrated that it could enable combat arms of all kinds to be much more lethal than otherwise would be the case. Space power showed that it enhances the fighting power of all military elements prepared technically, doctrinally, and organizationally to exploit its services.[30]

Today, space operations serve as a force multiplier in support of terrestrial combat units; given technological realities, it would be unrealistic to devise a grand strategy for space that incorporates space combat operations as its centerpiece.

Toward a Mahanian grand strategy for space

A grand strategy for space that advanced the warfighting and war-winning potential for space operations is premature, and would appeal to only the most committed advocates of military space programs.[31] What is instead needed is a far more fundamental statement of the contribution of space access technologies and operational capabilities to the economic, scientific, and security interests of the United States.[32] Grand strategy might be best thought of as an inspirational vision of space exploration that reflects the best elements of American culture, offering an emotionally and politically compelling description of how activity in space contributes to the future success of the United States and its allies. If the public and elected officials become disenfranchised from space operations, space activity will be treated as a tertiary endeavor, of interest only to engineers, science fiction buffs, and military technicians

Writing in the late nineteenth century, Mahan confronted a situation that resembles the discouraging situation that faces today's space enthusiasts.

Mahan posited that global trade, colonies, and a blue-water navy were the key elements of sea power; the only problem was that with a coastal navy and no colonies, the United States lacked two out of three of these elements. Similarly, the United States today only possesses a minimal human presence in space aboard the International Space Station, has barely moved beyond first-generation economic exploitation of space, and has only "out-year" plans[33] but no compelling reason, or the requisite budgetary support, to conduct combat operations in space. Mahan recognized, however, that without the political will to exploit the world's ocean highways, the elements of sea power would remain beyond the reach of naval advocates in the United States. Mahan thus posited a series of conditions that governed any nation's ability to exploit the seas to their full potential: size; population; national character; nature of the government; geographic setting; and access to the sea. In other words, a coastline did not automatically make a country a great seafaring nation. A nation's population and political leaders had to embrace the idea that their future prosperity and security was best secured by taking and exploiting the economic and military potential offered by the sea.

Today, space theorists lack a sweeping vision of how the exploitation of space can fundamentally contribute to national security and grandeur, despite the fact that we increasingly rely on space assets for a host of military and economic applications. They might be forgiven for this short-coming because of the novelty of space travel. After all, Mahan had the luxury of writing about mature technologies and he could rely on hundreds of years of economic and military history to illustrate his thinking about the relationship of sea power to national greatness. Nevertheless, members of the George W. Bush administration seem to understand the political basis of space policy by their repeated efforts to link US prosperity and scientific progress to a robust space program. In a policy statement issued in 2004, entitled "A Renewed Spirit of Discovery," the White House noted that "the fundamental goal of this vision is to advance US scientific and economic interest through a robust space exploration program." In advancing this vision, the administration proposed four objectives for the US space program:

- implement a sustained and affordable human and robotic program to explore the solar system and beyond;
- extend human presence across the solar system, starting with a human return to the moon before the year 2020, in preparation for human exploration of Mars and other destinations;
- develop the innovative technologies, knowledge, and infrastructure both to explore and to support decisions about the destinations for human exploration; and
- promote international and commercial participation in exploration to further US scientific, security, and economic interests.[34]

What is missing from the administration's vision of the US space program, however, is a compelling vision of exactly how the program fosters US scientific, security, and economic interests. By contrast, the administration has provided a detailed and compelling description of its immediate operational goals in space: complete the International Space Station by 2010; develop the new crew exploration vehicle and associated booster; and return to the moon by 2020 as an enabler for missions to other planets.

Conclusion: a grand strategy for space

The July 4, 2006 launch of the Space Transportation System, more commonly known as the space shuttle, carried with it a sense of purpose that had seemed to be lacking in the US space program. The mission was the first step in the long road back to the moon and the even more ambitious objective of exploring Mars in the decades ahead. With the decision to soon abandon the space shuttle for more reliable launch vehicles, the US space program seems to be back on track after years of indecision and disarray.

The scientific program adopted by the US space program is well defined, although some may object to its emphasis on enhancing the human presence in space at the expense of less costly and more technologically feasible robotic space vehicles. Defense advocates also might object that little official effort has been made to integrate the likelihood of future space warfare into an overall vision of the US space program, a problem that is becoming increasingly important in the aftermath of the January 11, 2007 Chinese test of an anti-satellite weapon. In many respects, however, US space policy better captures the elements of space power, not the conditions that will guarantee US leadership in space in the decades ahead. What is missing is a clear statement of how space exploration and activity contributes to American scientific and economic gains, or how space exploration is opening new vistas for all humankind. Although the imagery of a July 4 space shuttle launch might be especially appealing to Americans, one can only compare the picture of bravery, determination, and technological achievement demonstrated by the launch with the images that competed for attention on the world stage in early July 2006. One was a failed North Korean missile launch, accompanied by threats of nuclear war, which was apparently intended to intimidate regional actors and to extort further concessions for what appear to be empty promises of cooperation on the part of the lunocracy in Pyongyang. The other was the release of the last statement of Shehzad Tanweer, one of the suicide bombers that attacked the London transportation system in July 2005, which threatened continued violence in support of al Qaeda's objectives.

At its core, a grand strategy for space would have to offer not only a hopeful, but an exciting vision for all of humankind to abandon what amounts to petty squabbles over insignificant issues to turn our collective attention toward something more worthwhile. A space grand strategy might in the end have little, if anything, to do with military matters, an issue that hopefully would demand less attention in the decades ahead.

Echoing a call made by Gray nearly a decade ago, the world still awaits a Mahan for the space age, a theorist who can articulate how space power contributes not only to national security and prosperity, but creates a lasting and significant benefit for all of humankind. If such a strategy were clearly articulated and accepted by the American people, political support and funding for a space force would follow in its wake.

Notes

1 Edwin Mead Earle, "Political and Military Strategy for the United States," *Proceedings of the Academy of Political Science*, Vol. 19, No. 2 (1941): 2–9.
2 Christopher Layne, "From Preponderance to Offshore Balancing: America's Future Grand Strategy," *International Security*, Vol. 22, No. 1 (1997): 86–124.
3 Jack S. Levy, "Declining Power and the Preventive Motivation for War," *World Politics*, Vol. 40, No. 1 (1987): 82–107.
4 X (George Kennan), "The Sources of Soviet Conduct," *Foreign Affairs*, Vol. XXV (1947): 575–576.
5 Colin Gray, *Modern Strategy* (New York: Oxford University Press, 1999), pp. 206–207.
6 Carl Builder, *The Masks of War: American Military Styles in Strategy and Analysis* (Baltimore: Johns Hopkins Press, 1989).
7 Daniel J. Moran, "Geography and Strategy," in John Baylis, James Wirtz, Colin Gray, and Eliot Cohen (eds.), *Strategy in the Contemporary World* (Oxford: Oxford University Press, 2006).
8 Peter Paret, "Clausewitz," in Peter Paret (ed.) *Makers of Modern Strategy* (Princeton: Princeton University Press, 1986), pp. 186–213.
9 Stephen Biddle, *Military Power: Explaining Victory and Defeat in Modern Battle* (Princeton: Princeton University Press, 2004).
10 Giulio Douhet, *The Command of the Air*, trans. Dino Ferrari (New York: Coward-McCann, 1942).
11 Alfred Thayer Mahan, *The Influence of Seapower upon History 1660–1783* (New York: Dover Publications, 1987).
12 Martin van Creveld, *Supplying War: Logistics from Wallenstein to Patton* (Cambridge: Cambridge University Press, 1977).
13 Martin van Creveld, *Command in War* (Cambridge, MA: Harvard University Press, 1985).
14 Stephen Biddle, "Victory Misunderstood: What the Gulf War Tells Us about the Future of Conflict," *International Security*, Vol. 21, No. 2 (1996): 139–179; Stephen Biddle "Allies, Airpower, and Modern Warfare: The Afghan Model in Afghanistan and Iraq," *International Security*, Vol. 30, No. 3 (2005): 161–176; Bruce I. Gudmundsson, *Stormtroop Tactics: Innovation in the German Military* (New York: Praeger, 1989).
15 Williamson Murray and Allan Reed, *Military Innovation in the Interwar Period* (New York: Cambridge University Press, 1996).
16 B.H. Liddell Hart, *Strategy: The Indirect Approach* (New York: Praeger, 1968).
17 Anthony Kemp, *The Maginot Line: Myth and Reality* (New York: Stein and Day, 1982).
18 Perry M. Smith, *The Air Force Plans for Peace, 1943–1945* (Baltimore: Johns Hopkins Press, 1970); H.R. McMaster, *Dereliction of Duty: Lyndon Johnson, Robert*

McNamara, the Joint Chiefs of Staff, and the Lies that Led To Vietnam (New York: Harper Collins, 1997).

19 Thomas Schelling, *Arms and Influence* (New Haven, CT: Yale University Press, 1966).

20 Lawrence Freedman, *The Evolution of Nuclear Strategy* (New York: St. Martin's Press, 1981), p. 6.

21 Robert Pape, *Bombing to Win: Air Power and Coercion in War* (New York: Cornell University Press, 1996).

22 Benjamin S. Lambeth, *The Transformation of American Air Power* (Ithaca, NY: Cornell University Press, 2000).

23 Alfred Thayer Mahan, *Interest of America in Sea Power Present and Future* (Port Washington, NY: Kennikat Press, 1970); Mahan, *The Influence of Seapower Upon History 1660–1783*; Philip Crowl, "Alfred Thayer Mahan: The Naval Historian," in Peter Paret (ed.) *Makers of Modern Strategy* (Princeton: Princeton University Press, 1986), pp. 444–477.

24 John J. Mearsheimer, *The Tragedy of Great Power Politics* (New York: W.W. Norton and Co., 2001).

25 Wayne Hughes, *Fleet Tactics: Theory and Practice* (Annapolis: Naval Institute Press, 1986).

26 Everett C. Dolman, *Astropolitik: Classical Geopolitics in the Space Age* (London: Frank Cass, 2002), p. 157.

27 Gray, *Modern Strategy*, pp. 260–262.

28 Joel R. Primack and Nancy Ellen Abrams, "Star Wars Forever? A Cosmic Perspective," available at: http://physics.ucsc.edu/cosmo/UNESCOr.pdf.

29 Combat capabilities, however, have been and are currently available. Nuclear weapons have been tested in space as well as co-orbital and ascent anti-satellite weapons and missile defense systems. See Paul B. Stares, *Space and National Security* (Washington, DC: The Brookings Institution, 1987); Kenneth N. Luongo and W. Thomas Wander, *The Search for Security in Space* (Ithaca, NY: Cornell University Press, 1989). I would like to thank James Clay Moltz for offering this observation.

30 Gray, *Modern Strategy*, pp. 262–263.

31 Advocates of this sort of position, especially in the realm of missile defense, remain active. See: Independent Working Group on Missile Defense, the Space Relationship and the Twenty-First Century, "2007 Report," available at: www.ifpa.org/publications/IWGReport.htm.

32 Dolman takes this notion one step further, suggesting that the United States should lead the human race itself into space: "As the great liberal democracy of its time, the United States is preferentially endowed to guide the whole of humanity into space, to police any misuse of that realm, and to ensure an equitable division of its spoils," Dolman, *Astropolpolitik*, p. 181.

33 United States Space Command, *Vision for 2020*, available at: http://middle-powers.org/gsi/docs/vision_2020.pdf.

34 President of the United States, "A Renewed Spirit of Discovery," available at: www.whitehouse.gov/space/renewed_spirit.html.

2   Space, science, and technology

*Timothy Lawrence, William W. Saylor,
Ken Siegenthaler, and Thomas G. Ward, Jr.*

Introduction

Virtually all enterprises involving space are heavily dependent on modern technology. Space technology is developed mainly in response to the demands of missions that are deemed worthy and feasible, thus creating a "technology pull" dynamic. In some instances space technology is unique, having limited uses for other applications. In other cases, non-space technology is adapted for space use. The overall success of space technology development and application is strongly affected by a country's "technological climate," which includes political support, funding levels, educational opportunities, tolerance for failure, commercialization potential, remuneration packages, and scientific prestige.

While not always readily apparent, policy issues arise regarding space technology just as they do with all the other aspects of space utilization and exploration. In some cases, technology offers multiple paths to accomplishing a given space objective, and decision-makers must choose which path or which combination of paths to follow. In other instances, technology defines the feasibility of specific space capabilities, but actually pursuing them is fraught with political implications that must be resolved both nationally and internationally.

The space missions that technology supports may be classified in numerous ways. A common taxonomy comprises three sectors – military/defense, scientific, and commercial. Another breakdown differentiates near-Earth missions from lunar, interplanetary, and deep-space exploration, the former capturing military/defense, commercial, and some scientific missions, the latter encompassing the remaining scientific undertakings. Another approach is to separate manned from unmanned missions, and still another is to examine national missions versus international ones.

Regardless of how one chooses to classify space missions, all of them depend to varying degrees on at least five key technologies – propulsion, power, communications, lightweight high-strength materials, and computers. We will now explore these technologies and highlight some of the policy issues associated with them.

Propulsion technology

Propulsion technology is an essential contributor to every space mission, providing the means for putting any spacecraft into space. Propulsion is the means of providing acceleration, which means exerting a force on a body for a specified period of time. The most common method of accomplishing this in space is with rockets, which push mass in one direction to gain momentum in the opposite direction. The greater the mass pushed out of the rocket's nozzle or the greater the velocity of this mass, the greater the desired accelerative force. A common measure of performance of a rocket is *specific impulse,* which is defined as the amount of accelerative force delivered over a unit of time per unit of propellant mass expended. The common unit for specific impulse is seconds. Specific impulse turns out to be directly proportional to the effective exhaust velocity of the mass exiting the nozzle. The higher the specific impulse, the more efficient a rocket is, and thus a key research driver for propulsion technology over the years has been finding ways to achieve higher values for this key parameter. The actual feasibility of some space missions comes down to whether sufficiently high values of specific impulse can be achieved or not. Figure 2.1 shows that the best systems offer high thrust, high specific impulse, and low specific mass. Unfortunately, systems that exist today generally cannot meet all of these requirements.

Thermodynamic rockets

The most common rockets used for propelling spacecraft into Earth orbit or beyond the Earth's gravitational field are *thermodynamic rockets,* which transfer thermodynamic energy (heat and pressure) to a propellant and convert the energized propellant to high-speed exhaust by directing it through a nozzle. Depending on the source of the heat and pressure, thermodynamic rockets may be classed as *chemical, cold gas, solar thermal, thermoelectric, and nuclear thermal.*

Chemical rockets have been the workhorse of the world's launch vehicles, offering specific impulses as high as 477 seconds for the liquid oxygen/liquid hydrogen fuel combination that powered the United States' Saturn V moon rocket.

When intense, direct sunlight is continually available, solar heating of a hydrogen propellant can provide specific impulses on the order of 800 seconds. Electric arc heating of a propellant, assuming adequate electric power availability, can raise specific impulse as high as 1,000 seconds.

Nuclear propulsion

Nuclear propulsion, where the heat to expand the propellant through a nozzle is provided by a nuclear fission reactor, offers the advantages of

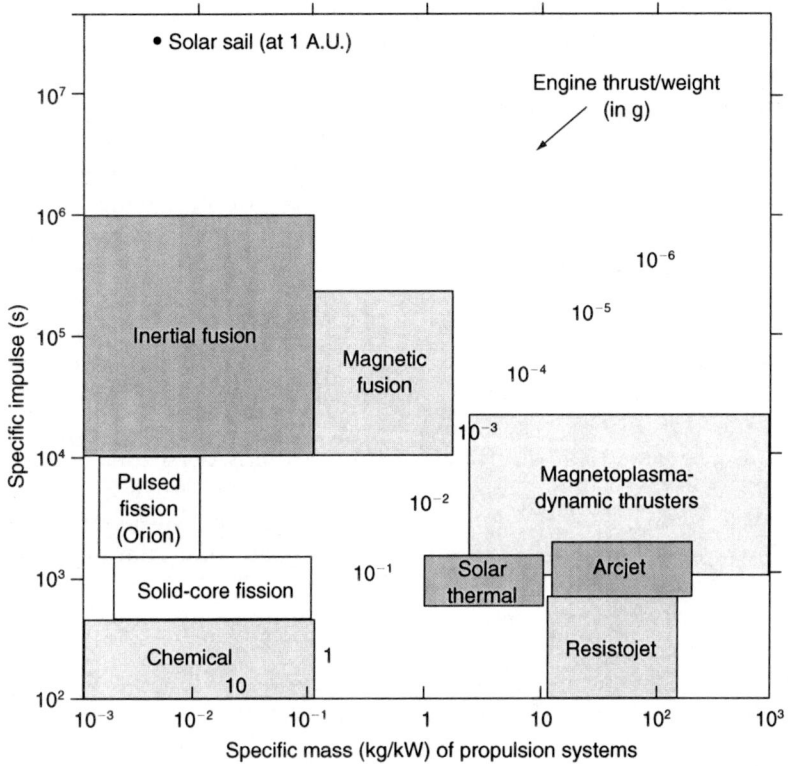

Figure 2.1 Performance of propulsion systems (source: Humble, Ronald, Gary Henry, and Wiley Larson (1995) *Space Propulsion Analysis and Design*. NY: McGraw-Hill, figure 8.2. Copyright by the McGraw-Hill Companies, Inc).

both high specific impulse (on the order of 1,000 seconds) and high thrust. The United States' Nuclear Engine for Rocket Vehicle Applications (NERVA) program tested nuclear-thermal rockets from 1947 until 1972, at which time environmental and political concerns about safe ground-testing and launching a fully fueled nuclear reactor forced sharp reductions in funding for this promising technology.

This stand-down came despite the proven safety record of developing and operating nuclear reactors in the United States. The US Navy's extensive nuclear reactor program for submarine and ship propulsion, for example, boasts a zero nuclear accident record. One mitigating feature of launching a nuclear reactor for use in deep space is that, assuming it is fueled with enriched uranium, it is totally inert until it is put into operation, presumably after lift-off from Earth on a conventional launch system. Once in operation, the radiation it gives off is based on the amount of time it operates.

The combination of performance parameters offered by a nuclear-thermal rocket is essential for some missions – for example, sending humans to Mars and back in a timeframe sufficiently short that they are not exposed to unacceptable levels of solar and cosmic radiation. No other form of propulsion currently known can accomplish this. Use of chemical rockets would result in a travel time to Mars of hundreds of days, whereas a nuclear rocket could shorten the trip to 40 days. For a round trip with a stay-time on Mars of 30 days, the total trip time entailed with the use of chemical rockets would be 433 days. A nuclear rocket would shorten it to 316 days. The radiation exposure of the space travelers would drop from 60 rem to 45 rem. Other missions such as sending heavier payloads to distant planets would fall into the realm of possibility were nuclear rockets used. See Table 2.1 for a breakdown of the change in impact on overall rem due to the use of nuclear propulsion.

There are recent signs that nuclear propulsion may be emerging from under the blanket of fear thrown over it. As part of NASA's 2005 budget, $10 million was identified for nuclear thermal propulsion work at the Marshall Spaceflight Center as part of the Prometheus Nuclear Systems and Technology Program. A promising potential application was the Jupiter Icy Moons Orbiter, which would orbit the Callisto, Ganymede, and Europa moons of Jupiter, suspected of harboring vast oceans beneath their icy surfaces, and collect data pertinent to their make-up, history, and potential of sustaining life. More recently, however, NASA canceled the Prometheus program to focus on lunar missions, but the concept is currently being pursued by a joint commercial/DOE program called IOSTAR. This program is looking at possibly developing a nuclear-propelled orbital transfer vehicle.

Electric propulsion rockets

Electric propulsion rockets, which entail the acceleration of charged particles by electrostatic or electromagnetic fields to produce thrust, offer the

Table 2.1 Radiation profile: chemical versus nuclear propulsion systems

Radiation source	433-day Mars mission using a chemical rocket	316-day Mars mission using a nuclear rocket
Van Allen belts	2 rem	2 rem
Mars surface	1 rem	1 rem
Galactic radiation	31 rem	22 rem
Solar flares	26 rem	15 rem
Reactor	–	5 rem
Mission total	60 rem	45 rem

Source: Ronald Humble, Gary Henry, and Wiley Larson, *Space Propulsion Analysis and Design.* (New York: McGraw-Hill, 1995), Table 8.1. Copyright by the McGraw-Hill Companies, Inc.

principal path to higher efficiencies unattainable with thermodynamic rockets. The two principal types of electrodynamic rockets are *ion thrusters* and *plasma thrusters*. Specific impulses up to 10,000 seconds may be achieved, but at a cost of low levels of thrust. Thus, electrodynamic propulsion is most suitable for long-duration interplanetary flights.

Ion thrusters use electric fields to accelerate ions, while plasma thrusters use electromagnetic fields to accelerate plasmas. NASA's Deep Space 1 mission, launched in 1998, was the first to rely on an ion thruster, operating at 2.3 kW and producing a thrust of 0.09 N with a specific impulse of 3,100 s. The most common plasma thruster is the Hall effect thruster, which employs a radial magnetic field to accelerate positive ions within a plasma. Hall effect thrusters produce somewhat higher levels of thrust at lower specific impulse for a given amount of power than ion thrusters. Electrodynamic propulsion has achieved a satisfactory level of maturity, and a robust variety of devices are available for various low-thrust applications as long as sufficient electrical power is available.

Table 2.2 below compares nuclear-thermal and nuclear-electric systems. Although nuclear-thermal propulsion can send a 500 kg payload to the outer regions of our solar system faster, it will require a large payload faring due to the high volume of the hydrogen propellant. The electric system will require prolonged testing to validate its ability to accommodate long periods of continuous thrusting due to the spiral transfer trajectory.

Low-cost propulsion

In addition to achieving higher efficiencies in propulsion systems – such as with nuclear reactors – propulsion engineers, entrepreneurs, and government decision-makers have all come to realize that drastically lowering the cost of putting objects into space is a compelling objective. The high cost of putting mass into Earth orbit, currently about $20,000 per kilogram (if one used conventional US launch vehicles), limits the scope and structure of many space applications. Moreover, because lowering the costs of launch encompasses simplifying system designs, manufacturing processes, and launch procedures, a reduction in launch preparation time

Table 2.2 Comparison of nuclear-thermal and nuclear-electric systems

	ΔV (km/s)	Initial mass (kg)	Height (m)	TOF (years)
Jupiter nuclear-thermal	7.6	15,250	12.2	4.1
Jupiter nuclear-electric (ion)	15.5	15,515	–	19.4
Pluto nuclear-thermal	13.4	41,680	34.4	19.0
Pluto nuclear-electric (ion)	31.5	17,285	–	57.5

Source: Timothy Lawrence.

can be expected as well, itself a money saver. This combined cost–time reduction opens the door to new military and commercial strategies that were heretofore seen as unacceptable, mainly having to do with spares on the ground instead of in space and less reliance on preemption in the arena of space defense. Alternatively, the cost–time reduction may be exploited to put greater quantities of mass into orbit without increasing costs. This also opens new vistas for military and commercial consideration. Shielding, for example, to harden satellites against attack or to protect humans from radiation becomes a more attractive option.

Although techniques for lowering launch costs have been explored to some extent, it has been capability, rather than cost, that has been the overriding consideration, with government programs leading the way. Only in recent years, with entrepreneurs playing a more active role, eyeing profits to be won from space tourism, satellite imagery, personal communications and the like, has the quest for cheaper launch capabilities intensified. A fundamental dichotomy in this quest is between trying to lower the cost of *expendable* launch vehicles versus switching to *reusable* vehicles, technology playing a very important underlying role.

If it were possible to construct a launch vehicle that could fly into orbit and return as a single stage – a sort of space airplane – great cost and time advantages would result. However, this would require high-thrust, high-specific impulse propulsion systems and lighter weight materials than currently exist, and a robust low-maintenance reentry protection system to return safely time and again. Getting into space with a single stage will probably entail a combined cycle propulsion system, exploiting jet and scramjet techniques in the atmosphere, and transitioning to rocket propulsion in the exosphere. Private investors have not yet concluded that single-stage-to-orbit (SSTO) approaches are safe for their investments, and funding from the government has dwindled since there has been no trade-off between chemical rocket engine design and lightweight materials advances to ensure an SSTO system would work with quick turnaround and reduced maintenance costs. Studies have shown that a nuclear-thermal propulsion SSTO system is feasible, but development costs and political concerns have prevented any significant progress to date.

At the present time, the more promising option for lowering the cost per kilogram of getting into space lies with the use of cheaper expendables, as evidenced by commercial endeavors in this arena. Current efforts such as DARPA's Falcon program hold promise for reducing the cost to orbit through its work with air-launch techniques, such as those pioneered by Orbital Sciences Corporation, and with SpaceX, a relatively new company determined to find ways to lower the cost of manufacturing ground-based expendable launchers. To date, SpaceX has made more progress than the new air-launched system. This company has developed a two-stage liquid propulsion system and has promised to deliver 1,500 kg to low Earth orbit (LEO) for one-third the price of current launch systems.

Completely privately financed, SpaceX's approach has been to use commercial off-the-shelf (COTS) technology as much as possible and sacrifice performance and mass to keep the costs down. Unfortunately, due to a procedural error, the maiden launch of SpaceX's Falcon-I out of Kwajalein Atoll was unsuccessful, but the company's later attempt, financed by itself and DARPA, succeeded in March 2007.

A substantial number of companies have tried to solve the low-cost spacelift conundrum through an SSTO approach, including Kistler, Beal, Conestoga, Roton, Scorpios, and even the larger corporations Boeing and Lockheed Martin. The main reason that all of these development efforts failed was that the companies relied on developing new technology as part of their new vehicle concepts. Unfortunately, they all got bogged down with new technology development and found themselves unable to develop their system. Conestoga attempted to use existing solid rocket engines, but they used too many and floundered on integration issues.

The lowest cost systems in the world today are Russian. The Russians have kept their costs down by refraining from modifying their designs since the 1960s. New designs require new analysis, new manufacturing procedures, and substantial testing, and the Russians' strategy holds that for cheap expendable systems, it is not worth the extra mass that can be saved through design improvements. Their Proton rocket, for example, has just seven engineers involved in its manufacture, mainly due to keeping the manufacturing simple with as little change as possible over the years.

A recent new system in the United States and an existing one in Europe is a secondary payload adapter ring that permits the launch of multiple small satellites with a medium-lift expendable launcher. The Europeans have flown their ring in the *Ariane IV* and call it the ASAP ring, while the new US system (ESPA) made its maiden voyage on the Atlas V in March 2007. The Europeans have charged $200,000 per satellite or $2 million for the entire ring. This could become a low-cost alternative for 150 kg, 1.5 m³-class payloads – ideal for distributed network constellations, for example.

Hypersonics technology holds promise for contributing to a cheaper, more reliable route to space, either via an SSTO or an air-launched approach. A committee of the National Research Council, in their *Evaluation of the National Aerospace Initiative* (2004) performed a very detailed evaluation of hypersonics applications to space and the critical technologies that needed to be matured. Four critical technologies they highlighted are: (1) air-breathing hypersonic propulsion and flight test; (2) material thermal protection systems and structures; (3) integrated vehicle design and multidisciplinary optimization; and (4) integrated ground test and numerical simulation and analysis. The committee's prediction was that a two-stage-to-orbit, reusable Mach 12 vehicle could be developed in the 2015–2018 timeframe.

Power technology

A reliable source of electrical power is essential for the operation of every craft that goes into space. Payload sensors and spacecraft housekeeping modules are invariably powered by electricity, and, as noted above, some propulsion schemes also require this form of energy. A key performance parameter for space power systems is the power produced per unit mass, often expressed in W/kg, since the mass of a system is such a critical factor in its suitability for space launch (for terrestrial systems, where mass is hardly relevant, the key parameter is energy conversion efficiency, although greenhouse gas and pollutant emission levels are becoming increasingly important).

Solar panels, which convert sunlight directly into electricity with efficiencies on the order of 25 percent, are the preferred source of electrical power for long-lived spacecraft in Earth orbit. Performance levels approaching 140 W/kg are realizable for some designs. In the future, development of "rainbow" multiple-band gap photovoltaic cells may well double the amount of power produced by these systems. Additionally, manned spacecraft take advantage of fuel cells to generate electricity, since the byproduct of those using hydrogen and oxygen is potable water. They can produce powers on the order of 1,000 W, but their main limitation is the amount of fuel that can practically be carried into space to power them.

Inescapably, the energy per unit area emitted by the Sun falls off as the square of the distance away, which means that an additional 93 million miles out into space from the Sun, the solar flux is only one-quarter of what it is in Earth orbit. Thus solar panels are not practical for use beyond Mars, and other sources of electricity must be considered for deep-space missions. Here, nuclear energy again comes into play, in the form of radioisotope thermoelectric generators (RTGs). RTGs use the heat generated by radioactive decay to heat thermocouples to produce electricity. Their most important advantage is their long life, a result of the long half-life of the radioisotope used, often uranium or plutonium. On the downside, RTGs are about five times more expensive than solar cells per unit of power delivered, and, like nuclear fission reactors for space propulsion, they prompt political concerns centered on possible accidents involving radioactive materials. Fortunately, disposal of the radioactive RTG is not a problem for deep-space missions, since they are not designed to return to Earth. In the future, it would be possible to adopt a nuclear fission reactor instead of an RTG to generate large amounts of electrical power in space or on another planet or moon in the solar system. Again, it could be designed such that radioactive emissions would not begin until it reached its destination and began operating.

One of the more intriguing possibilities involving power in space, perhaps coming to fruition 20 or 30 years from now, is the deployment of one or more solar power satellites in Earth orbit. According to one concept, sunlight would be concentrated by thin Fresnel lens membranes and focused on "rainbow" multiple-band gap solar arrays, each lens–array combination generating nominally 1 megawatt of power. The power from multiple arrays would be combined, converted to microwaves, and beamed to different locations on Earth by means of kilometer-sized antennas. Five-kilometer "rectennas" (antenna–rectifier combinations) at different locations on the ground would receive the microwaves and convert them to electricity for insertion into the local power grid. Power levels as high as 400 mW might be produced by a 5 km array of 1 mW modules in orbit.

Clearly, undertakings of this nature must get their start at government expense before commercial interests take over. Pursuit of this concept is currently in the hands of NASA under its Space Solar Power Exploratory Research and Technologies (SERT) program. The overriding consideration, of course, is the ultimate cost per kilowatt of power from solar satellites compared with more conventional sources. A major hurdle is the high level of investment required before any power is produced. One design, a non-tracking geosynchronous satellite with an integral phased array, offers an eightfold reduction in initial investment costs over earlier designs. Other factors besides cost play a role however, such as the "cleaner" nature of the power and the greater ease of distribution to lesser-developed regions, such as those in the Third World. Although solar power satellite energy production will progress as a civilian enterprise, once in place, it is likely that its availability in remote locations will be a feature of great interest to the military. It is not difficult to envision deployed military forces employing this source of energy in lieu of transporting heavy generators and large quantities of fuel to the field.

Communications technology

The most pervasive space application today is communications, being central to the military and civilian communities alike. Military communications satellites and leased channels on commercial satellites permit unparalleled amounts of information to flow in secure fashion to and from military commanders and their subordinate echelons. To the extent desired, military campaigns may now be monitored and controlled from afar. Remotely piloted vehicles collecting intelligence or delivering weapons in hostile territory can be controlled from a separate continent. Costly and sometimes vulnerable ground stations on foreign territory can be eliminated through the use of satellite-to-satellite communications.

Commercial satellite communications, relying mainly on long-lived spacecraft in geosynchronous orbits, is now a well-established, profitable enterprise, supporting extensive television services and high-capacity data and voice channels. Constellations of lower-orbit satellites, such as Iridium, provide worldwide communications access through hand-held transceivers.

Communications satellite technology continues to advance, driven in part by NASA's Space Communications program. Objectives of this program include higher power spacecraft transmitters that reduce the size and complexity of ground equipment, antennas with greater gain and directionality yet lighter in weight, use of different bands of the spectrum such as the Ku and Ka bands that help relieve spectrum congestion, and more extensive use of digital components to exploit the advantages of digital communications.

While a principal driver for this work is the desire to equip NASA's scientific satellites with greater communications capabilities, the results are also available to the commercial sector to help improve the capabilities of its satellites as well. Commercial applications include global long-distance telephone services; wireless communications and wireless data links; private wireless networks for voice, data, and multimedia; point-of-sale data gathering; news gathering; information distribution; video conferencing and employee training; entertainment services (high-quality television transmission); digital audio broadcasting; Internet services; tele-education; telemarketing; crime-prevention networks; intelligent highway services; and high-performance, special-purpose radiofrequency components, antennas, digital electronics, and digital signal processing. With fiber optic communications networks growing rapidly, both on land and under the sea, integration of satellite communications with fiber optic networks into a seamless system is a key requirement and is being pursued actively.

A particularly promising avenue of research underway at NASA's Glenn Research Center in Cleveland is the development of ways to make connections between aircraft and spacecraft as easy as Internet connections. The goal is to transform the United States' current 1960s voice-dominated air transportation system into a twenty-first century global system that combines communications, navigation and surveillance systems to provide a ground-to-air-to-space network where all users can connect with each other.

From a political point of view, it may be argued that communications satellite technology, with the systems it has spawned, has brought the world community closer together than any other single achievement. There are the obvious services that carry vast quantities of pictures, commentary, data, and messages to every corner of the globe, increasing awareness and understanding among governments, corporations, and private citizens alike. But, somewhat hidden, is an extraordinary degree of international cooperation that has occurred to enable many countries to obtain their own satellite communications capabilities or to cooperatively exploit the capabilities of other countries. In 2004, for example, 16 communications satellites were launched for customers comprising the United States, Japan, the European Union, Russia, the United Nations, China, Argentina, Canada, and Spain. There is every expectation such levels of cooperation will continue.

Lightweight, high-strength materials

Given the high cost of boosting mass into space and the payload limitations of current launch vehicles, there are strong incentives for attempting to reduce the mass of the structural materials used both in launch vehicles and their payloads, without sacrificing performance. For liquid-fuelled launch vehicles, propellant tanks present a major structural challenge that is customarily addressed with either aluminum alloys or filament-wound fiberglass composite materials. The Atlas launcher employs thin aluminum for its tanks, which are subjected to a modest positive pressure during launch to maintain structural integrity, whereas the US shuttle carries its liquid launch fuel in a large fiberglass tank structure that drops back to Earth following launch. The shuttle also employs two strap-on solid propellant boosters to provide additional thrust during launch, the casings of which are also made of composite materials.

If there is to be a breakthrough in structural materials for space applications it is likely to come in the form of carbon nanotubes, sometimes called "buckytubes," as they are a variant of "buckyballs," a novel allotrope of carbon named after the geodesic structures invented by Buckminster Fuller (graphite and diamond are the most common forms of carbon). Nanotubes of carbon comprised of carbon atoms arranged in long, thin tubes whose diameter is approximately 1 nanometer, or 10^{-9} m. They exhibit strengths as high as 100 times that of steel at one-sixth the weight, a result of their pure carbon–carbon atomic bonds. Currently still in the research stage, their initial applications are occurring in the electronics industry, taking advantage of their unique electrical properties, where small quantities are highly useful. Ways must be found to manufacture large quantities at low cost before they become feasible for structural applications, which Ivan Bekey estimates will occur in the 2010–2030 time period. The current forecast is that production levels will reach 1,000 lbs per day by 2008 at a cost of $100–$1,000 per pound. Further cost reductions are likely as demand and production levels continue to increase. It is envisioned that initial space applications will be reducing the weight of payloads, resulting in launching larger payloads with current launch vehicles or reaching higher orbits with current missions. Bekey predicts that other space applications, currently seen as impractical or impossible, will be viewed in an entirely different light with the advent of structural nanotube materials.

Computer technology

Computers are ubiquitous in space systems. Were it not for steady advances in computer technology, both for use in space and on the ground, during the late 1950s and through the 1960s, the Apollo program that landed men on the Moon multiple times would not have been possible. Whether it be for communications, guidance and control, or other requirements of data handling, the need for continual improvements in speed, reduction of size and power levels, and increased storage capacity of digital electronics keeps marching on at a staggering rate. Fortunately, this technology has continued to advance at a rapid rate and promises to continue doing so. Moore's Law asserts that the number of transistors on a semiconductor chip doubles about every two years. It has proven true for the past 35 years and shows no signs of slackening off, thanks to nanotechnology and other technical advances, although eventual theoretical limitations are well known. The tremendous commercial potential of small computer-controlled devices, such as GPS receivers, cell phones, laptop computers and the like, is a key driving force behind this progress, as it results in exponential reductions in the cost of the digital electronics they exploit.

Digital electronics for space use, however, must be protected from radiation, as they must be for many defense applications as well, where there is the perceived need for operations in a nuclear weapons environment. This can, in theory, be accomplished either by shielding the circuitry or making it intrinsically radiation-resistant by using semiconductor materials that themselves are less prone to radiation damage. Shielding, which requires pure mass to accomplish, is much less attractive for space use because of the high cost of putting mass into space as described earlier. Thus, the US government defense and space sectors have steadily supported research and development of radiation-hardened electronics and, there being little commercial incentive for this, will likely be forced to continue doing so, always chasing the improved performance of COTS devices.

Satellite constellations

The Iridium constellation of 66 communications satellites in medium Earth orbit and the 24-satellite constellation of global positioning satellites, also in medium Earth orbit, illustrate the importance of constellations in providing specific capabilities. In the case of Iridium, the desired use of hand-held transceivers to communicate directly with satellites, coupled with the requirement for continuous availability from any point on Earth, requires a constellation of satellites considerably closer to Earth than geosynchronous altitude. With the GPS system, the position-finding algorithm requires receipt of signals from four different satellites, and its use worldwide at any time again requires a robust constellation. The need for many copies of the same satellite in both cases led to a successful push to use innovative mass-production techniques in spacecraft production. This, in part, has stimulated a harder look in recent times at the pros and cons of satisfying other space needs, heretofore addressed by single high-cost, long-life satellites or using a few such satellites by larger numbers, i.e. constellations of simpler lower-cost satellites.

Technology plays in important role in satellite constellation considerations. Lower cost, simpler launch systems render the multiple launches needed to put a constellation into orbit more affordable and permit more rapid replacement of individual satellites that fail or expire. Given less dire consequences of failure of a satellite that is part of a constellation, there is less risk in using newer technology, such as more advanced solar panels or more powerful traveling wave tubes, thus quickening the pace of spacecraft improvement.

Further down the road there is the possibility that advancing technology will permit the creation of responsive space systems (RSS), comprising constellations of satellites that have reconfigurable capabilities and permit collaborative operations among them. Anticipated capabilities include intensive data processing and information fusion, precision satellite positioning and deep-space navigation, a robust, survivable capability for sensing and surveillance, and new capabilities that protect against hostile threats from space or the ground.

Constellations that incorporate reconfigurable spacecraft technologies can be expected to exhibit many advantages. Foremost among them is that the system as a whole will be much more tolerant to faults, damage to individual satellites, and the inevitable degradation of performance due to age. Second, such systems, as a result of dynamic interconnection, can be readily applied to multiple, and in some cases, novel problems. The individual elements will have lower cost and risk, since it will be feasible to accomplish multiple builds of much smaller platforms, which will lead to faster flight qualification. This will allow the more rapid integration of the latest technology into systems in a controlled, tested, and responsive manner. Additionally, expected advances in astrodynamics and satellite control can be tested and integrated such that the ability to prepare for and respond to the unexpected will be greatly increased. Intelligence collection from space is a key mission that can be expected to profit considerably from cooperative and collaborative capabilities among multiple space platforms.

Examples of key technologies that will underpin spacecraft reconfigurability include reconfigurable radar antennas and feed structures, docking and other physical connectors, on-board robotic arms, integrated multispectral imaging, precision pointing, and on-board data processing and information extraction architectures. The critical capability of converting data to information in space rather than on the ground will minimize communication bottlenecks, allowing, for example, rapid dissemination of location-critical intelligence to multiple ground receivers.

Still further in the future lies the possibility of the virtual satellite multi-element system (VSMES), which is a fully distributed embodiment of advanced technologies in any of a multiplicity of satellite configurations. Natural orbits may be used for composing constellations; sensor information from multiple elements may be fused to synthesize complex functions; and built-in redundancy will enhance reliability and prolong mission duration. An illustrative concept employing the VSMES approach is that of visible or multi-static radar imagery from distributed aperture elements, based on very accurate inter-satellite ranging capability and precise position control. Other possibilities exploiting this "virtual satellite" concept, include:

- removing multipath ambiguities
- three-dimensional imaging
- hyperspectral overlays of information
- event correlation within a military theater
- vastly improved signal qualities
- long-baseline interferometry
- medium to high resolution of the Earth's surface from geosynchronous altitude.

One promising approach for accomplishing high-precision station-keeping at geosynchronous altitude is to use electrostatic (Coulomb) forces activated by adjusting the electric charge on each spacecraft making up the array. Free-space laser relay crosslinks are envisioned as a critical subsystem of distributed arrays, supporting data rates in excess of 100 Gbps.

Decision-makers must weigh the costs and risks versus the benefits of pursuing satellite constellations for multiple applications as they parse out the limited funds for space systems research in the years ahead. Over time promising architectures will evolve. The principal question is, over *how much* time?

Defense applications

Technological advances stand to benefit civil–military, scientific, and commercial activities in space alike. It is in the military arena, however, that controversy within the international community is most likely to arise, principally as launch costs drop and access to space increases. On a positive note, to the extent the governments of the world, especially the space-faring ones, can agree on ways to avoid arms competition in space and continue to collectively exploit its advantages, the greater the overall net benefit they can achieve in the ongoing struggle against terrorism. Terrorists have no meaningful space capability and don't have the wherewithal for securing it. Nor do they have any hopes of achieving the ability to attack any country's space assets, with the possible exception of a ground station or two, after which intense reactive security measures would preclude any sort of repetition.

In the United States, however, there are currently several areas of potential controversy with respect to space operations by the military and national security components of the government. Current US policy is essentially to ensure unfettered access to space and operational use of space. Assets of strategic importance to many countries operate in space, and to date unregulated operations by all in the space community has been the accepted practice – even during times of conflict. As the only global superpower, the United States generally has an asymmetric advantage with respect to both terrorists and other nations, and there is a strong incentive to protect this advantage. Since the on-orbit random failure of spacecraft can mask deliberate interference by an adversary, either from the ground or from space, the United States views it as absolutely critical to be able to differentiate between random component failures; inadvertent interference (e.g. ground-based commercial transmitters); deliberate denial-of-service attacks; and intentional attacks meant to permanently disable space assets so as to avoid unsettling ambiguity in nation-to-nation relations regarding space.

As a consequence, some of the most critical areas of technology development are those techniques and means that provide space situational awareness (SSA). This includes sensors that define the environment around a satellite and detect the presence of electromagnetic radiation (RF, optical wavelengths, gamma and x-rays, etc.) and sensors that detect and identify other objects in space that could represent a threat. While historically the threat has often been perceived to be from kinetic energy impactors or covert space mines (particularly nuclear), the future threat may involve highly sophisticated attacks from relatively small and inexpensive satellites that have the ability hide in close proximity to national assets. Therefore, another important SSA capability is to be able to perform autonomous rendezvous and proximity operations (RPO) in the vicinity of space assets – particularly at geosynchronous altitudes – where objects cannot be detected by ground-based observers.

The inability to remove ambiguity associated with the failure of on-orbit assets or temporary loss of service has the potential for very negative consequences. Failure to understand a problem could lead to the temptation for the United States to seek greater military advantage in that realm, through a combination of offensive and active and passive defensive measures.

Least alarming to the international community would be passive defenses, which could involve hiding, hardening, evasion, and replacement strategies, facilitated by the ability to employ additional mass for coatings, shielding, and divert propellants and to quickly launch a replacement for a negated satellite. Active defenses would entail applying force to attacking weapons, such as "killer" satellites, to negate them, but such a capability is not far removed from the ability to attack other types of satellites. The next step, a "no-holds-barred" approach to military advantage in space, would bring in offensive capabilities, such as the United States' own killer satellites. An international view articulated by Nancy Gallagher is that such a policy of coercive prevention

> could provoke a major international policy confrontation in which the United States would be isolated unless it restores a diplomatic

dimension to its space security policy and considers more collaborative steps to protect its own space assets without threatening other countries.

(Logsdon and Schaffer, 2005: p. 24)

The counter argument is that as the only superpower willing to expend lives and resources to police the planet and attempt to provide some semblance of the rule of law in disparate parts of the Earth, the United States will take whatever measures it feels are essential to national security. Fortunately, more efficient launch systems lessen the potential mass advantage of offensive and active defensive approaches to space warfare and tilt the scale more toward the neutral point where passive defenses become more competitive and the likelihood of alarming the international community declines.

The ever-growing equities of the commercial sector in space must not be overlooked in this context. Concerns would undoubtedly arise among commercial satellite owners in the face of an all-out struggle for space supremacy by competing powers. Would commercial satellites become targets? Who would defend them?

Hopefully the governments of the world's space-capable nations will collectively realize the serious disadvantages of striving to achieve offensive weapons dominance in space and will refrain from doing so. Their comforting alternative may well be prudent passive defenses, facilitated by advancing space launch technologies.

Interplanetary space exploration

Interplanetary space exploration (ISE) traditionally responds to mankind's quest for knowledge about the solar system, though it is expanding to include objects around nearby stars in our galaxy – other solar systems as it were. It is funded primarily by government. In the United States, NASA's FY-2004 budget identifies approximately $4 billion for space science, of which about half is for solar system exploration, including Mars but excluding Sun–Earth connections.

In theory, ISE can be accomplished by unmanned flyby, one-way lander, round-trip lander spacecraft, and manned spacecraft. Additionally, Earth orbiting and ground-based telescopes can provide valuable data. Objects for study, besides the planets, include planetary rings and moons, asteroids, comets, objects residing in the Kuiper Belt and the Oort Cloud farthest from the Sun and, as noted above, planets orbiting nearby stars.

Given the non-defense, non-commercial nature of ISE, it is not a powerful technology driver. It exploits basic space technologies developed for more compelling missions to a great extent, such as launch systems, information processors, and communications. The innovations of ISE focus on the scientific sensors that collect the key mission data and on a few specialized technologies such as long-life, non-solar power sources and robotics. Additionally, in an overall sense, great attention is paid to ensuring that all components, aside from the launch system, function either continually or intermittently over long periods of time in the harsh deep-vacuum, low-temperature, high-radiation environment of space and planetary surfaces.

Robotics is a technology that has wide application in unmanned interplanetary exploration and, perhaps somewhat surprisingly, may have extensive application in future manned space missions back to the Moon and possibly to Mars. Robotics profits from a high commercial demand, and thus requires less investment from the space community than might otherwise be the case. Currently over 160,000 robotic devices are in use in US manufacturing, which places the United States second in robotic use, after Japan. China is rapidly expanding its use of robots. This extensive commercial development, combined with the previous work by NASA and JPL, implies that advances in robotic technology are likely to well match the requirements of future space missions. Current spacecraft are basically robots on orbit, and their autonomy is increasing every day. Obviously, robots are better suited than humans to function in the harsh environment of space and on the Moon, Mars, and other planets. Robotics has already served as a key enabler in the detailed exploration of Mars, and it is sure to be used extensively in future exploration missions of this nature.

One concept is to combine the advantages of manned and unmanned missions on the same mission. For example, on a mission to Mars, spacecraft with supplies and robots would be launched well in advance of the manned spacecraft. Since the supplies and robots are not affected by the radiation of the trip, their spacecraft could use a more efficient rocket engine. The manned spacecraft, in turn, would carry less instrumentation and travel to Mars in a faster orbit to reduce the radiation exposure time. The humans on the mission, in addition to their own exploratory tasks, would each act as an "element leader" of 5–10 robots. When the robots needed repairs or redirection, the humans would step in. Such a mission could accomplish the equivalent of a large manned mission.

The 2006–2015 New Horizons flyby mission to Pluto and beyond readily illuminates the challenges of ISE. Launched in January 2006, the New Horizons observatory carries seven scientific instruments and will traverse three billion miles of space over a period of ten years or more, and arrive at Pluto within a 120-mile circle at precisely the right time for an alignment of two planets, a moon, the Sun, and an Earth-bound network of antennas. The spacecraft's instruments comprise an ultraviolet imaging spectrometer, a high-resolution optical telescope, a combination optical/ infrared imager, two particle detectors, a radio transmission analyzer, and a dust particle counter. These sensors will take photographs, map terrain, analyze the atmosphere, sample space dust, and measure the solar wind. The data they collect will be relayed back to Earth over a nine-month period as the spacecraft continues onward toward the Kuiper Belt, where additional data collection will be attempted. Clever use of a gravity-assist trajectory past Jupiter will shorten the travel time to Pluto by at least three years. Nevertheless, while traveling between Jupiter and Pluto, the spacecraft will be in a dormant state lasting eight years during which it must monitor and regulate itself in a carefully structured manner that avoids self-destructive processes when addressing anomalies. Propulsion limitations preclude New Horizons going into orbit around Pluto.

Few would argue that only manned missions are capable of answering the questions about our solar system. Nevertheless, a steady stream of studies continues to focus on returning man to the Moon and, for the first time, sending him to Mars. The radiation environment in interplanetary space is very hazardous to living organisms, comprising galactic cosmic rays (GCR), made up of protons, alpha particles, and high-energy, high-charge (HZE) particles, as well as solar particle events. Such radiation differs drastically from the x-rays and gamma rays reaching the Earth's surface, producing more complex types of DNA and cellular damage that may lead to adverse health risks, even at low doses. Without mitigating steps, the radiation exposure experienced by astronauts going to Mars, exploring the surface, and returning might well be close to or above acceptable limits. Under NASA's Space Radiation Health project research is underway to investigate radiation-caused chromosomal damage and mutations; damage to eye, skin, and brain tissues; and radiation effects on the formation of cataracts and tumors. On the side of mitigation, work is underway on designing improved shielding for spacecraft and spacesuits and on pharmaceutical radioprotectants, gene therapy, and dietary regimens to counteract harmful radiation effects. Much of this work is supported by refined mathematical models that address the space radiation environment, radiation transport, and DNA damage and repair.

Other technological innovations required to send humans to Mars and back include adequate propulsion for the round trip, which could require a nuclear-thermal approach, as described earlier, and a spacecraft capable of providing artificial gravity.

Interstellar space travel

Will humans ever travel to a distant solar system hosted by a distant star? The answer may be yes, but not anytime soon. To accomplish such a monumental feat, advances in power and propulsion capability well beyond anything even researched today will be required. Breakthroughs comparable to the discovery of relativity and quantum mechanics in the early twentieth century must be achieved if humans are to attempt to reach just the nearest star to Earth, Alpha Centauri, which is 4.3 light years (25 trillion miles) away. More promising destinations, i.e. those having high probabilities of solar systems similar to Earth's, are more distant, however. Two examples are Epsilon Eridani, which is 10.8 light years away, and Tau Ceti, at 11.8 light years distant. Even using the most efficient propulsion system known to man today (specific impulse of 50,000 s) requires the mass of the Sun in propellant to get to Alpha Centauri in 50 years.

The speed of light (186,000 miles per second) is the upper limit on how fast a spacecraft can travel toward a distant star. Traveling at this speed to Alpha Centauri and sending data back would entail a total mission time of 8.6 years. A more realistic scenario employs a propulsion system capable of sustaining an acceleration of 0.1 g for one year to reach a velocity of 0.1 c (one-tenth the speed of light) and coasting for 43 years to reach Alpha Centauri. Counting the 4.3 light years for the information to return to Earth, a fly-by mission to Alpha Centauri would take about 48 years. A comparable mission to Tau Ceti would require 133 years to complete.

It is generally agreed that to plan a mission of longer than 50 years to Alpha Centauri may mean that a later mission with more advanced propulsion technologies might arrive before the original one. The implication, of course, is that it would have been better to invest the time and money into improving the technologies instead of rushing the mission as soon as it became feasible. Only in the case of distant space travel do such conundrums arise!

Conventional thermochemical rocket propulsion is impractical for interstellar travel because of the combined thrust–propellant requirement for a one-year burn. NASA's former Breakthrough Propulsion Physics (BPP) program attempted to find new technical concepts that could overcome this limitation, such as propulsion systems requiring no propellant mass, propulsion technologies that attain the maximum transit speeds physically possible, and breakthrough methods of energy production to power these technologies. Some of the concepts are:

- Nuclear-electric rocket: a nuclear reactor combined with a thermal-to-electric generator powering an ion thruster. The current level of technology applied in this manner would result in a travel time to Alpha Centauri of 12,500 years.
- Nuclear-pulse rocket: small nuclear explosions (fission or fusion) impart thrust through a momentum-conditioning unit (shock absorber) to a momentum absorber (pusher plate) on the vehicle. Development of this technology is on hold, as it would violate the Nuclear Test Ban Treaty.
- Antimatter rocket: propulsive energy extracted from the annihilation of matter and antimatter. A key challenge is keeping the two types of matter apart until they are used.
- Beamed-pellet propulsion: a long electromagnetic mass driver in the solar system accelerates small pellets in a path toward the star. The interstellar vehicle rides the pellet beam using a strong magnetic field to reflect the pellets and convey momentum to the spacecraft.

- Beamed-microwave propulsion: reflected microwaves off a very large sail convey momentum to the spacecraft. This is only good for light-weight probes. Microwaves are only practical at close ranges.
- Beamed-laser propulsion: reflecting laser beams off a very large sail conveys momentum to the spacecraft.

These and other concepts are addressed in *Advanced Space System Concepts and Technologies* by Ivan Bekey and other sources.

Interestingly, some of the breakthroughs necessary for interstellar travel would totally change the quality of life on Earth. For example, a manned interstellar mission to Epsilon Eridani based on beamed-laser propulsion would require a 43,000 terawatt laser running for 1.6 years to complete the mission in 50 years. Supplying that quantity of energy would require the achievement of controlled nuclear fusion. If nuclear fusion is solved, all civilization will witness a profound increase in the availability of energy.

The magnitude of the technical problems to be solved is so great that it will probably be at least 100 years before human interstellar travel is attempted. The financial investment will also be formidable.

Asteroid defense

Defense of the Earth against near-Earth objects – asteroids and occasional comets – also demands powerful technological capabilities, both for detection and, should the need arise, effective action. An estimated 1,000–1,500 asteroids 1 km or greater in size exist in the inner solar system. Relatively small asteroids, about 100 m in size, are thought to hit the Earth every few hundred years with potentially deadly effects. One such impact unleashed the force of 1,000 atomic bombs when it struck an unpopulated area of Siberia in 1908.

The goal of the Spaceguard Survey, funded mainly by NASA, is to find 90 percent of the 1 km or greater objects by 2008. Data from this and other sources are fed into the Minor Planet Center at the Smithsonian Astrophysical Observatory in Cambridge, Massachusetts, which operates on behalf of the International Astronomical Union and serves as the central repository of information on sub-planet objects. The Jet Propulsion Laboratory's Sentry System assesses the impact potential of detected near-Earth asteroids, classifying those that exhibit a minimum orbit intersection distance of 7.5 million kilometers or less and a size greater than 150 m as potentially hazardous asteroids (PHAs). In January 2006, there were 760 PHAs being monitored. For the period from January 1 to March 11, JPL posted 24 upcoming close approaches to Earth, including object 2000 PN9, estimated to be 1.6–3.6 km in size, traveling at a relative velocity to Earth of 31.3 km/s, and predicted to pass at a distance of 7.9 lunar distances, or about three million kilometers. At any given time, JPL makes available its current estimated cumulative probability of impact of each threatening NEO and the period of time for which it applies. For example, in the period 2036–2037, object 2004 MN4, 320 m in diameter, will have two chances of impacting the Earth with a cumulative probability of about one chance in 6,000.

In 2003, NASA issued a study report addressing the feasibility of detecting near-Earth objects smaller than 1 km in size, recognizing that such objects striking Earth could cause massive regional damage, including that from tsunamis generated by ocean impacts. The study team recommended a goal of a search and detection program sufficient to eliminate 90 percent of the expected damage from sub-kilometer objects greater than 140 m in size. The envisioned sensors comprised a mix of ground-based and space-based telescopes, with no breakthroughs in technology anticipated.

The other half of the asteroid defense equation is the action to be taken against truly threatening objects. According to NASA's Bill Cooke, the notion of the populace that NASA or some other governmental entity has a means of preventing a collision by an asteroid or comet is false. Cooke states that

> Despite the certainty that an impact will occur sometime in the future, practically nothing has been done to establish the infrastructure that would be required to mitigate the calamities caused by such events. Inaction has been justified by the assumption that there will be decades, if not centuries, of warning, plenty of time to prepare a defense.

Still, various concepts have been advanced for preventing an asteroid from colliding with Earth, including nuclear explosions, kinetic impact, mass ejection, ablation, and solar pressure. For example, lasers or giant space mirrors could evaporate ice on its surface, creating jets that propel it away from Earth. Alternatively, half-painting a threatening asteroid could make it radiate heat differently on each side, slowly nudging it off course. However, techniques of this nature require several years of advance warning in order to move the asteroids into safe orbits.

If an overlooked asteroid or comet were suddenly discovered on a collision path with Earth, presenting much less time to react, a faster means of action would be required. One concept that could offer somewhat more rapid response is the space tug as conceived by Schweickart *et al.* that employs low-thrust plasma engines powered by electricity from a nuclear fission reactor to move an asteroid into a different orbit. The proposed pathfinder for a full-blown space tug is the B612 mission, which is designed to demonstrate the ability to attach an engine assembly to an asteroid and apply a low thrust over several months, sufficient to alter its velocity and spin to a convincing degree. Assuming the B612 spacecraft can use off-the-shelf power and propulsion systems and that launch can be accomplished by a single existing launch vehicle, proponents project the cost of the mission to be about \$1 billion. There is no indication, however, that this experiment is in line to receive funding from NASA anytime soon.

Another concept, described by Didier Massonnet and Benoît Meyssignac of France's National Centre for Space Studies, entails capturing a small 40 m asteroid and "parking" it at a stable Lagrange point 1.5 million kilometers from Earth, where the gravity of the Earth and the Sun balance. When an object is found to be on a collision course with Earth, the small rock could be moved into its path within eight months. This "David's stone" would be too puny to cause any damage to Earth if things went awry, claim its advocates. But other experts say the plan is not realistic. It relies on using a small hopping robot to excavate rock at tens of meters per second from the little asteroid in order to provide the force to capture it and send it toward the larger rock. The capture would take a year of digging and would require the robot to remove 66 percent of the small rock's mass, a very difficult thing to do technically, according to Dan Durda, a planetary scientist at the Southwest Research Institute in Boulder, Colorado.

More direct impact techniques promise quicker results, and one such concept would entail crashing an impactor spacecraft into a dangerous asteroid. Dario Izzo, an aerospace engineer at the European Space Agency's Advanced Concepts Team in the Netherlands, is currently formulating an experiment associated with ESA's Don Quijote mission to do this. Don Quijote is a technology demonstration mission designed to put a spacecraft in orbit around an asteroid to watch as another is sent crashing into it. Izzo's team has been working on ways to have this impactor spacecraft demonstrate its ability to deflect a threatening asteroid. As a test case, the team used the orbital parameters of Apophis, a 400 m asteroid that will pass by Earth in 2029. During that pass, Apophis may change course enough to hit Earth when it returns again in 2036 – a possibility currently seen as having a one in 5,000 chance of happening. The team found that Apophis could be deflected by a 700 kg (1,540 lbs) spacecraft launched in 2026, whose impact would change its speed by only 0.01 mm per second – a tiny change, but enough to prevent it from colliding with Earth a decade later.

NASA has not yet funded the Congressionally-mandated program recommended in its 2003 study cited earlier to eliminate 90 percent of the threat from objects greater than 140 m in size. Neither has the US government assigned any agency responsibility for dealing with any impact threat that may be identified. So the unsettling words of Bill Cooke chastising decision-makers for their assumption that there will be plenty of time to react when a real threat is finally identified seem to hold true.

Conclusions

Space exploration and exploitation are enterprises that depend very heavily on technological advances, some of which occur in response to non-space related incentives, but many of which come only as a result of substantial government investment in space. The current climate for government investment in space technology is not encouraging, presumably a result of greatly increased expenditures by the government in other sectors, such as the pacification of Iraq and Afghanistan, the war on terrorism, disaster prevention and relief, and the growth in the cost of entitlement programs. In May 2006, for example, the Space Studies Board of the National Research Council reported that NASA had not been provided with the resources necessary to carry out the tasks it had been assigned and that it was particularly concerned that the shortfall in funding for science had fallen disproportionately on small missions and on basic research and technology. Priority shifts within the defense budget also do not bode well for vigorous space technology efforts. All indications point to a continuation of this austere fiscal climate for some years to come.

The main challenge, then, is for the government to make the wisest choices in how to allocate level or declining sums of money year to year among the numerous space technology projects and proposals competing for support. This is extraordinarily difficult, in part because it requires comparisons of value, feasibility, and risk across an extremely wide spectrum of technical possibilities. How important is it, for example, to confirm the presence of water on one or more moons of Jupiter compared to, say, preparing to divert a sizable asteroid that has one chance in 5,000 of striking the Earth ten years from now? In a zero-sum or declining-sum game, each new initiative comes only at the expense of one or more existing initiatives.

To meet this challenge, scientists and technologists on the one hand and politicians on the other must continue striving to comprehend each others' interests, objectives, and obligations. Sometimes there is the happy coincidence that a politician *is* a scientist or engineer, but not often. Technical people must accept political realities, and politicians must try to overcome any aversions to technical issues and explanations. Advisory panels comprising experts with diverse interests and experience have served usefully in the past in bridging this "culture gap," and their continued use is commended. Prestigious universities, often private, with large endowments and generous alumni have the privilege of acting on farsighted proposals that excite their faculty and students, unimpeded by consideration of commercial potential or political pressures. Hopefully, the call of space exploration and space technology will continue to captivate and energize some of these institutions.

Not to be overlooked are the benefits of international cooperation in space technology, even though past experiences have not been without difficulty. As with the high-energy physics community, dividing up the research pie means that more frontiers can be explored, with more progress likely being made. Fortunately, cooperation in space has always been accepted as an international objective, serving, for example, as a stabilizing influence during some of the bleakest days of the Cold War.

Over the long-term, technology for space will continue to advance, hopefully for the benefit of the whole world. It is important in the meantime, as other priorities dominate, that space activities not lose their luster, their infusion of capable, young, highly educated enthusiasts, and thus their momentum.

Bibliography

Barter, Neville J. (ed.) (2003) *Space Data*. Fifth edition. Redondo Beach, CA: Northrup Grumman Space Technology.

Bate, Roger R., Donald D. Mueller, and Jerry E. White (1971) *Fundamentals of Astrodynamics*. New York: Dover Publications, Inc.

Bekey, Ivan (2003) *Advanced Space System Concepts and Technologies: 2010–2030+*. El Segundo, CA and Reston, VA: The Aerospace Press and the American Institute of Aeronautics and Astronautics, Inc.

Collard-Wexler, Simon, Jessy Cowan-Sharp, Sarah Estabrooks, Amb. Thomas Graham Jr., Robert Lawson, and William Marshall (2004) *Space Security 2004*. Waterloo, ON: Space Security Organization.

Humble, Ronald W., Gary N. Henry, and Wiley J. Larson (1995) *Space Propulsion Analysis and Design*. New York, NY: McGraw-Hill Companies, Inc.

Keiger, Dale (2005) "Mission: Pluto," *The Johns Hopkins Magazine*, November, pp. 26–33.

Logsdon, John M. and Audrey M. Schaffer (eds.) (2005) *Perspectives on Space Security*. Washington, DC: Space Policy Institute, Elliot School of International Affairs, the George Washington University.

Nadis, Steve (2003) "Keeper of the Objects," *Scientific American*, August, pp. 84–85.

NASA budgets. Available at: www.NASA.gov.

Near-Earth Object Program website. Available at: http://neo.jpl.nasa.gov.

Schweickart, Russell, Edward T. Lu, Piet Hut, and Clark R. Chapman (2003) "The Asteroid Tugboat," *Scientific American*, November, pp. 54–61.

Sellers, Jerry Jon (2000) *Understanding Space: An Introduction to Astronautics*. Second edition. New York, NY: McGraw-Hill Companies, Inc.

Space Radiation Health Project. Available at: www.haco.jsc.nasa.gov/projects/space_radiation.htm.

Wertz, James R. and Wiley J. Larson (eds.) (1999) *Space Mission Analysis and Design*. Third edition. El Segundo, CA and Dordrecht: Microcosm Press and Kluwer Academic Publishers.

3   Space and the economy

Bruce Linster

Although spaceflight has been predominantly a government activity since its earliest days, the commercial sector has become an increasingly important part of the space industry, beginning in the early 1980s and continuing to the present. The commercial sector of the space industry, as it currently exists, consists primarily of firms that supply telecommunication and remote sensing services via satellites. It would be difficult to overstate how important commercial space activity is to the US position as the world's premier military and economic power. American consumers rely heavily on the communications and remote sensing services that the commercial space sector makes possible. These same products and services are important to US civil and military space capabilities.

Before we can appreciate the importance of commercial space to national security, we must first understand the size and scope of the activities we have in mind. We will begin with a brief history of satellite communications and remote sensing. Next we will look at the three segments of the commercial space industry – satellite manufacturing and services, launch vehicle manufacturing and services, and ground equipment manufacturing. Then we will be able to discuss the role commercial space activities play in national security along with US space transportation policy and industrial policy in the launch vehicle manufacturing sector. We will conclude the chapter with some ideas about the future of commercial space activities.

Commercial space activities in retrospect

From the earliest days of the space age, satellites have been used for communications and remote sensing. In 1946 the concept of using satellites for communications was tested by bouncing radio signals off the Moon, and in 1954 the US Navy used the Moon as a reflector for Earth-to-Earth voice radio communications. President Eisenhower's broadcast of a Christmas message via the US Army's Project SCORE in 1958 marked the beginning of communication via man-made satellite. These telecommunications satellites are generally launched into geosynchronous Earth orbits (GEO) to maintain a constant position in the sky relative to the Earth's surface.

Private sector satellite communication was born with the design and launch of Telstar 1 by an international collaboration of Bell Labs, NASA, and French and British Telecom. By 1975 Home Box Office (HBO) began live satellite television feeds to cable affiliates, and in 1980 Ted Turner began the Cable News Network (CNN) using satellite communications to broadcast live news 24 hours a day. NASA went a step further in 1993, enabling satellite-based Internet communication by launching the Advanced Communications Technology Satellite using Ka-band frequencies that provide the foundation for such capabilities as e-learning and video conferencing.

Today, satellites routinely provide telecommunications services that include direct-to-home (DTH) television, telephony, satellite radio, and broadband Internet. Since the late 1990s new markets for commercial satellite services like mobile telephones and data messaging have appeared. In 2001 satellite radio services were first offered in North America. Remote sensing constitutes another segment of the commercial space market that is important and becoming more significant with time. The earliest remote sensing from space began in the early 1960s when Soviet cosmonauts and American astronauts took photographs from their spacecraft. Today, satellite remote sensing employs much more sophisticated optical, radar, and infrared sensors to observe, measure, and record electromagnetic signals that are either emitted or reflected from the Earth's surface.

Despite excellent and useful imaging capabilities, demand has not grown as rapidly as predicted for satellite imagery. Currently, three US companies operate remote sensing satellites – Space Imaging, Digital Globe (formerly Earthwatch), and Orbimage. These companies compete with aircraft imaging companies that are older and have a broader customer base. Internationally, these companies have to compete with commercial remote sensing from the likes of Russia, China, Brazil, Israel, India, and Turkey.

The data obtained from remote sensing satellites have been important for acquiring environmental information that has proven extremely valuable in the management of renewable and non-renewable resources. For example, satellite data have been used effectively in mineral exploration, map-making and verification, agricultural planning and management, urban planning, and natural disaster assessment.

Commercial space activities

Just like the other sectors of national space activity, the commercial sector includes satellite manufacturing and services; launch vehicle manufacturing and services; and ground equipment manufacturing. The commercial, civil, and military space sectors have worked together closely from the very beginning. Advances in the space industry have been fueled primarily by government investment. Industry and government alike expected that the commercial sector would bear a greater burden for future developments as that sector expanded. However, a downturn in the commercial satellite industry caused by the bursting of the telecom bubble of the late 1990s, along with technological advances in satellite capacity, has kept this goal from being realized. Cooperation among the sectors has allowed the industry to leverage the total research and development budget for maximal effect, and this interdependence of the defense, civil, and commercial sectors has aided the commercial sector's survival.

Five companies dominate the commercial satellite manufacturing market. Figure 3.1 shows that in 2004, US companies controlled approximately 70 percent of the global market – Lockheed Martin controlled 23 percent, Boeing had 12 percent, and Space Systems/Loral provided 23 percent of the market. Two companies, EADS Astrium (European) and NPO PM (Russian), supplied the remaining 30 percent of the market with 18 percent and 12 percent respectively.[1]

While demand for commercial launches increased steadily from the early 1980s until the late 1990s, the number has decreased substantially since then. Table 3.1 shows the overall level of space launch activity, commercial space launch activity, and the US, Russian, and European market shares for commercial launches. Note the steep decline in launch activity following the softening in the telecom market in 2000. As Table 3.1 shows, the number of commercial launches peaked at 36 in the late 1990s and has since declined to 15 in 2004.[2]

The commercial space sector continues to be plagued with overcapacity in both satellite manufacturing and launch services. Overly optimistic demand forecasts for both non-geosynchronous orbit (NGSO) and GEO

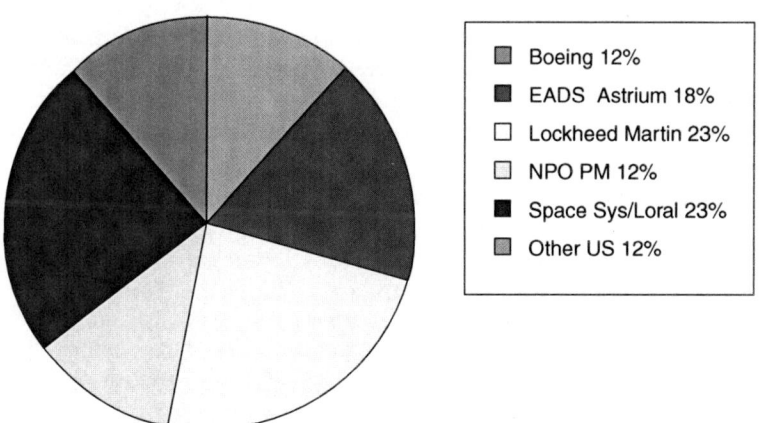

Figure 3.1 Satellite manufacturer market share (2004) (source: Futron Corporation (Bethesda, MD). Satellite Manufacturing Report (January 2005). Available at www.futron.com/resource_center/friends_of_futron/report_ archives/ satellite_manufacturing_report.htm). All rights reserved by Futron Corporation.

Table 3.1 Commercial launch market shares

	1997	1998	1999	2000	2001	2002	2003	2004
Total launches	89	82	78	85	59	65	63	54
Commercial launches	35	36	36	35	16	24	17	15
US market share	14 (40%)	17 (47%)	13 (36%)	7 (20%)	3 (19%)	5 (21%)	5 (29%)	6 (40%)
Russian market share	7 (20%)	5 (14%)	13 (36%)	13 (37%)	3 (19%)	8 (33%)	5 (29%)	5 (33%)
European market share	11 (31%)	9 (25%)	8 (22%)	12 (34%)	8 (50%)	10 (42%)	4 (24%)	1 (7%)

Source: Federal Aviation Administration, *Commercial Space Transportation Year in Review.* Years 1997–2004. Available at: www.faa.gov/library/reports/commercial_space/year_in_ review.

launches led to surplus of launch capability in the early 2000s. This surplus drove prices to as low as $50 million for a commercial GEO launch. Both Boeing and Lockheed Martin have needed government subsidization to keep production lines open.

The launch vehicle manufacturing and launch services sectors of the commercial space market are characterized by a small number of firms on both the demand and supply sides of the market. In 2004 only three companies provided launch services for US commercial launches. These were Sea Launch (SL), International Launch Services (ILS), and Orbital Sciences Corporation (OSC). The five Russian launches in 2004 were accomplished by the Russian Space Agency (RKA), and the European launch was conducted by Arianespace.

ILS is a joint venture between Lockheed Martin and Khrunichev State Research and Production Space Center. Nine of the 15 commercial launches in 2004 were conducted by ILS. Five of these launches used variants of Lockheed Martin's Atlas launch vehicle, while the other four involved Russian Proton M launch vehicles. Sea Launch is a multinational (US, Russian, Ukrainian, and Norwegian) consortium that launches from floating platforms near the equator. This allows a launch vehicle to put a given payload into a higher orbit or put heavier payloads into GEO. Sea Launch employs the Zenit 3SL for its GEO launches, and Orbital Science Corporation employs its Taurus XL to launch small payloads into low Earth orbit (LEO). Arianespace used its Ariane 5G, and ISC Kosmotras launched a Dnepr 1, which is a variant of a Soviet-era SS-18 ICBM.

Although the increase in demand for telecom, data, and entertainment services provided by commercial space activities has not translated into economic prosperity in the satellite manufacturing and launch sectors, ground equipment manufacturing has shown increasing revenue growth over the years. Ground equipment includes gateways, network operations centers, satellite news-gathering equipment, flyaways (portable satellite communications stations that can easily be transported on commercial aircraft), very small aperture terminals (VSATs), direct broadcast satellite (DBS) dishes, digital audio radio services (DARS) equipment, and satellite phone equipment. In current dollars, world ground equipment revenues have increased from $9.7 billion in 1996 to $23.3 billion in 2004. Despite declining prices for ground equipment, total revenue has continued to grow because of the large increase in the number of units sold.

In 2002 commercial space transportation and enabled industries generated more than $95 billion in economic activity, producing more than $25.5 billion in earnings and providing more than 576,000 jobs. This is up from $61.3 billion in economic activity, $16.4 billion in earnings, and employing more than 497,000 people in 1999. This growth came primarily from satellite services like DTH television, satellite radio, and broadband Internet. The shift in the composition of the commercial sector can be seen most clearly by looking at employment figures. From 1999 to 2002, employment in "launch vehicle manufacturing and services" fell from 28,617 to 4,828, reflecting the decreased demand for commercial launches, while employment in "satellite services" jumped from 186,984 in 1999 to 278,287 in 2002. This reflects a surge in the number of satellite TV subscribers, despite overcapacity in other areas like telecoms and remote sensing.

Commercial space and national security

In addition to being economically important, commercial space activities are vital to US national security, primarily in three ways. First, commercial space activities provide products and services to the Department of Defense (DoD) and intelligence resources. Second, satellite telecommunications support vital activities like transportation and banking. Finally, commercial satellites are essential to daily commerce. The United States, indeed the global economy, has become very dependent on commercial space activities, and along with dependence comes vulnerability.

Perhaps the most vulnerable of US commercial space assets is the satellite infrastructure. Like most equipment, it is subject to a variety of threats that must be addressed. For example, ground stations and equipment are targets that could prove lucrative to enemies of the United States. Additionally, computer and other communications networks may be subject to cyber attack. It is not very difficult to come up with examples of cyber attacks that have caused considerable damage. A coordinated cyber attack could be devastating if conducted by either state or non-state actors seeking to harm US interests.

The globalization of the world economy, along with new multinational alliances, raises national security questions. US firms that enter foreign markets or merge with foreign companies pose national security issues. Products may now be produced in many countries and by companies with many owners worldwide. Additionally, financial capital moves around the globe almost instantaneously as a result of the communications available because of commercial space activities. Generally speaking, these relationships are beneficial. However, it doesn't take much creativity to imagine a situation in which national security considerations arise. When implementing space policy, the US government must carefully balance the needs of the commercial sector with those of national security. The ability of US firms to compete globally must certainly be considered.

These national security threats should not be overstated though. There is certainly an upside to the globalization process that has been fostered by commercial space activities. Thomas Friedman summarizes the phenomenon in *The Lexus and the Olive Tree* when he points out that two nations that both have McDonald's restaurants have never gone to war against each other. The interconnectedness resulting from the global economies makes war almost unthinkable. The effect one problem economy will have on others was well documented during the global financial crisis that began in Thailand in 1997. Unfortunately, the major national security threat today does not come only from other states. Although many states are integrated into the global economy, some are not. States that are not vested in globalization have little to lose and may pose a serious threat by sponsoring various terror groups or allowing them to operate freely inside their borders.

The national security implications of commercial remote sensing are profound as well. Access to real-time remote sensing data can have a major impact on world events, and such information in the hands of our foes could adversely affect US national security. The security threat was obvious early on. In 1992 the Land Remote Sensing Act, along with Presidential Decision Directive 23 Foreign Access to Remote Sensing Space Capabilities, in March 1994 established the legal framework for commercial remote sensing activities. By 2000 the United States, Canada, France, Russia, China, India, Brazil, Japan, and the European Space Agency had space-based remote sensing satellites capable of high-resolution imagery. In April 2003, President Bush established a new US Commercial Remote Sensing Policy. This new guidance established policies for licensing and regulation of commercial remote sensing activities and foreign access to US commercial remote sensing capabilities. US satellite manufacturers can now sell high-resolution capable satellites to foreign entities. Clearly, national security concerns must be carefully considered. As an added safeguard, the United States has reserved the right to impose shutter controls on commercial remote sensing satellites during a national crisis. The availability of high-resolution satellite imagery means a new period of global transparency and security risks.

US policy and commercial space

On December 21, 2004 President Bush authorized a US Space Transportation Policy that outlined six goals to ensure the US ability to maintain
access to space for national and homeland security as well as civil, scientific, and commercial purposes. The commercial space sector will play an
important role in achieving these goals. Just as the commercial air transportation network is essential to achieving US national security objectives,
the commercial space transportation network will be vital to achieving the
same goals. The potential for future profit in commercial space activities
provides an incentive to develop the technologies necessary to achieve the
President's goals.

As part of the policy implementation guidelines, US government agencies are directed to allow commercial space transportation providers to
compete with government providers on a level playing field. Exactly how
this will play itself out is another question since another part of the implementation guidelines directs the Secretary of Defense to pay the annual
fixed costs for the Evolved Expendable Launch Vehicle (EELV) program.
Recognizing the importance of space launch bases and ranges, the guidelines direct the Secretary of Defense and the Administrator of NASA to
operate the facilities to accommodate users from the military, civil, and
commercial sectors.

In order to better meet the goals of the US Space Transportation
Policy, the US government is committed to encourage and facilitate a
viable commercial space transportation industry in the United States.
The policy goes on to direct agencies and departments of the US government to purchase commercially available products and services to the
maximum extent possible consistent with US law and mission requirements. These agencies and departments are further directed to refrain
from competing with US commercial space interests, unless required by
national security.

Licensing and regulation of commercial space activities are important
areas of government involvement, and the space transportation policy is
committed to providing the best regulatory situation possible. The regulatory environment facing the commercial space industry has had a stifling
effect. In 1998 Congress transferred export licensing jurisdiction to the
State Department (DoS) from the Department of Commerce (DoC)
because of technology leaks to China. While the DoC states that maximizing US competitiveness is one of its strategic goals, the DoS makes no such
claim. The result of this transfer of jurisdiction has been to hinder the
growth of the satellite and space launch services industries.

Prior to this transfer of licensing jurisdiction, US satellite manufacturers
controlled as much as 75 percent of the world market. In 1998 US producers
supplied 64 percent of the global market, but this decreased to 36 percent by
2002. In 2003, *Business Week* reported that the United States had become a

net importer of commercial satellites, and European satellite-makers are eager to exploit the perceived licensing problems, gaining at US manufacturers' expense.[3]

Jerry Grey, the Director of Aerospace and Science Policy for the American Institute of Aeronautics (AIAA), suggested in Congressional testimony that more effective implementation of the current export control legislation on satellites and space launch services would be beneficial. Since much of the space equipment and technology is "dual-use" (they can be used for military or commercial purposes), their export is heavily regulated. Grey pointed out that these export controls have virtually terminated interactions with companies in NATO nations, reducing US ability to maintain global competitiveness.[4]

One issue that requires government involvement if we are to maintain a globally competitive commercial space sector is liability risk-sharing for launch and reentry activities. This is another area identified by the Space Transportation Policy. There are two types of risks faced by satellite operators. First, there is the primary risk that a launch will fail or a satellite will not operate properly in orbit. Second, there is third-party liability risk. Here, commercial space launch companies face the risk of injury or damage to a third-party by a launch or reentering debris. Restrictions on technology transfers have resulted in reluctance on the part of foreign underwriters to participate in the launch insurance market. Space insurance accounts for an increasing share of launch costs, now about 8–15 percent. According to Futron Corporation, insurance rates increased by 129 percent in the period 1997–2001. Congress has assisted the industry's risk management efforts through the Commercial Space Launch Act which indemnifies commercial launch providers against any third-party claims above the required insurance coverage. This protection is set to expire in 2009.[5]

Another commitment made to the commercial space sector by the President in the US Space Transportation Policy is to provide "stable and predictable access" to space launch and recovery facilities on a direct-cost basis for commercial launches. In addition, the policy document directs US agencies and departments to encourage state and local governments as well as the private sector to invest and participate in the development and improvement of the space infrastructure. An excellent example of such a government–private partnership is the Southwest Regional Spaceport (Spaceport America), which is under construction near Upham, NM. In addition to the commercial prospects, the new spaceport will be able to support US military and civil space activities.

Finally, the US Space Transportation Policy directs that the private sector become more involved in designing and developing capabilities to meet US government needs. As an incentive to achieve this end, expanded retention of technical data rights in acquiring space transportation capabilities will be provided for the private sector.

Industrial policy for commercial launches

One of the most contentious issues in commercial space right now is the role of the EELV program in the market for space launch vehicles. The EELV program is the result of the space shuttle program's failure to reach its goal of 50 launches per year. The US government originally planned to use the space shuttle as the launch vehicle of choice for any type of launch. Of course, private firms could build their own launch vehicles and compete with the space shuttle, but the high level of funding and subsidies for the space shuttle program made success for a competitor very unlikely. The Commercial Space Launch Act (CSLA) of 1984 leveled the playing field somewhat, but the shuttle program was still required for government payloads and the possibility of successful commercial competition with the space shuttle was still very remote. Also in 1984, the Air Force began the Complementary Expendable Launch Vehicle (CELV) program. The vehicle was to be complementary in the sense that it would provide an alternative launch system for ten flights each year.

The tragic loss of the *Challenger* in early 1986 changed the landscape dramatically. The Air Force and NASA initially joined forces to develop a cheaper alternative for space launches. The Advanced Launch System and the National Launch System were proposed but cancelled in their early stages. The EELV program began in 1994 with the intention of developing a new booster for military, civil, and commercial space launches. As initially conceived, the EELV program would result in one company winning the competition and providing all launch services. The predicted growth in the commercial launch industry appeared able to support two companies rather than one, so the competition ended with McDonnell Douglas (now Boeing) and Martin Marietta (now Lockheed Martin) producing the Delta 4 and Atlas 5 rockets, respectively. Stiff competition from Europe, China, and Russia, along with the overall decline in demand for commercial launches, has continued to strain the EELV program, and now Lockheed Martin and Boeing intend to form a 50–50 joint venture to solidify control of the launch market. Although the proposed joint venture has been approved by the European Union, the US Federal Trade Commission has not ruled on the case yet.

The US government has paid dearly for the EELV program to ensure access to space. In an effort to maintain the viability of the launch market in this time of overcapacity, the government has had to increase payments to Lockheed Martin and Boeing by approximately 50 percent. In order to cover fixed costs that are now spread over fewer launches, the payments will increase to "as much as $135 million per launch from $91 million," according to the Air Force Deputy Director of Space Acquisition, Richard McKinney.[6] The DoD sees these increases as necessary to keep the companies viable and in the military launch program. Interestingly, the DoD has reversed course from the initial plan for only one EELV provider to making multiple launch vehicle providers part of DoD policy. One of the lessons from the *Challenger* disaster is that relying on one space launch vehicle is risky.

In October 2005, Space Exploration Technologies (SpaceX) Corporation filed a lawsuit charging Lockheed Martin and Boeing with antitrust and racketeering law violations in response to what it sees as abuses in relation to the EELV program. SpaceX, founded in 2002 by Internet entrepreneur Elon Musk, filed the suit after the Air Force decided to scrap its price-based competition for a recent round of EELV launches in favor of a division of the market between Lockheed Martin's Atlas 5 rocket and Boeing's Delta 4.

Issues in commercial space

The satellite services sector has been plagued with overcapacity problems along with the launch services sector. While demand for satellite bandwidth increased by 31 percent from 1995 to 2003, the supply of bandwidth available increased 54 percent. This has resulted in a substantial excess of supply. Despite forecasts of bandwidth demand increasing at a rate of more than 4 percent per year, the overcapacity gap is not predicted to close until 2011 according to a Futron Corporation report.[7] The rapid growth of bandwidth demand will not translate into corresponding increases in satellite manufacturing and launch activity, however. Because of an increased number of transponders per satellite and longer satellite service life, communications satellites are now nine times more capable than those produced and launched 15 years ago.

The satellite communications industry is changing rapidly. Three trends have become apparent in the industry according to a Futron Corporation report. First, the share of total satellite capacity dedicated to broadcast services is predicted to increase to 60 percent in 2010, from a baseline of 15 percent in 2002. Second, while satellite communications capacity was about evenly split between C-band and Ku-band in 2002, by 2010 they will make up less than half of the overall capacity. Finally, the proportion of fixed satellite services carrying digital high definition (HD) signals will increase from about 10 percent in 2002 to more than 60 percent in 2010.

All three DTH satellite television providers are increasing their capacity dramatically. EchoStar has acquired the use of five more satellites to enhance its capacity. DirecTV launched two new Ka-band satellites in 2005. These additions gave DirecTV the capacity for more than 500 local HD channels. DirecTV completed a successful launch in July 2007, adding the capacity for more than 1,000 local HD channels and more than 150 national HD channels. These new satellites will give DirecTV a tenfold increase in bandwidth from what they currently use. The third DTH provider, Cablevision's VOOM HD satellite television service, ordered five new satellites that will give it coverage of the entire United States including Alaska and Hawaii. With relatively flat predictions for fixed satellite service capacity over the next five years, these increases in DTH services will result in DTH capacity over the United States increasing to about 60 percent.

A Futron Corporation report predicts this growth in the DTH television capacity will largely be in Ka-band frequency satellites. The increased capacity for both EchoStar and DirecTV will come from Ka-band satellites. While there were very few Ka-band frequency satellites available in the past few years, by 2010 more than half the capacity over the United States will be Ka-band. This increase will be important in the transformation of DTH television services from analog to digital television. Futron Corporation predicts that by 2010 more than 60 percent of the demand for broadcast and cable capacity will be for digital signals.[8]

The outlook for commercial space

The commercial space industry has a number of problems it must overcome in order to reach its full economic potential. First, there is a need for industry-generated standards in order to facilitate design, development, and space operations. Also, the industry must seek customers from outside the industry itself or traditional groups – NASA, DoD, and NOAA. Some new markets – space passenger travel, microgravity processing, and solar power generation – are currently being explored. Expansion of the market for commercial space services is perhaps one of the most promising avenues for developing commercial space transportation to its full potential.

In June 2004 the first privately funded manned vehicle was launched into space. The first commercial astronaut wings were presented to Mike Melville, pilot of SpaceShipOne, after he achieved an altitude of 337,500 feet. In October the second set of commercial astronaut wings was earned by Brian Binnie after a flight on the same vehicle. This was the flight that earned Burt Rutan the $10 million Ansari X-Prize, which was awarded to the first company to launch a vehicle that carried the equivalent of three people to space and returned twice in a two-week period. This accomplishment marked the beginning of a new era in commercial space transportation.

What was once thought to be only a distraction from the real issues facing the space industry is now seen as an important area of potential growth for the industry. So far, three people have paid $20 million each to travel to the International Space Station and back, and two more have paid for the trip and are awaiting their opportunities. Space Adventures, the company that owns the right to sell the open third seat on Russian Soyuz spacecraft is negotiating with other potential space tourists. Space Adventures also has plans to send passengers who pay $100,000 to the edge of space for 5–10 minutes of weightlessness aboard a five-passenger spacecraft from a spaceport it plans to open in the United Arab Emirates by 2009. Additionally, Virgin Galactic is accepting $20,000 deposits for

$200,000 tickets for sub-orbital space rides in 2008. Virgin Galactic has reported that up to 38,000 people from 126 countries have reserved seats. It has been estimated that 40 percent of people would like to fly in space, and with approximately 800 billionaires and 20 million millionaires, expected space passenger demand of 5,000 per year does not seem unreasonable. Other companies set to enter the market are Rocketplane Kistler – who expect to fly in late 2007 – and PlanetSpace – who expect to enter the space tourism market in 2008. All signals indicate growth in the sub-orbital space tourism market as entrepreneurs around the world attempt to open space to passenger travel.

In addition to space passenger travel, private industry has contributed to the development of new launch vehicles. SpaceX Corporation claims that its Falcon 1 subscale technology testbed rocket has the lowest cost per flight in the world for a production rocket, and it is entirely American built. The Falcon 1 is also the only semi-reusable rocket in the world other than the space shuttle. Follow-on models Falcon 5 and Falcon 9 are designed for medium and heavy lift to carry people or large satellites. Both the Falcon 1 and Falcon 9 are scheduled for launches through 2008.

Despite many challenges, research continues on the economic viability of commercial space missions providing platforms for research and development in certain industries. The possibility of creating and manufacturing new materials in space's microgravity environment is exciting. Currently, launch costs are too high for such a plan to be commercially viable.

Space could also provide virtually unlimited clean and cheap sources of energy. One plan is to allow solar arrays in orbit to collect energy and send it to Earth using microwave lasers. This would involve very complex and difficult problems that require substantial technological progress. Although progress has been made, there are many hurdles yet to overcome. Once solved, however, these technological issues will have applicability in many areas of space transportation. Right now, whether or not such a plan is possible is an open question.

Summary

The commercial sector of the space industry continues to play an ever-increasing role in national security policy. Increasing globalization made possible through satellite communications and increased transparency through the proliferation of remote sensing capabilities present security planners with new problems and opportunities. Despite the current over-capacity in launch vehicle and satellite manufacturing and services, the commercial satellite sector has continued to grow. While launch vehicle and satellite manufacturing have been stagnant or shrinking in terms of economic importance, the satellite services sector has experienced extremely rapid growth. Services like DTH television and satellite radio have generated increasing revenues despite much lower prices. The interconnectedness of the global economy that has made US business more dependent on foreign governments and companies has both generated rapid growth and increased vulnerability. The development of remote sensing capability has also been a two-edged sword. National security planners will be faced with a delicate balancing act trying to assure national security while respecting commercial interests.

Government policy changes are necessary for the commercial space sector to achieve the kind of growth that is possible. Regulation and licensing have been stumbling blocks in further development of markets for US commercial space companies. Also, since there are extraordinary benefits to the US government from successful American space companies, some form of risk sharing is essential for the industry to grow as rapidly as possible. Removing sunset clauses on third-party indemnification rules would go a long way in ensuring the stability of the market. Finally, policy-makers should encourage commercial space companies to expand their markets. Passenger space travel, microgravity environment research and manufacturing, and the possibility of clean and cheap energy are promising areas that should be explored and promoted.

Notes

1 Futron Corporation, *Satellite Manufacturing Report* (Futron Corporation).
2 The data used to populate Table 3.1 were extracted from the FAA's publications, *Commercial Space Transportation Year in Review* for the years 1997–2004.
3 Stan Crock, "Commentary: What's Shooting Down Satellite Sales," *Business Week*, August 4, 2003.
4 Jerry Grey, "Statement to Subcommittee on Space and Aeronautics Committee on Science U.S. House of Representatives on Barriers to Commercial Space Launch," June 10, 1999.
5 Futron Corporation, "Satellite Insurance Rates on the Rise – Market Correction or Overreaction?" July 10, 2002, available at: www.spacedaily.com/news/futron-insure-2002.pdf.
6 Tony Capaccio, "Payment Rates Rise for Boeing Launches," *Seattle Post-Intelligencer*, February 26, 2004, p. 1.
7 Philip McAlister, "The State of the Space Industry," Futron Corporation briefing, January 8, 2004 in Industrial College of the Armed Forces, *Spring 2004 Industry Study Final Report: Space Industry*.
8 Futron Corporation, "The Transformation of the Satellite Services Industry," White Paper, January 2005, available at: www.futron.com/pdf/resource_center/white_papers/Transformation_SatelliteServicesIndustry.pdf.

4 Understanding space law
Legal framework for space

Jonty Kasku-Jackson and Elizabeth Waldrop

The desire to define and codify legal principles applicable to Man's activities in space began in the context of the Cold War, during which time the governments of the Soviet Union and the United States developed and operated many military satellites and dominated the world's space activities. During these early years of the space age, satellites were mainly useful in maintaining peace and stability through reconnaissance, intelligence-gathering, early warning, and as the national technical means (NTM) of verification for monitoring arms control compliance. Viewed in this context, it is thus not surprising that space law is a relatively new, specialized body of international law that is very permissive for national security space operations. This chapter will first summarize the main principles of space law as reflected in the major international space treaties, examining these concepts in the context of other general international law principles that may impose additional restrictions on the use of space for national security purposes, particularly the use of force in space. The second part of this chapter will discuss US domestic law and policy that further shape how the United States cooperates with others on the use of space. Included in this section is an overview of US domestic laws associated with commercial space activities, since the US domestic commercial space industry is essential to meeting national security requirements as well as international legal obligations. It should first be noted that although US domestic commercial space-related legislation is necessarily restricted to what is allowable within the broad architecture of the previously discussed international laws, domestic laws are primarily enabling – not restrictive – in the sense that they promote domestic investment and cooperation on a scale that would not otherwise occur. Finally, the chapter will identify some space law-related issues about which nations do not agree and which are or may be the source of international and legal conflict in the future.

Fundamental principles reflecting agreement among states

General international law

International law reflects many space law principles generally accepted by the international community. Before we examine the four major international law treaties that apply specifically to space, we must briefly outline some general principles of international law.

There are two primary sources of international law: customary law (consensual principles that have evolved from the practices and customs of nations over time), and international agreements (those things which nations have explicitly agreed to in a convention, treaty, or agreement). Under international law, the terms "treaty" and "international agreement" are synonymous, although the terms do have different meanings within the US Department of Defense (DoD).[1] For the United States, treaties are concluded under the authority of the Constitution, Article II, which states that the president has the power, by and with the advice and consent of the Senate, to make treaties, provided that two-thirds of the senators present concur. As such, treaties are part of the "supreme law of the land" under the Constitution, Article VI. Generally, treaty terms take precedence over conflicting US statute terms. The major exception to this is when Congress explicitly intends for a later statute to override the conflicting treaty provision.

Since this is not a detailed chapter on international law, we can only summarize general international law principles important to our review of space law:

1   During time of conflict, treaty terms that are inconsistent with a state of armed conflict may not apply between belligerents, unless the terms of the treaty itself are specifically intended to apply during conflict (for example, the Geneva Conventions)[2]. Thus, many space law treaty provisions might not apply between belligerents during armed conflict.
2   States assume legal obligations only by affirmatively agreeing to do so or, arguably, by acquiescing by silence to activities of another state (lack of protest to known activity).[3]
3   Generally, and with some limits, activities are presumed to be allowed unless prohibited by law. This is the US view, and it is admittedly controversial in the international arena.[4]

The four major space treaties

There are four main treaties that make up the specialized body of space law: the Outer Space Treaty (1967), the Rescue and Return Agreement (1968), the Liability Convention (1972), and the Registration Convention (1975).[5] The United States and all other major space powers are party to all four of these treaties. Most of the principles in these treaties are generally accepted as customary international law binding on all nations, even those nations that are not party to them.

The Outer Space Treaty is the cornerstone of space law and sets out its major guiding principles: the common interest principle (Article I); the freedom principle (Article I); and the nonappropriation principle (Article II). These principles taken together establish the general idea that outer space (including the Moon and other celestial bodies) is not and cannot be owned by anyone, but that everyone is equally free to use it.[6] Another powerful provision of the Outer Space Treaty is the statement that international law, including the UN Charter, applies in outer space (Article III).

As noted previously, in the earliest years of the space age satellites were mainly useful in maintaining peace and stability through reconnaissance, intelligence-gathering, early warning, and as the NTM of verification for monitoring arms control compliance.[7] In part to assure the continued availability of satellite reconnaissance (especially of the very-secretive Soviet Union during the Cold War), the United States had a strong interest in establishing early on that the law of space is different from the law of the air, with perhaps the most important distinguishing aspect being a "right of overflight" by satellites over the territory of other sovereign nations (the opposite of existing air law, which recognizes sovereignty over a state's territory).[8] This concept of an outer space "right of overflight" was effectively established through US and Soviet satellite operations with no formal opposition from other states, and the concept was formally recognized in the 1967 Outer Space Treaty.[9]

Some of the greatest misconceptions about space law, however, concern limitations on weapons in space. In fact, the Outer Space Treaty only provides two "arms control" provisions limiting military uses of space:

1   nuclear or other weapons of mass destruction will not be placed in orbit around the Earth, on the Moon or any other celestial body, or in outer space; and
2   the Moon and other celestial bodies will be used exclusively for peaceful purposes; establishing military bases, testing weapons of any kind, or conducting military maneuvers on the Moon and other celestial bodies is forbidden.[10]

Consequently, ICBMs carrying nuclear warheads can traverse space without violating the treaty – they don't go into orbit, and they aren't installed or stationed in space or on celestial bodies. In addition, there is no prohibition against anti-satellite weapons (ASATs).

However, there has been much debate about the Outer Space Treaty's statement that the Moon and other celestial bodies must be used only for "peaceful purposes." It is from this language that other states and scholars have argued that space is a "sanctuary" that should be protected against weaponization. In reality, recent years have seen a continuous escalation of the uses of space for national security purposes. As space powers reiterate their commitment to the use of space for "peaceful purposes," they also now routinely and overtly use satellites and space systems in direct support of military operations, stating that this direct support is "peaceful."[11] Such direct support includes the use of satellites for: communications between forces engaged in armed combat; intelligence-gathering for selection of targets; precision guidance systems to accurately steer weapons to their targets; and data-collection by remote sensing for battle damage assessment. These uses, coupled with a lack of formal protests regarding them, have led some experts to conclude that all military uses of space other than those specifically prohibited by treaties are lawful, so long as they do not violate other international law provisions.[12]

Thus, the definition of "peaceful" seems to be expanding according to state practice. For example, for over 40 years the United States has defended the position that "peaceful" means "non-aggressive," so that any military use is lawful so long as it does not violate either Article 2(4) of the UN Charter, which prohibits "the threat or use of force," or Article IV of the Outer Space Treaty.[13] Under this interpretation the development and deployment of weapons in space, as long as they are not weapons of mass destruction prohibited under Article IV, and if they are used for "peaceful purposes," would not violate the Outer Space Treaty.[14]

Further, the Outer Space Treaty contains a provision that "[i]n the exploration and use ... parties ... shall conduct all their activities ... with due regard to the corresponding interests of all other states." In the exploration of outer space (including the Moon and other celestial bodies), states must "avoid harmful contamination." In addition, there is an obligation for international consultation if a state's space activity could potentially cause harmful interference with the space activities of other states (Article IX).

The other space treaties expand on concepts introduced in the Outer Space Treaty. The Outer Space Treaty and the Liability Convention make states responsible and liable for all activities that occur in outer space, even those conducted by civilians and private entities. Thus, for example, if a foreign country or its nationals are damaged by the space activities of the fictional US corporation "Space Bus," that country would file its claim against the United States, not "Space Bus." The United States maintains control over this responsibility by imposing licensing requirements on commercial entities, and protects against its governmental liability through insurance requirements.

The Liability Convention further expands on the idea that "launching states" are liable for damage caused by space objects (including debris). If damage is caused to another space object *in outer space*, liability is based on fault. In other words, State A is liable to State B for damage by State A's space object to State B's space object *only if* State A was at fault. On the other hand, if damage is caused by a space object *on Earth* or to *an aircraft in flight,* liability is absolute. For example, if State A's space object causes damage on Earth to State B, State A is liable *regardless of whether State A was at fault.* However, states are liable only for direct damage caused by a space object (e.g. loss of life; personal injury or other impairment of health; or loss of or damage to property).[15] Notably, there can be more than one "launching state" – a launching state is any state that launches an object, procures the launch of an object, or from whose territory or facility an object is launched. If there is more than one launching state, the states may apportion liability between them.

While space law was first being established, astronauts were often returned to Earth in capsules that landed in the ocean and were recovered. Accordingly, it was important to the spacefaring states that provisions be made to ensure the safe return of astronauts (and the spacecraft) to the launching state. In this context, the Rescue and Return Agreement established some key principles. It requires proactive, prompt, and safe rescue and return of spacecraft personnel who land in international waters and in foreign countries. The treaty also prohibits taking such persons hostage or imprisoning them.[16] Presumably, the term "spacecraft personnel" would cover space tourists in the future, but likely not combatants in a future conflict since they would likely be governed by laws of war (such as the Geneva Conventions).

While still protected by the treaty, space objects receive less protection than spacecraft personnel. If State A's space object lands in a foreign country, State A must request its return. If State A does so, the foreign nation must take steps to recover the object, if practicable, and return it. It is important to note that there is no requirement to return an object in the same condition in which it was found; therefore, the foreign country can inspect the object, reverse engineer it, take it apart, etc., prior to returning it. The launching state is responsible for the costs of the recovery and return. If State A learns that a space object has returned to Earth in its own territory or the high seas, or anywhere not under the jurisdiction of any state, State A must inform the launching state and the UN.

The Registration Convention sets up a UN registry for space objects and also requires states to establish their own national registries. This Convention has been criticized for its "loopholes" that enable states to avoid providing detailed information about their space objects:

1   States are not required to mark the space objects with the registration number; therefore, it is not always obvious to whom an object belongs.
2   States are only required to notify the UN "as soon as practicable" after launch. The treaty does not define "as soon as practicable" – therefore, the country decides for itself when it's practicable to notify the UN, which could be years after the launch or maybe never.
3   Because the treaty only requires a general description of the function of the satellite, countries do not often provide a very helpful description of

the function of the objects (for example: USSR entry "to explore the cosmos"; US entry "to conduct practical applications such as weather or communications"). Military satellites must be registered, as well as civil and commercial satellites.

4   States are only required to provide notice on the initial orbital parameters of the object. Therefore, if the object is moved later, there is no requirement to amend the initial notification or provide the updated information to the UN.

There are also a number of UN resolutions dealing with space activities. While UN resolutions are non-binding, in some cases they reflect international consensus on international law principles or are an attempt to contribute to the formation of customary international law. Due to the brevity of this chapter, the principles will not be discussed, but we mention in passing that there are resolutions governing the following space activities: direct TV broadcasting; remote sensing; and the use of nuclear power sources.

This brief summary of major space law principles illustrates that space law is quite permissive for national security space operations. However, since international law generally applies to outer space under the terms of the Outer Space Treaty, it is important to look at other areas of international law (as well as domestic law) that may further affect or limit space operations.

Other international law impacting national security space activities

Given the backdrop of relatively permissive international space law, it is important to look at other constraints on US national security uses of space imposed by other treaties and bodies of international law. First and foremost, the UN Charter,[17] which explicitly applies to space operations under the terms of the Outer Space Treaty, contains limitations on the use of force and the right to self-defense against an armed attack, which in the US view includes anticipatory (or preemptive) self-defense. As complicated as the analysis of these terms and issues is on the ground, when these principles are applied to space, satellites, and computer networks there are many more unresolved issues.

Article 2(4) of the UN Charter prohibits "the threat or use of force against the territorial integrity or political independence of any state."[18] Thus, the first question for national security space operations is whether an action against a satellite or its communications links is a "use of force." There are different approaches to the analysis of whether an act is a "use of force" under Article 2(4) of the UN Charter. The approach most likely to be taken by US national security decision-makers is that the *effect* of the attack is what matters (i.e. whether the damage done is equivalent to that done by actual force), not the actual means by which the attack was made.

On the other hand, many in international circles, particularly academia, argue that the *means* of attack governs the issue, and that "use of force" means exactly what the plain terms indicate, using actual force.[19] A third approach would combine these two approaches in a case-by-case analysis.

There are two exceptions to the UN Charter's prohibition on the use of force: first, an action taken pursuant to a UN Security Council mandate under Article 42; and second, an action taken in self-defense under Article 51. Article 51 of the UN Charter states in part: "nothing in the present Charter shall impair the inherent right of individual or collective self-defense if an armed attack occurs against a member of the United Nations." The next obvious issue in space operations, then, is whether an attack on a satellite or space system is an "armed attack" that would trigger the right of self-defense. As guidance, most international lawyers look to the definitions of the phrases "use of force" (from Article 2(4)) and "armed attack" (from Article 51) as given by the International Court of Justice in the famous Nicaragua case.[20] Under these definitions, a "use of force" is not always an armed attack (it could be lesser acts or indirect force, such as arming and training rebels),[21] but an "armed attack" would most likely require property damage or injury to humans. For national security space operations, then, it could be argued that providing information (such as satellite imagery) to rebel forces is not a "use of force," since it is more like providing money to rebels than equipping them with weapons.

Further, most states interpret Article 51 of the UN Charter to be much more limited in its coverage than the broader right of self-defense granted to states under customary international law – the right of preemptive self-defense. The United States, however, has long maintained that so-called "anticipatory" self-defense is authorized under both customary international law and the UN Charter.[22] This view is controversial and not accepted by many UN member states.[23] Essentially, the same unresolved controversies about using force and responding to armed attacks in self-defense (whether or not preemptively) will exist in space as they do on Earth.

The law of armed conflict (LOAC, also called the "law of war") is the branch of international law regulating the use of force in armed hostilities.[24] Under the US military's standing rules of engagement (SROE), "US forces will comply with the Law of War during military operations involving armed conflict, no matter how the conflict may be characterized under international law."[25] In other words, the United States does not have to be in a declared war for LOAC principles to be binding on its military forces. Although a detailed discussion of LOAC is beyond the scope of this chapter, it is important to briefly outline its sources and general principles to understand how they may apply to space operations.

Like the rest of international law, LOAC is derived from two main sources: customary international law and treaty law. The treaties regulating the use of force were concluded at conferences held at The Hague, the Netherlands and Geneva, Switzerland and can be divided into two main areas: the "law of The Hague" and the "law of Geneva."[26] In general terms, The Hague treaties deal with the behavior of belligerents and the methods and means of war (for example, lawful and unlawful weapons and targets), while the Geneva agreements address the protection of personnel involved in conflicts (e.g. prisoners of war, civilians, the wounded). LOAC sets boundaries on the use of force during armed conflicts through the application of several principles:

1   Necessity: only that degree of force required to defeat the enemy is permitted. In addition, attacks must be limited to military objectives whose "nature, purpose, or use make an effective contribution to military action and whose total or partial destruction, capture, or neutralization at the time offers a definite military advantage."
2   Distinction or discrimination: military objectives must be distinguished from protected civilian objects such as places of worship and schools, hospitals, and dwellings.
3   Proportionality: military action must not cause collateral damage which is excessive in light of the expected military advantage.
4   Humanity: the use of any kind or degree of force that causes unnecessary suffering is prohibited.
5   Chivalry: war must be waged in accordance with widely accepted formalities, such as those defining lawful "ruses" (e.g. camouflage and mock troop movements) and unlawful treachery (e.g. misusing internationally accepted symbols in false surrenders).[27]

The combination of these LOAC principles, as implemented on the US domestic level by the SROE, imposes a legal and moral obligation to reduce non-combatant civilian casualties. In application, this can be difficult as military and civilian systems, particularly space systems, become more and more intertwined.[28]

While maintaining its own space assets and capabilities, in the past few years the US military has increasingly relied on commercial and civilian space assets, owned and operated by foreign, domestic, and even international entities. As part of a larger general trend toward military "outsourcing," such non-military organizations may provide cheap, technologically advanced space commodities in a number of areas, e.g. launch, communications, remote sensing, and weather. Even in situations in which the military relies on its own space assets (such as navigation, launch, and surveillance), partnerships with and investment in non-military (and even non-domestic) entities are common and openly encouraged. Thus, the United States must consider the LOAC implications of using its own civilian space systems for military purposes; such dual uses may turn these systems, under LOAC principles, into legitimate military targets. Likewise, the United States must be concerned with targeting adversary civilian space systems which are used for military purposes, and must consider such factors as collateral damage to civilian space users. On the other hand, space systems also provide an enhanced ability to meet these LOAC requirements (particularly necessity, distinction, and proportionality), since military use of space systems enables accurate targeting and a reduction in unnecessary civilian collateral damage.

Under LOAC principles, legitimate military targets must be distinguished from protected civilian objects. Anticipated collateral damage must be weighed against expected military advantage, and excessive civilian damage avoided. However, force may lawfully be used against objects which an adversary is using for a military purpose, if negation of the object would offer a definite military advantage.[29] The analysis becomes even more complex, however, when the object being used by the adversary belongs to a "neutral" third party.

Nonparticipants in a conflict may declare themselves to be neutral.[30] As long as the neutral state does not assist either belligerent party, it is immune from attack by the belligerents. However, if one of the belligerents uses the territory of a neutral nation in a manner that gives it a military advantage and the neutral nation is unable or unwilling to terminate this use, the disadvantaged belligerent has the right to attack its enemy in the neutral's territory.

Traditionally, the laws of neutrality did not require a neutral state to prevent its private entities from trading with belligerents.[31] However, increasing governmental control and involvement in trade led to the practical erosion of the distinction between private and governmental actors, and it is now commonly accepted that neutral states have an obligation to prevent acts of supply to belligerents by their private entities.[32] Since space law accords states responsibility over their private entities involved in space operations, an even stronger argument can be made to hold a neutral state responsible for the actions of its private entities.[33] In addition, when a state issues a license authorizing a private entity to provide certain services, there can be little argument that the state should be held responsible for the subsequent conduct of the private entity. Accordingly, if a neutral state permits its space systems to be used by a belligerent military, the opposing belligerent would have the right to demand that the neutral state stop doing so. If the neutral state is unwilling or unable to prevent such use by one belligerent, it would seem reasonable to authorize the other belligerent to prevent the offending use. In the context of space systems used in a time of conflict, before resorting to force a belligerent could (or should) demand a neutral nation not to provide satellite imagery, navigation services, or weather information to its adversary.[34]

However, belligerents may have no similar right to limited self-defense in neutral territory when the use of satellite communications systems is involved. Articles 8 and 9 of The Hague Convention V (which notably was concluded in 1907, decades before satellite communications systems were even envisioned) provide that a neutral state is not required to restrict a belligerent's use of "telegraph or telephone cables or of wireless telegraph apparatus belonging to it or to Companies or private individuals" as long as these facilities are provided impartially to both belligerents.[35] An argument can be made that these Articles would apply to modern-day satellite communications as well, but this remains an open question. In any event, scholars point out that the law of neutrality is heavily influenced by pragmatic factors such as power differentials between the parties to a conflict and nonparticipants; the intensity, time duration, and geographical scope of a conflict; and other available coercion techniques, including economic pressure.[36] There is no reason to believe that the application of the law of neutrality to space uses will be any different.

Military uses of outer space may also be limited by specific disarmament and arms control agreements. In addition to the Outer Space Treaty already discussed, the following merit mention:[37]

1   The 1963 Limited Test Ban Treaty prohibits "any nuclear weapon test explosion, or any other nuclear explosion" in the atmosphere, underwater, or in outer space.[38]
2   The Biological and Toxins Convention of 1972 and the Chemical Weapons Convention of 1992 prohibit development, production, stockpiling, and acquisition of biological agents, weapons containing toxins, and chemical weapons for hostile purposes.[39]
3   The 1980 Environmental Modification Convention prohibits all military or hostile environmental modification techniques that might cause long-lasting, severe, or widespread environmental changes in Earth's atmosphere or outer space.[40]
4   A series of bilateral agreements between the United States and the former Soviet Union (now binding on Russia) prohibit interference with early warning systems and NTMs of verification (reconnaissance and communications satellites) to reduce the risk of nuclear war and monitor treaty compliance.[41] Also, these agreements carry additional notification requirements for launches and reentry of unidentified objects from space into the Earth's atmosphere.

The United States is also party to numerous bilateral or multilateral agreements that, although not traditional "arms control" agreements, may restrict space activities by limiting certain activities from being performed in or from the territory of a state. For example, in the US pursuit of a ballistic missile defense system, it is entirely foreseeable that states could impose additional restrictions on US space activities (or use of data there from) in exchange for the US right to base ground- or link-segments in that state. The existence of such agreements and potential limitations on space activities thereby imposed should not be ignored in a discussion on national security uses of space.

Satellites require the use of communications links between space and the Earth, both for commanding, controlling, and monitoring them, as well as to get data to and from them. The International Telecommunication Union (ITU), a UN specialized agency which governs the use of the radio frequency spectrum, is therefore important to consider in an examination of national security space operations. The ITU member states, which include the United States, have established a legal regime for the radio frequency spectrum in order to avoid harmful interference among users of the spectrum. This regime is detailed in the ITU Constitution, Convention, and the Radio Regulations and is based on the main guiding principles of efficient use of and equitable access to the radio frequency spectrum and the geostationary satellite orbit (GSO).[42] To meet these goals, the ITU allocates different parts of the radio frequency spectrum to different types of radio communication services, allows member states to allot an assigned spectrum to specific users, and records the resulting frequency assignments and orbital positions. Recognizing the special importance of certain high-demand frequency bands and the GSO, the ITU regulates them slightly differently to allow more equitable access to these limited resources. For national security purposes it is important to note that while the ITU is mainly concerned with radio frequency interference, for many satellites (notably those in the GSO) the ITU also assigns physical slots, and satellites must stay within their assigned physical slots. Notably, although the ITU has no jurisdiction over the use of the spectrum for military purposes,[43] the United States implements ITU rules by domestic law and applies the ITU rules to the military.[44]

US domestic law and the regulation of commercial space

As is evident from the previous review of international law, there are few restrictions on national security space activities at the international level. This section will examine domestic law, including law associated with commercial space activities. In such a brief introductory chapter into the law affecting national security space operations, it would be impossible to discuss all relevant domestic laws. Therefore, we will discuss only some key provisions in this chapter, focusing on those that implement international obligations, those that most impact the competitiveness of the US commercial space industry in the international market, and those that impact international cooperation. In other words, we will focus mainly on those commercial space law provisions that have the greatest impact on national security policy.

In the previous section we introduced the controversial issue of weapons in space. There are no domestic laws that would further prohibit space weapons; however, US policy will drive whether the United States will pursue them in the future. It is important to note, though, that all proposed "space weapons" being considered by the United States are not for "shooting down" adversary satellites. Some new and planned space weapons systems are designed to be capable of incapacitating a satellite temporarily by degrading, denying, or disrupting its signal (in essence, jamming them). In fact, US DoD Space Control Policy states that the preferred US approach to negating space systems or services hostile to US national security interests is such "tactical denial." Tactical denial means that the denial or negation of the hostile system will have localized, reversible, and temporary effects.[45] In September 2004 the Air Force declared a new system with such capabilities operational – the Counter Communication System (CCS), which is a "ground-based deployable system designed to deny a potential enemy the use of a satellite communications system employing temporary and reversible methods."[46] The CCS can be considered a space weapon under a broad definition, even though it is ground-based, since it is designed to negate (albeit temporarily) a satellite. That the CCS has not received extensive negative media coverage is likely due to the fact that it is ground-based, an Earth-to-space weapon, and therefore "less provocative" than a space-based system.[47]

In light of this new CCS capability, it is important to discuss a US code provision that makes it a crime to interfere with satellite signals. Title 18 of the United States Code, section 1367 (18 USC §1367) prohibits interference with satellite communications. Despite the fact that military activities are not excepted from this prohibition, there is general recognition that this provision would not prohibit authorized military activities using systems such as CCS. This is because:

1   if the system is deployed overseas this provision would not apply at all, since generally US law is presumed not to apply extra-territorially unless it is explicitly intended to;
2   if the operator of the satellite consents to such interference, the activity is not prohibited; and
3   there is an implied exception for properly approved, "national security" activities.

Other US law may impact national security space activities as well, including intelligence oversight rules (under Title 50 of the United States Code) and environmental law (especially for launch and launch ranges). These areas of the law are beyond the scope of this introductory chapter. Instead, this section will now examine the laws associated with commercial space activities.

It should first be noted that although US domestic commercial space-related legislation is necessarily restricted to what is allowable within the broad architecture of the previously discussed international laws, domestic laws are primarily enabling – not restrictive. Second, although it may not be easy to see at first glance, the domestic commercial space industry is essential to meeting US national security requirements. Therefore, the laws and regulations that govern that industry are as important as international law. The following section will first examine how domestic laws reveal assumptions about the importance of the commercial space industry in meeting US national security requirements. It will then examine how the commercial space-related statutes work within the architecture established by international law and how they facilitate the US government's responsibility to meet its international treaty obligations. This covers domestic laws and regulations governing licensing and liability requirements, provision of resources by the US government, and arms control laws and regulations that impact the commercial space industry. An examination of these domestic laws will highlight the delicate balance between protecting US national security interests through regulation and facilitating a robust commercial industry, necessary to meet the US national security requirement of assured access to space. This delicate balance becomes even more complex when economic and arms control agreements are considered, as they may further impact international competition and cooperation. This, too, will be briefly explored in the next two sections of this chapter.

The two most widely known statutes which are specifically related to the commercial space industry are the Commercial Space Launch Activities statute (CSLA) and the Commercial Space Competitiveness Act (CSCA).[48] The statute, which established the National Aeronautic Space Administration (NASA) and delineated its roles and responsibilities, also discusses the commercial space industry; however, only portions of that statute (found in Title 42) that directly address commercial and private sector activities will be examined in this chapter.

National Security requirements and the commercial space industry

It is obvious from the language of these statutes that US space-related laws are in place, at least in part, to facilitate the growth of the commercial space industry. One of the stated purposes of the CSLA is to promote economic growth and entrepreneurial activity through use of the space environment for peaceful purposes and to facilitate the strengthening and expansion of the US space transportation infrastructure.[49] The CSCA finds that a robust US space transportation capability remains a vital cornerstone of the US space program and that the availability of commercial launch services is essential for the continued growth of the US commercial space sector.[50] Further, the CSLA acknowledges that private entities offer potential for growth in space-related technologies.[51] This language does seem to focus on "growing" the commercial space sector. Other language in these same laws also recognizes the importance of the commercial space industry to national security.

For example, the language of the CSLA recognizes that the private sector can complement launch/reentry services already available from the

US government (primarily NASA and the US Air Force).[52] Those complementary commercial services could also be used by the US government to launch its national security payloads. Further, the CSLA acknowledges that the development of commercial launch and reentry services helps the United States retain its competitive position internationally, "contributing to the national interest and economic well-being of the United States."[53] Additionally, it clearly indicates the United States values the commercial space industry's ability to provide launch and reentry services, as these commercial capabilities are "consistent with the national security and foreign policy interests of the United States."[54] Finally, the CSLA states that the participation of state governments in facilitating private sector involvement in establishing a space transportation infrastructure is in the national interest and of significant public interest.[55]

One of the key national security interests referred to in the above statutory language is assured access to space for US government national security payloads. The importance of maintaining assured access to space is of enough concern to the US government that it has been codified in US law,[56] which actually requires sufficient resources to sustain two families of commercial space launch vehicles which can be used to launch national security payloads. Such payloads can directly support national security requirements when they are used to support Homeland Security, the Global War on Terrorism, and theater operations. Assured access to space may also indirectly support national security requirements by allowing launch and test of demonstrator technology that could later be used on a national security payload.[57]

The statute, similar to language found in the CSLA and the CSCA, not only recognizes the importance of assured access to space, but also facilitates active support of the commercial industry that will provide that access. By requiring the US government to acquire space transportation services from commercial providers when possible,[58] these statutory provisions reflect the importance of maintaining a robust commercial launch industry. These same laws actively protect the US space launch industry in the context of international competition, mandating that NASA notify the public prior to implementing any agreement that involves foreign launch or satellite providers.[59]

Further, the CSLA and CSCA allow US government agencies to provide resources to the private sector as a means to facilitate and encourage the commercial launch industry.[60] However, even under these laws there are constraints on what a US government agency may provide to a commercial company. Property and services must be "launch" property or "launch" services. Under the CSLA "launch property" means an item built for, or used in, the launch preparation or launch of a launch vehicle. "Launch services" are defined as activities involved in the preparation of a launch vehicle and payload for launch, and the conduct of a launch.[61] In addition to the requirement that the property or services in question be closely associated with the conduct of a launch, they must be excess (launch services) or not otherwise needed (launch services and launch property) to prevent interference with the mission of the US government agency providing the property or service and to prevent a potential misuse of appropriated funds. Finally, the US government agency may only provide launch property or services for direct cost.[62] Similarly, the CSCA allows use of "space-related facilities" for direct costs only, and there are five restrictions placed on the use of "space-related facilities":

1   the facilities will be used to support commercial space activities;
2   such use can be supported by existing or planned Federal resources;
3   such use is compatible with Federal activities;
4   equivalent commercial services are not available on reasonable terms; and
5   such use is consistent with public safety, national security, and international treaty obligations.[63]

Reflecting the belief that a robust, commercial US space industry is essential to meeting US national security requirements, these statutes actively support the US commercial space industry. While such US government "facilitation" of the growth of its commercial space industry may be a good idea from the US government point of view, others may see such facilitation as a subsidy that allows US companies to compete unfairly in the international market. Surprisingly, that view is expressed by both US and international space companies.[64] This issue will be addressed in more detail later in this chapter, as we examine international economic and technology control agreements.

Domestic implementation of international obligations

Although the stated purpose of these statutes is the facilitation of a robust commercial space industry and recognition that the commercial industry is essential to US national security, the statutes also recognize that the US has certain international obligations which must be met. The previously discussed international laws establish the constraints under which a nation may conduct space activities as well as the responsibilities and liabilities a nation assumes while conducting space activities. US laws implement and reflect those constraints, responsibilities, and liabilities.

First, the "peaceful uses of outer space" discussed previously is one of the central tenets of international space law. Similarly, US laws (particularly the CSLA, the CSCA, and the statute establishing NASA[65]) restate the "peaceful uses" principle. In the US view, the use of space in furtherance of national security objectives is peaceful.

In addition, US laws reflect the national duty to comply with liability obligations under international law. US laws contain mechanisms intended to ensure compliance with these international liability obligations. For example, the CSLA regulates private entities wishing to conduct launch or reentry activities involving US citizens (including private corporations or entities) or occurring from US territory. Such entities are required to obtain insurance or demonstrate financial responsibility sufficient to pay third-party (those not involved in the launch activity) claims for damages to people or property in the event of a launch or reentry mishap. The specific insurance provision requires commercial launch/reentry providers to carry sufficient insurance or demonstrate sufficient financial responsibility to cover maximum probable loss from all claims resulting from space launch/reentry activities occurring under a US-granted license. Section 70112 of the CSLA thus recognizes the necessity for the US government to comply with its obligations under the Liability Convention and creates a mechanism which facilitates that.

However, the US government also realizes that a mishap might occur in which claims for damage exceed the level of insurance required of the commercial launch company. The indemnification clause found in §70113 of the CSLA provides a mechanism for the Department of Transportation (DoT) to request additional funding from Congress to cover those third-party claims. This statutory provision ensures funding will be available to fully compensate third-party claims for damages as is required by the Liability Convention regardless of the amount of insurance a company has purchased. While §70112 transfers a reasonable amount of financial responsibility to the entity actually carrying out the space launch/reentry activity, §70113 thus tempers a commercial launch company's financial risk. This provision preserves the commercial companies for the sake of the economy and maintaining international competitiveness. In fact, this mechanism is also recognized as essential to the competitiveness of the commercial industry in the CSCA.[66] More importantly, it also preserves assured access to space for US government national security payloads by ensuring that commercial companies remain robust and internationally competitive.

A second similar mechanism is found in Title 42, which permits NASA to provide insurance to users of its space assets on a reimbursable basis or to indemnify such users.[67] (Indemnification means to secure from harm or to reimburse for damages incurred.) The major difference between the CSLA insurance provisions and the NASA insurance provisions is that NASA may use appropriated dollars to provide insurance to users of its space vehicles and then seek reimbursement from the user. Under the CSLA the commercial or private sector entity must purchase its own insurance.[68] There are limits to this, however; NASA will not pay for any actions which arise from actual negligence or willful misconduct of the user.[69]

To implement the international law requirement that states also take responsibility for space activities, including the activities of its private entities. US law enables governmental supervision of commercial launch and reentry activities through licensing requirements. The CSLA ensures the US government has sufficient oversight and control of the launch/reentry activity to ensure the "public health and safety, safety of property, or national security or foreign policy interest of the United States."[70] Section 70104 of the CSLA requires a license for US citizens (including commercial companies) conducting launch/reentry activities, and for any launch/reentry activities that occur on US territory. Thus, companies such as Sea Launch, which launches in international waters, must still be licensed because it is 40 percent US-owned.[71] Similarly, International Launch Services, a US–Russian company, requires a license whether it launches out of Cape Canaveral Air Force Station in Florida or the Baikonur Cosmodrome in Kazakhstan.[72]

The enforcement agency for the licensing provision of the CSLA is the DoT which, in 1995, delegated licensing authority to the Federal Aviation Administration, Office of Commercial Space Transportation (FAA/AST). FAA/AST is responsible for drafting the regulations that govern the issue of commercial launch and reentry licenses and US launch site licenses.[73] However, FAA/AST relies heavily on the US Air Force to develop the safety standards in those regulations and to oversee commercial launch and reentry activities that occur on federal launch sites (Vandenberg Air Force Base, CA and Cape Canaveral Air Force Station, FL). The FAA/AST regulations apply to commercial launches that occur on non-federal launch sites[74] as well. This system ensures that some US government agency has an appropriate level of oversight and control of commercial space activities, which permits the US government to comply with its international obligation of responsibility for space activities.

Historically, the DoD was the primary developer or purchaser of both satellites and launch vehicles. To lift national security payloads into space, the DoD either developed its own launch vehicles or purchased launch vehicles from commercial companies under contracts. Liability clearly belonged to the US government, since it owned the launch vehicle; it therefore had the responsibility to indemnify launch activities conducted on its behalf by the commercial company. Accordingly, in such cases the US government exercised great supervision and effectively directed the launch activities. Therefore the US government, through DoD oversight and indemnification, reduced its potential liability exposure and ensured there was sufficient funding to pay third-party claims in the event of a mishap. Because the DoD owned the launch vehicle, indemnified the launch/reentry activity, and conducted high levels of oversight, there was no reason for any other US government agency to be involved in a DoD launch of national security payload. There is an exception, in fact, to the licensing requirement for any launch which has sufficient government supervision or oversight; such launches need not be licensed by the DoT under the CSLA.[75]

However, as the commercial launch market began to emerge in the mid-1980s, it became necessary to consolidate a myriad of government approvals and assign one US government agency the responsibility to ensure public health and safety for non-government launch/reentry activities. The licensing requirements of the CSLA cover launches in which the US government does not have interest enough to indemnify the launch/reentry activity (e.g. no US government ownership of the launch vehicle) and does not have sufficient control or supervision to be considered as essentially directing the activities. It is interesting to note that a US government agency may choose to buy a purely commercial launch/reentry service that must be licensed under the CSLA rather than supervising the launch. Many US Navy launches fall into this category. The Navy routinely chooses to "buy on orbit" – only taking control of the satellite once it is in orbit – and chooses not to be involved in the actual launch activities. Therefore, another US government agency must provide the oversight and control necessary to ensure unwarranted risk is not being assumed. For the Navy launches in this example, this oversight is accomplished through the mandatory licensing of commercial launches by the DoT under the CSLA.

Licensing also allows a state, in this case the United States, to exercise control over space activities and restrict them if necessary in the interests of national security. This is illustrated in the Remote Land Sensing Act[76] licensing provisions. In order to obtain a license, US providers of commercial imagery are constrained in the resolution of imagery they may sell on the open market due to national security concerns.[77] Commercial providers complain that such restrictions hurt their competitiveness in the international market. However, the stated purposes of the same regulation are also to encourage the development of the commercial remote sensing industry in the United States, preserve US national security, and ensure availability of collected unenhanced data to the countries that were sensed (an obligation under the non-binding remote sensing UN resolution that the United States abides by).[78]

While the licensing provisions of the CSLA ensure the United States can comply with its international obligations, those provisions do have the potential to be burdensome to the private space sector – which would be contrary to the stated purpose of facilitating a robust commercial space industry. Accordingly, in the CSLA Congress requires that commercial launch and reentry services be regulated only to the extent necessary to ensure compliance with international obligations.[79] The United States must continually balance the regulatory burden, the requirement to protect US national security interests, against facilitation of a robust commercial industry, necessary to meet the US national security requirement of assured access to space. This delicate balance becomes even more complex when international economic and arms control agreements are considered, which will be explored in the next two sections of this chapter.

Economic agreements – the World Trade Organization

While allowing the US government to "facilitate" the growth of its commercial space industry may be a good idea from the government point of view, others may see such facilitation as a subsidy that allows US companies to compete unfairly in the international market. Because the economies of the world have become so interdependent, the United States must address this concern.

The World Trade Organization (WTO), established in 1994, has the stated goal of encouraging smooth, predictable, fair, and free trade. This is accomplished through international negotiations aimed at lowering trade barriers and eliminating discriminatory treatment in international trade relations.[80] One of the primary means to meet those goals is by "substantial reduction of tariffs and other barriers to trade and to the elimination of discriminatory treatment in international trade relations."[81] The WTO "umbrella" covers trade in goods and services. Since commercial telecommunications (including those provided by satellite), remote sensing, space-based navigational aids, and space launch services are "services," they fall under the General Agreement on Trade in Services (GATS).[82] Trade in "goods," on the other hand, is addressed by the General Agreement on Tariffs and Trade (GATT). Therefore, government subsidies for the development of launch vehicles, spacecraft, satellites, and infrastructure would fall under GATT.[83]

GATS has three principles that could impact companies providing space-related services. They are: (1) most favored nation (MFN); (2) market access; and (3) national treatment. The MFN principle is a "general obligation" principle that applies to a service as soon as it is offered in a national market. For general obligations such as MFN, a state must affirmatively make an exemption for a specific service if it doesn't want the principle to apply to it. In essence, a state must "opt out" a specific service for a general obligation such as MFN to not apply. The MFN principle requires states to offer the same "deal" given to one state to all other states on a non-discriminatory basis.[84] This exemption option was the route the United States chose when it entered into each of the bilateral launch services agreements between the United States and the states of China, Russia, and Ukraine in the late 1980s and early 1990s, which set launch quotas and limited pricing. The United States specifically exempted launch services from the application of the MFN principle so that it did not have to offer the same "good deal" to everyone.[85] Those bilateral agreements have expired, but any similar subsequent agreements would still be subject to the MFN principle unless specifically exempted.

The other two GATS principles guarantee access to a domestic market regardless of how a service is provided (market access)[86] and require states to treat foreign service providers no differently from domestic providers (national treatment).[87] Unlike the MFN principle, these two principles are not general obligations and thus do not automatically apply to all services. They instead require a specific commitment by a nation to apply the principles to a particular service. Essentially this means parties must explicitly "opt in" specific services to have these two principles apply to those services. In 1997 69 WTO member states, including the United States, representing over 90 percent of the world's basic telecommunications revenues, signed the Fourth Protocol to the GATS and made specific commitments relating to basic telecommunications, including satellite telecommunications.[88] Significantly, no state has made a specific commitment for any of the other space services mentioned herein.[89] The ability to exempt a particular service from the MFN principle, and the necessity to affirmatively apply the market access and national treatment principles minimize the impact of the GATS on US commercial space industry.

Similarly, any impact to the commercial space industry under GATT may be minimal because GATT allows 75 percent of industrial research costs and 50 percent of pre-competitive development costs to be "non-actionable subsidies."[90] As previously noted, under the CSLA and the CSCA, US government agencies may subsidize the launch/reentry activities by providing property and services to commercial companies conducting launch/reentry activities. However, that could only be considered a subsidy to the launch service, which would therefore be subject to GATS, and not development of a spacecraft or launch vehicle which would be subject to the GATT cost restrictions. The United States has subsidized the development of the new Evolved Expendable Launch Vehicle (EELV) by providing $500 million each to two separate companies (Lockheed Martin and Boeing), which were expected to pay for the rest of the development costs. However, since the development costs of the EELV for at least one provider were closer to $2.5 billion,[91] the US government did not exceed the GATT thresholds. It is not likely the US government will cross that threshold in the near future for items that are intended to be offered in the international market. There is another principle that will most probably prevent GATT/GATS from having significant impact on US commercial space companies. There is a "national security exception" under both GATT and GATS, stating that nothing in the agreements shall be construed "to prevent a party from taking any action which it considers necessary for the protection of its essential security interests."[92] Not uncommon in multilateral treaties, such exceptions free states of restrictions otherwise imposed by international agreements and give them latitude to spend on their armed forces to protect the nation from foreign threats.[93] As the United States increasingly relies on services provided by its commercial space industry, the chances of using this exception increase. However, the exact definition of "essential security interests" has not yet been defined and may cause controversy when applied in the future. Indeed, some have charged the United States with protectionist actions based in part on legitimate national security concerns.[94]

For example, the United States has invoked this exception twice – once to defend the boycott against Cuba and once to defend selective purchasing against Burma – in part claiming that unilateral sanctions served US security interests by responding to human rights violations committed by the two regimes (e.g. resulting in a heavy influx of refugees from Cuba).[95] In addition to the lack of clarity in the terms themselves, most industrialized nations take the position that a state's determination of the existence of a national security interest is "self-judging," based exclusively on the discretion of the party invoking the exception, and therefore inherently non-justiciable.[96] Under this view, "[W]ithout a mechanism for a review of such actions, each nation has the sovereign right to define its own national security interests without foreign interference. In effect, it is impossible for a nation to violate article XXI."[97]

Implementation of the Defense Authorization Act of FY 2004 (modifying 10 USC §2273), which requires resources for two separate families of launch vehicles, may also cause difficulty with other domestic space companies that believe the two current EELV manufacturers are receiving an unfair subsidy.[98] The US commercial industry is likely to be further impacted as the US government decides how it will further "facilitate and encourage" the commercial space industry and domestic and international companies express their concerns about fair competition.

Arms-related agreements

In addition to purely economic agreements, arms-related technology transfer agreements also impact the US commercial space industry. Two key international agreements are the Wassenaar Arrangement on Export Controls for Conventional Arms and Dual-Use Goods and Technologies (Wassenaar Arrangement) and the Missile Technology Control Regime (MTCR), which complement each other. The Wassenaar Arrangement was created to prevent transfers of conventional arms and dual-use technologies to nations that pose security risks. The MTCR is an informal agreement among 33 nations to control the spread of missile technology that can be used to proliferate weapons of mass destruction (WMD). Enforcement of both agreements is left to participating states through national laws.[99]

Although it is not the intent of the MTCR or the Wassenaar Agreement to negatively impact national space activities, there is great potential for that to occur. For example, space launch vehicles are considered "missiles" and are therefore covered by the MTCR. Space launch vehicles may be transferred to other MTCR partner states if sufficient assurances are given about the proper use of the launch vehicle by the recipient state. However, in the past such transfers have been the source of great controversy, with some partner states criticizing others for transferring technology based on suspicion that the recipient country is trying to get launch vehicle technology to use for ballistic missile development.[100]

The United States complies with these agreements via two sets of domestic laws and implementing regulations that are overseen by two different agencies. Often criticized for its complexity, the current US export control regime reflects the climate in which it has evolved – the climate of conflict between "pro-business" and "pro-national security" advocates.[101] At the national level in the United States, the Department of Commerce (DoC) and the State Department (DoS) are primarily responsible for licensing the export of strategic goods, including space technologies. The DoS deals with those technologies which are inherently military,[102] while the DoC is concerned with dual-use items. The Arms Export Control Act (AECA) and the Export Administration Act (EAA) are the main statutes in the US export control regime.[103]

Through the AECA, the DoS licenses the commercial export of exclusively military items and related technical data. Promulgated by the DoS, the International Traffic in Arms Regulations (ITAR) are the implementing regulations for the AECA. Items such as weapons, ammunition, and civilian articles designed, adapted, or modified for military or intelligence uses are monitored and controlled if they are included on the United States Munitions List (USML).[104] Effective March 15, 1999 commercial satellites were placed on the USML (under the AECA) and therefore require DoS approval for export. The 1999 law also requires the DoD to approve any satellite export.[105]

Again, the ITAR implements the Arms Export Control Act (AECA).[106] These regulations specifically regulate the export of military or "defense" articles and services. Defense articles are defined broadly, as any specifically designed, developed, configured, adapted or modified for a military application *and* that have neither predominant civil applications or performance equivalent to that of an article or service used for civil applications, unless the item has significant military or intelligence applications such that control under ITAR is necessary.[107] A defense service includes furnishing of technical data to foreign persons in the United States or abroad. Under ITAR, software is considered to be technical data.[108]

The DoC oversees the Export Administration Regulations (EAR)[109] which implement the EAA.[110] The technologies covered under the EAA include many difficult-to-classify, dual-use items (which may or may not also be regulated under the AECA), listed on a lengthy, very technical Commerce Control List (CCL) which covers such items as high-speed computers, navigation devices, and other items which have potential military and civilian application with little or no modification.[111] Items subject to the EAR include commodities, software and technology within the United States, as well as US-originated items that are not within the United States.

The EAR does have exceptions for items exclusively controlled for export by another US agency, publicly available software and technology, and foreign-origin items that have less than a specified percentage of US content.[112] It is important to note that the EAA has a "sunset clause" (expiration date) and has not always been renewed. During those times where it has lapsed the president has typically invoked his emergency powers under the International Emergency Economic Powers Act.[113] It is also important to note that the EAA seems to be moving away from its original purpose of denying Warsaw Pact countries and China strategically important technology, and is now moving toward prevention of the proliferation of WMD and limiting the ability of certain countries to conduct or support terrorist activities.[114]

Launch technology and services and satellite technology and services may be regulated under both regimes. Since launch vehicles are in essence modified missiles they are logically covered as part of the USML and require DoS approval for export. Although the ITAR is intended to regulate defense articles that do not have predominant civil application or civil performance, satellites and associated technical data are controlled under Category XV of the ITAR. However, for the purposes of ITAR, the launch of a launch vehicle or a payload is not considered an "export," although the vehicle or payload may be regulated to prevent transfer of technology to prohibited countries.[115] Again, since 1999 purely commercial satellites have been on the USML. They are therefore subject to the ITAR and require DoS approval for export. Prior to 1999, commercial satellites and associated data were regulated under the EAR and were licensed by the less restrictive DoC. However, due to alleged transfers of sensitive technology and data to China, Russia, and the Ukraine in the 1990s, Congress transferred licensing authority to the more restrictive DoS.[116]

Before it is possible to fully understand the impact these laws and regulations may have on the commercial space industry it is essential to understand exactly what "export" means. Under the EAR, an export includes Internet or fax transmission to an addressee abroad, posting on the Internet without limiting foreign access, and release of specified technology or software to a foreign national who is not a permanent resident of the United States (such a release to a foreign national is a "deemed export").[117] Under ITAR, an export also includes performing a defense service for the benefit of a foreign person.[118] Both ITAR and EAR require a license for export to specified countries. EAR also requires an export license for transfer to specified companies/individuals, whenever the exporter "knows" or has been informed by the DoC's Bureau of Industry and Security of a link to weapons proliferation.[119] In contrast, the intended end use does not matter as to whether ITAR applies.[120]

These far-reaching restrictions on a broad range of items under EAR and ITAR mean that US commercial space companies are greatly impacted by export control regulations. Added to the fact that both the DoS and DoC may impose financial penalties as well as imprisonment for violations, it is clear that these regulations may have a chilling effect on the ability of US space companies to do business in the international market. Impacts occur across the gamut of space-related business opportunities: from research and development to partnerships, to mishap investigations. Because the definition of export includes the transfer of any information to a national of that country, whether or not the information actually reaches the government of that country, US businesses must be extremely careful in how they set up and manage any partnerships. This creates a resource impact to companies with international partners, because they must develop and manage firewalls to prevent any EAR- or ITAR-controlled information from being transferred to a foreign national. Companies must also develop and execute manufacturing license agreements and technical assistance agreements that contain ITAR-prescribed clauses and specifically describe the assistance and technical data given.[121] Further, companies must be extremely careful about the nationality and residency status of the personnel they hire.

In addition to affecting a company's business management practices, "deemed exports" also impact partnerships with universities. Commercial space companies have been known to create partnerships with university science and engineering departments. A significant number of students in these departments may be foreign nationals studying abroad and could be from countries that are on a restricted export list under ITAR or EAR. This adversely affects the participation of those students and their ability to contribute meaningfully. There is a limited research exemption to ITAR and EAR, but it only applies to basic research. Basic research is restricted to a systemic study directed toward greater knowledge or under-standing of the fundamental aspects of phenomena and observable facts, without specific applications toward processes or products in mind.[122] Additionally, the foreign student must be a legal US resident and must not be from a country on a list of prohibited countries. Because of these limitations, this exemption is probably of limited use to a commercial company interested in partnering with a university.

Probably one of the most potentially harmful side effects of ITAR and EAR is lost business opportunities. This was recognized in 1999 by William A. Reinsch, Under Secretary for Export Administration, DoC, who stated in his testimony before Congress:

> The ability of U.S. manufacturers to penetrate foreign commercial markets is dependent on their ability to secure export licenses for pro-posed sales. Industry has reported that export licensing procedures limit their ability to respond rapidly to overseas sales opportunities, putting them at a disadvantage compared with their non-U.S. com-petitors. Many of the U.S. companies in these product lines are small businesses and are thus particularly vulnerable to the effects of licens-ing delays caused by a jurisdictional transfer. In sum, interfering with the ability of U.S. firms to compete in international markets can ulti-mately threaten U.S. production capabilities in these critical, dual-use product lines and thereby hurt our national security.[123]

This expressed concern was borne out as the early 2000s saw potential satellite customers moving from US companies to foreign companies. In 2003, Arabsat awarded two new satellites to Astrium (a European company) instead of Lockheed Martin, due primarily to fear that export regulations would delay delivery. Similarly, Telesat Canada chose to award the Anik F1R satellite to Astrium.[124] China has successfully marketed its DFH-4 bus to other countries fearing US export policies, including Nigeria and Venezuela.[125] In addition to losing satellite manufacturing opportunities to foreign businesses, the US is also losing launch opportunities. China has launched numerous foreign satellites (including US satellites) from its launch sites.

Restrictive export control regulations may also potentially impact the ability of a commercial company to obtain insurance. The insurance pool for satellite launch and operations consists of a number of multinational underwriters, and there is no single company or underwriter that can underwrite the launch of a satellite. Also, the pool of underwriters available to insure space activities grew smaller after the September 11, 2001 attacks in the United States, because the same pool of insurance underwriters covers both air and space policies. The smaller pool, combined with reluctance from commercial space companies to provide information that might lead to a violation of export control regulations, has made it more difficult to adequately insure commercial space activities.

Clearly, EAR and ITAR, as they implement international export control agreements, could significantly impact US commercial space companies. While these regulations may directly stifle competitiveness of the US commercial space industry in the international market, such restrictions may also ultimately harm national security through reduced international cooperation, resulting in a less effective "engagement strategy."[126] It is not surprising that affected companies would request relief from the regulations in order to prevent economic impacts to their industry. However, the United States must balance the potential harm to national security resulting from proliferation of WMD and sensitive technologies against potential harm to national security resulting from decreased international cooperation and economic harm to US domestic space industry due to export controls.

Unresolved issues for possible future international and legal conflict

From a legal perspective, then, it is clear that space law is very permissive for national security space activities. There are also few or no enforcement mechanisms to punish violators, at least at the international level. Although this legal and regulatory permissiveness is seen as positive for the United States (at least to most of those in the defense community), many in the international community are trying to close these perceived "loopholes" in international and domestic law. These are areas which are ripe for legal conflict in the future. This section will outline some of these areas.

Weaponization of space

The 2002 withdrawal of the United States from the 1972 Anti-Ballistic Missile (ABM) Treaty[127] and recent US ballistic missile defense efforts have prompted many states and international non-governmental organizations to urge a ban on arms in outer space and/or a strengthening of space law in a new, overarching convention or treaty. The United States opposes these efforts, based on its belief that the "existing multilateral arms control regime adequately protects states' interests in outer space and does not require augmentation."[128] The United States has long refused to consider any negotiations on the creation of a comprehensive space treaty or one on space weapons. Recently, even the United States' closest allies have begun to criticize this refusal even to negotiate.

The United States has pushed space weaponization issues into the Conference on Disarmament (CD), rather than discussing them in the UN Committee on the Peaceful Uses of Outer Space (COPUOS). Two items that have been on the CD agenda for years are efforts toward the prevention of an arms race in outer space (PAROS), which would prevent weapons in space, and the Fissile Material Cutoff Treaty (FMCT). These issues are significant, because the United States has prioritized FMCT while China (with the support of Russia and Canada) has prioritized PAROS, with the result being an impasse in both. Proposals by other nations to break the deadlock over these two issues in the CD have failed.

Despite persistent objection by the United States, on June 28, 2002 China and the Russian Federation (in conjunction with the delegations of Vietnam, Indonesia, Belarus, Zimbabwe, and Syria) submitted a joint working paper titled "Possible Elements for a Future International Agreement on the Prevention of Deployment of Weapons in Outer Space, the Threat or Use of Force against Outer Space Objects."[129] The proposal, based on an earlier Chinese version, contained proposed elements for an international legal agreement to prohibit deployment of weapons in space. It would also generally prohibit the threat of use of force against space objects.

In addition to the repeated PAROS calls for an international convention to ban space weapons outright, there have also been more moderate middle-ground proposals such as those that would encourage unilateral restraint in developing or deploying all or certain types of space weapons, establish a "code of conduct" or "rules for the road," governing behavior in space and the use of weapons in space.[130] Key provisions of one such proposed code of conduct would include rules for

> avoiding collisions and dangerous maneuvers in space; creating special caution and safety areas around satellites; developing safer

traffic management practices in space; prohibiting simulated attacks and anti-satellite tests in space; providing reassurance through information exchanges, transparency and notification measures; and adopting more stringent space debris mitigation measures.

Even US allies have begun to call for negotiations on the space weapons issue, if not an outright ban on weapons. For example, on June 30, 2005 the UK Ambassador in Geneva stated

> Given the difficulty of verifying or agreeing on further legal treaties, we suggested last year in an informal setting that it might be a good idea to think about adopting "rules of the road" in space, similar to those that already exist at sea. These would not be easy to reach agreement on, but they might have immediate benefits such as reducing the risk of accidental collisions, preventing incidents, and promoting "safe passage" for satellites.

Desiring safety in commercial space operations and a strengthening of commercial space markets, even US commercial entities have entered into preliminary discussions with the US government on the mutual benefit they perceive in adopting such "rules of the road" for space operations, mainly focused in collision avoidance. China's destruction of one of its aging, yet orbiting, satellites by an anti-satellite weapon on January 11, 2007 has further focused international attention on the issue of space weapons. Accordingly, based on these widespread efforts at the national and international levels, and with the involvement of government and commercial entities in the debate, it is obvious that the controversial issue of space weapons will remain in the forefront in the international arena.

Data-sharing and space surveillance

International and commercial entities have begun to call for increased data-sharing by the United States, particularly for space situation awareness (SSA) in support of collision avoidance. A perception that the United States alone has such data and refuses to share it has spurred some in the international community to propose creation of an alternative space surveillance system to the SSA system. In addition, there have been proposals to create an international space traffic management authority that would rely on internationally created, maintained, or distributed space surveillance data. Again, safety for commercial space operations, as well as safety for national space assets, are concerns behind such proposals for data-sharing and collision avoidance.

Effective November 2004, a new law in the US Defense Authorization Act (Commercial and Foreign Entities or CFE) switched control of the distribution of US space surveillance data, orbital characteristics of spacecraft and debris, from NASA to Air Force Space Command (AFSPC). In the new legislation (which created CFE as a pilot program), AFSPC, through the CFE Space-Track website, distributes two-line elements (TLEs), satellite catalog messages, satellite decay messages, and most of the miscellaneous messages previously offered by the NASA Orbital Information Group (OIG) website. Although the data is provided with the same latency that was provided by the NASA OIG website for many years, the international community has viewed the switch from NASA to AFSPC control as further restriction on data access by commercial and foreign entities. There was strong reaction from amateur astronomers and scientists whose work depends on this data. In particular, there has been sharp criticism of restrictions on redistribution of data and analyzed data without Secretary of Defense approval, and criticism over the legislated US option to charge for the data in the future. Thus, there has been increasing concern in the international community about dependence upon the United States for such crucial information, as well as increased calls for an international collaborative effort to develop an international space-monitoring and data-distribution capability. As outlined in the previous section, there has even been discussion that such data should be used for some form of international space traffic management authority.

Near-space/high altitude operations

The term "near space" is not a legal term, since "air" and "space" are the only legally defined regions above the surface of the Earth with legal significance. Rather, "near space" or "high altitude" merely describes a new US Air Force mission area in which it is envisioned that extremely high altitude balloons or aircraft would operate to provide effects similar to those provided by satellites. "Near space" or "high altitude" may be loosely defined as that region above which most military and civilian aircraft are unable to fly, but still below the altitudes at which satellites and other space objects orbit.

There is no defined altitude where "air" ends and "space" begins. In fact, the United States has consistently resisted defining such an altitude, despite international (mainly Russian) proposals to define the boundary as 100 km. In fact, a proposal has been on the UN COPUOS agenda for the past 40 years to define this altitude without resolution, due to US opposition.

However, despite US opposition to defining the exact altitude dividing air from space, it is almost certain that current "near space" or "high altitude" operations being considered by the US Air Force will be governed by "air law" rather than "space law," due to the altitudes and technologies currently being considered. Accordingly, as with any aircraft, overflight by these "near space" technologies over the territory of another nation may be an issue, since air law recognizes sovereignty over a nation's territory.

Balloons and other lighter-than-air craft are defined as "aircraft" under Annex 2 of the Chicago Convention, which governs international air law. Annex 2 defines an aircraft as "[A]ny machine that can derive support in the atmosphere from the reactions of the air other than the reaction of the air against the Earth's surface." Since a balloon or other lighter-than-air craft derives its lift through the displacement of air, it clearly fits within this definition. This is underscored by the fact that Annex 2 promulgates rules of the air for "unmanned free balloons," which are defined as "non-power-driven, unmanned, lighter-than-air aircraft in free flight." That being the case, it would be extraordinarily difficult to contend that these technologies, which would clearly operate as aircraft, are operating in something other than airspace.

Property rights, ownership, and resources

There has been increasing interest in the international community about ownership of the Moon and exploitation of its resources, particularly as private entities consider the potential for the mining of resources in the future. In fact, private citizens and entities have begun to claim ownership of the Moon, purportedly selling acreage there. No state currently recognizes such sales as legal. However, most legal scholars recognize that the outer space property rights debate, including resource exploitation rights, should be addressed.

The basis for the legal debate originates in the terms of the Outer Space Treaty. The plain terms of the treaty only state that the Moon is not subject to claims of sovereignty by states; it does not specifically mention ownership by private parties. The Moon Treaty of 1982, which has not been signed by any space power for a number of reasons, would have explicitly clarified that the Moon and celestial bodies could not be owned by private entities.

It is generally accepted in the international community and among legal scholars, however, that the Outer Space Treaty does prohibit private ownership of the Moon and celestial bodies. First, the prohibition of national appropriation precludes national legislation that would form legal recognition of a private claim. Second, the Outer Space Treaty, by its terms, indicates that activities of a state's private entities are considered national activities for which states bear responsibility.[131] Third, the negotiating history of the Outer Space Treaty indicates that private entities cannot do what states are prohibited from doing. Further, allowing them to do so would defeat the very purposes of the non-appropriation and freedom principles. One thing is clear, that the topic of ownership and exploitation of celestial bodies will be addressed in international forums in the future. It may be that issues such as outer space property rights, which are important to civil and commercial entities, may open the door to future international space law negotiations, despite traditional resistance by the US government to opening the Pandora's box of unresolved issues that could ultimately restrict national security space activities.

Space debris mitigation

The United States is a major proponent in the international community for debris mitigation measures in outer space. NASA is the official US representative to the Inter-Agency Space Debris Coordination Committee (IADC), whose mitigation guidelines form the basis for US-proposed international debris mitigation guidelines in the UN. The current US position before the UN is that the IADC debris mitigation guidelines should form the basis of voluntary, non-binding international debris mitigation guidelines, with specific exceptions for national security or defense mission accomplishment. US national law (the CSLA specifically), national policy, and DoD policy is to minimize the creation of space debris. These requirements are consistent with the Outer Space Treaty's mandates (in Article IX) that states act with due regard for the interests of other states, "avoid harmful contamination," and engage in international consultation if a space activity could potentially cause harmful interference with the space activities of other states. As it focused international attention on the space weapons issue, the on-orbit space debris created by China's destruction of one of its aging satellites by an anti-satellite weapon on January 11, 2007 has also focused international attention on the space debris mitigation issue.

Conclusion

The legal framework of space law, both at the international and domestic levels, impacts US defense policy. At the international law level, space law is very permissive for national defense activities. Most restrictions on defense space activities come from domestic law and/or policy, although even domestic space law is primarily enabling in nature. There is a strong international movement, however, to have international law impose more restrictions on national security space activities. With the recent international and commercial focus on space weapons and debris mitigation in particular, as well as recent efforts to increase international cooperation on space activities, it remains to be seen whether the law will change to restrict national space activities, or whether the law will remain as permissive as it is today.

Notes

1 The DoD defines "international agreement" more broadly, as:

> Any agreement concluded with one or more foreign governments (including their agencies, instrumentalities, or political subdivisions) ... that: Is

signed or agreed to by personnel of any DoD Component, or by representatives of the DoS or any other Department or Agency of the U.S. Government; Signifies the intention of its parties to be bound in international law; Is denominated as an international agreement or as a memorandum of understanding, memorandum of agreement, memorandum of arrangements, exchange of notes, exchange of letters, technical arrangement, protocol, note verbal, aide memoire, agreed minute, contract, arrangement, statement of intent, letter of intent, statement of understanding or any other name connoting a similar legal consequence.

(DoDD 5530.3, International Agreements, June 11, 1987, Encl 2)

2 See infra, note 26.
3 As discussed later in this chapter, under international space law states are responsible for the actions of their citizens and private entities as regards space activities. In addition, despite general areas of agreement in international space law, there are few or no enforcement mechanisms (mainly diplomatic consultation) to punish violators, at least at the international level.
4 Permanent Court of International Justice (PICJ), Lotus case, Series A, No. 10, p. 18.
5 Treaty on Principles Governing the Activities of States in the Exploration and Use of Outer Space, including the Moon and other Celestial Bodies, January 27, 1967, TIAS. 6347, 610 UNTS 205, Articles VI and VII (Outer Space Treaty); Agreement on the Rescue of Astronauts, the Return of Astronauts and the Return of Objects Launched into Outer Space, April 22, 1968, 672 UNTS 119, 19 UST 7570, TIAS 6599, 7 ILM 151 (Rescue and Return Agreement); Convention on International Liability for Damage Caused by Space Objects, March 1972, 961 UNTS 187 (Liability Convention); Convention on the Registration of Objects Launched into Outer Space, January 14, 1975, 28 UST 695 (Registration Convention). Although there is technically a fifth space treaty (the Moon Treaty of 1979), it will not be discussed in this chapter in any detail, because it only has nine parties, none of which is a space power (however, France – which does has a viable space program – is a signatory). The United States is not a party or signatory to this treaty. Moon Treaty provisions imply requirements to transfer technology and equitably share profits from mining the Moon, for example, with developing nations: Agreement Governing the Activities of States on the Moon and Other Celestial Bodies, December 5, 1979, 1363 UNTS 3, 18 ILM 1434 (Moon Treaty).
6 See the last section of this chapter for a thorough discussion of ownership of the Moon.
7 Recognition of the important role played by NTMs was and is still evident in many arms control treaties, which explicitly prohibit interference with them. Thus, for example, the 1972 Anti-Ballistic Missile (ABM) Treaty provided for the use of NTMs (with satellite observation as a critical component) to verify compliance with strategic arms limitations. The ABM Treaty recognized the importance of the role played by NTMs and therefore prohibited interference with them. Treaty Between the United States of America and the Union of Soviet Socialist Republics on the Limitation of Anti-Ballistic Missile Systems, 23 UST 3435 (entered into force October 3, 1972, but is no longer in effect as of June 13, 2002 due to US withdrawal), Article XII (ABM Treaty); US White House press release, "Statement by the Press Secretary Announcement of Withdrawal from the ABM Treaty," (December 13, 2001) available at: www.whitehouse.gov/news/releases/2001/12/20011213–2.html. Many other treaties still in force today contain this same prohibition, for example START I, INF, CFE.

8 To complicate matters, the delineation of "airspace" from "outer space" has never been defined. See the last section of this chapter for a more thorough discussion. Throughout the space age, the US view has been that there is no real need to establish any boundary between airspace and outer space, since the absence of such a boundary has, thus far, not created any major problems, and the utmost freedom of action in the peaceful exploration and use of outer space is both necessary and desirable. However, there have been repeated attempts to define such a dividing line (for example, Russia proposes that 100 km be the dividing line). Convention on International Civil Aviation, December 7, 1944, 15 UNTS 295, ICAO Doc.7300/6 (Chicago Convention).

9 Outer Space Treaty, supra note 5, Articles VI and VII.

10 Outer Space Treaty, supra note 5, Article IV, which states:

> States Parties to the Treaty undertake not to place in orbit around the Earth any objects carrying nuclear weapons or any other kinds of weapons of mass destruction, install such weapons on celestial bodies, or station such weapons in outer space in any other manner.
>
> The Moon and other celestial bodies shall be used by all States Parties to the Treaty exclusively for peaceful purposes. The establishment of military bases, installations and fortifications, the testing of any type of weapons and the conduct of military maneuvers on celestial bodies shall be forbidden.

11 See for example the US White House National Science and Technology Council, National Space Policy (September 19, 1996), available at: www.ostp.gov/NSTC/html/pdd8.html, which states:

> The United States is committed to the exploration and use of outer space by all nations for peaceful purposes and for the benefit of all humanity. "Peaceful purposes" allow defense and intelligence-related activities in pursuit of national security and other goals.

12 See for example Ivan A. Vlasic (1991) "The Legal Aspects of Peaceful and Non-Peaceful Uses of Outer Space," in B. Jasani (ed.) *Peaceful and Non-Peaceful Uses of Space: Problems of Definition for the Prevention of an Arms Race* (New York: Taylor & Francis), p. 45.

13 Charter of the United Nations, June 26, 1945, 59 Stat. 1031, 145 UKTS 805, 24 UST 2225, TIAS 7739 (UN Charter).

14 Arguments have also been made that Article IX of the Outer Space Treaty, which allows each state party to request consultation if it believes the space activities of another state might cause harmful interference to the peaceful use of space, could be used to challenge and constrain a particular military activity. Outer Space Treaty, supra note 5, Article IX.

15 A launching state will not be liable if the state claiming damage was itself grossly negligent or committed an intentional act or omission causing the damage. Liability Convention, supra note 5.

16 Whether military personnel on board spacecraft will be protected by Rescue and Return Agreement will probably depend on the mission. If they were on a mission into space to repair a satellite, for example, they will probably be protected by the treaty. If they were engaged in military operations supporting combat or actually in combat, they probably will not fall under the treaty but would be lawful combatants, thus prisoners of war if captured by a belligerent and detainees if captured by a neutral party (governed by the international Law of War).

17 UN Charter, supra note 13.

18 UN Charter, supra note 13 at Article 2(4): "All members shall refrain in their international relations from the threat or use of force against the territorial integrity or political independence of any state, or in any other manner inconsistent with the purposes of the United Nations."

19 Thomas C. Wingfield (2006) "When is a Cyber Attack an Armed Attack? Legal Thresholds for Distinguishing Military Activities in Cyberspace," The Potomac Institute for Policy Studies.

20 *Military and Paramilitary Activities in and against Nicaragua* (1986) ICJ Rep. 14 (June 27) at 259 [*Nicaragua* v. *United States*]. In this case, the court decided against the US claim that its use of force against Nicaragua was a lawful act of collective self-defense of El Salvador. The United States had argued that Nicaraguan support (in the form of weapons and supplies) to rebels in El Salvador was an armed attack justifying self-defense. See also, Gregory M. Travalio (2000) "Terrorism, International Law, and the Use of Military Force," *Wisconsin International Law Journal*, 18: 145–158.

21 Ibid.

22 National Security Strategy of the United States of America, September 2002, available at: www.whitehouse.gov/nsc/nss.html (accessed November 3, 2004):

> The United States has long maintained the option of preemptive actions to counter a sufficient threat to our national security. The greater the threat, the greater is the risk of inaction – and the more compelling the case for taking anticipatory action to defend ourselves, even if uncertainty remains as to the time and place of the enemy's attack. To forestall or prevent such hostile acts by our adversaries, the United States will, if necessary, act preemptively.
>
> The United States will not use force in all cases to preempt emerging threats, nor should nations use preemption as a pretext for aggression. Yet in an age where the enemies of civilization openly and actively seek the world's most destructive technologies, the United States cannot remain idle while dangers gather.

23 W.A. Stafford (2000) "How to Keep Military Personnel from Going to Jail for Doing the Right Thing: Jurisdiction, ROE, and the Rules of Deadly Force," *Army Lawyer*, November: 1.

24 James C. Duncan (1998) "Employing Non-lethal Weapons," *Naval Law Review*, XLV: 43; Joint Pub 1-02 (1994) *Department of Defense Dictionary of Military and Associated Terms*.

25 Chairman of the Joint Chiefs of Staff Instruction (CJCSI) 3121.01B, Standing Rules of Engagement for US Forces (June 13, 2005), Enclosure A, para 1.

26 Ingrid Detter (2000) *The Law of War*, 2nd edn. (Cambridge: Cambridge University Press), p. 158. E.g. Geneva Convention (I) for the Amelioration of the Condition of the Wounded and Sick in Armed Forces in the Field, August 12, 1949, 75 UNTS 31, Article 13 (Geneva I); Convention (II) for the Amelioration of the Condition of the Wounded, Sick and Shipwrecked Members of Armed Forces at Sea, August 12, 1949, 75 UNTS 85; Convention (III) Relative to the Treatment of Prisoners of War, August 12, 1949, 75 UNTS 135; Convention (IV) Relative to the Protection of Civilian Persons in Time of War, August 12 1949, 75 UNTS 287; Protocol Additional to the Geneva Conventions of 12 August 1949, and Relating to the Protection of Victims of International Armed Conflicts (Protocol I), June 8, 1977, 16 ILM 1391; Hague Convention (V) Respecting the Rights and Duties of Neutral Powers and Persons in Case of War on Land, October 18, 1907, 36 Stat. 2310, UST 540 (Hague V). For a complete list, see Adam Roberts and

Richard Guelff (eds.) (2000) *Documents on the Laws of War*, 3rd edn. (New York: Oxford University Press).

27 Roberts and Guelff, p. 10 (noting that proportionality and discrimination are generally incorporated into the other principles); Duncan, "Employing Non-lethal Weapons," p. 50.

28 In fact, US space policy for the military goes further than mere recognition of the interdependence of the commercial and the government space sectors and openly encourages it. In addition, for mandating a "preference for commercial acquisition," DoD policy encourages military–industrial partnerships, outsourcing, and privatization of DoD space-related functions and tasks, and even extends a promise of "[s]table and predictable US private sector access" to DoD space-related hardware, facilities, and data. The goal of the U.S. government to promote commercial–governmental interdependence is furthered by requiring that government space systems be based on widely accepted commercial standards to ensure future interoperability of space services. U.S. DoD Directive (DoDD) 3100.10, Space Policy (July 9, 1999) p. 6 (Space Policy).

29 Duncan, "Employing Non-lethal Weapons," p. 50.

30 Hague V.

31 Myres S. McDougal and Florentino P. Feliciano (1961) *Law and Minimum Public World Order: The Legal Regulation of International Coercion* (New Haven and London: Yale University Press), p. 438, citing Hague V, Article 7.

32 Ibid., p. 443.

33 David L. Willson (2001) "An Army View of Neutrality in Space: Legal Options for Space Negation," *Air Force Law Review*, 50: 175–213 (referring to the Outer Space Treaty and the Liability Convention, March 29, 1972, 961 UNTS 187).

34 DoD General Counsel (May 1999) "An Assessment of International Legal Issues in Information Operations."

35 Ibid.; Hague V.

36 McDougal and Feliciano, *Law and Minimum Public World Order*, p. 435.

37 M. Lucy Stojak (August 2001) Excerpt from a report prepared for the Canadian Department of Foreign Affairs and International Trade entitled "The Non-Weaponization of Space," (copy on file with the author).

38 The Treaty Banning Nuclear Weapon Tests in the Atmosphere, in Outer Space, and Under Water, 480 UNTS 43 (entered into force October 10, 1963).

39 Convention on the Prohibition of the Development, Production, and Stockpiling of Bacteriological (Biological) and Toxin Weapons and on their Destruction (1976), 11 UKTS, Cmd 6397 (entered into force March 26, 1975) (Biological Weapons Convention); Chemical Weapons Convention 1992, 32 ILM 800 (entered into force April 29, 1997) (Chemical Weapons Convention).

40 Convention on the Prohibition of Military or any other Hostile Use of Environmental Modification Techniques, 31 UST 333 (entered into force October 5, 1978).

41 Agreement on Measures to Reduce the Risk of Outbreak of Nuclear War (1972), 807 UNTS 57 (entered into force September 30, 1971); Agreement on Measures to Improve the USA–USSR Direct Communications Link (1972) 806 UNTS 402 (entered into force September 30, 1971); Agreement Between the United States of America and the Union of Soviet Socialist Republics on the Prevention of Nuclear War (1973), UST 1478 (entered into force October 5, 1978); Agreement Between the United States of America and the Government of the Union of Soviet Socialist Republics on Notifications of Launches of Intercontinental Ballistic Missiles and Sub-Marine Launched Ballistic Missiles (entered into force May 31, 1988); Agreement Between the United States of America and the Government of the Union of Soviet Socialist Republics on the

Prevention of Dangerous Activities (entered into force January 1, 1990); Memorandum of Agreement Between the Government of the United States and the Government of the Russian Federation on the Establishment of a Joint Center for the Exchange of Data from Early Warning Systems and Notifications from Missile Launches. See Stojak, "The Non-Weaponization of Space."

42 ITU Constitution Article 44:

> radio freqs and GSO are limited natural resources and they must be used rationally, efficiently, and economically ... so that countries may have equitable access to both, taking into account the special needs of the developing countries and the geographical situation of particular countries.

43 Constitution and Convention of the International Telecommunications Union, December 22, 1992 (Geneva: ITU) Article 48(1) ("Members retain their entire freedom with regard to military radio installations"). Although the ITU regulations do not, therefore, apply to the military, armed forces must avoid harmful interference with other users as a practical matter. Further, Article 48(2) requires military radio installations to observe, to the extent possible, measures designed to avoid harmful interference.

44 In the United States, non-federal radio frequency users are managed by the Federal Communications Commission (FCC) and federal agencies are managed by the National Telecommunications and Information Administration (NTIA). The Air Force Frequency Management Agency implements rules for Air Force users. The governing domestic law is 47 USC 502.

45 DoD Space Control Policy, DODI S-3100.15 (January 2001). Notably, however, this policy also explicitly requires that the option for irreversible denial, including destruction, be retained.

46 John Hyten and Robert Uy (2004) "Moral and Ethical Decisions Regarding Space Warfare," *Air & Space Power Journal* 28(2): 51–60; Edmond Lococo (2004) "US Air Force Anti-Satellite Weapon is Operational," *Bloomberg.com* (September 30).

47 Bruce M. DeBlois, Lt. Col., US Air Force (1998) "Space Sanctuary: A Viable National Strategy," *Airpower Journal* 12(4): 41.

48 The CSLA is at 49 USC §70101 et. seq., and the CSCA is at 15 USC §5801 et. seq.

49 49 USC §70101(b)(1) and §70101(b)(4).

50 49 USC §70101(b)(1) and §70101(b)(4).

51 49 USC §70101(a)(2–3).

52 49 USC §70101(a)(5).

53 49 USC §70101(a)(5).

54 49 USC §70101(a)(6).

55 49 USC §70101(a)(9).

56 10 USC §2273

> (a) Policy. – It is the policy of the United States for the President to undertake actions appropriate to ensure, to the maximum extent practicable, that the United States has the capabilities necessary to launch and insert United States national security payloads into space whenever such payloads are needed in space.
>
> (b) Included Actions. – The appropriate actions referred to in subsection (a) shall include, at a minimum, providing resources and policy guidance to sustain –
>
> (1) the availability of at least two space launch vehicles (or families of space launch vehicles) capable of delivering into space any payload designated by the Secretary of Defense or the Director of Central Intelligence as a national security payload; and

(2) a robust space launch infrastructure and industrial base.

(c) Coordination. – The Secretary of Defense shall, to the maximum extent practicable, pursue the attainment of the capabilities described in subsection (a) in coordination with the Administrator of the National Aeronautics and Space Administration.

57 10 USC §2273 (b)(2).

58 42 USC §2475a.

The Federal Government shall acquire space transportation services from United States commercial providers whenever such services are required in the course of its activities. To the maximum extent practicable, the Federal Government shall plan missions to accommodate the space transportation services capabilities of United States commercial providers; 15 USC §5801 (8) the Federal Government should purchase space goods and services which are commercially available, or could be made available commercially in response to a Government procurement request, whenever such goods or services meet Government mission requirements in a cost effective manner.

59 42 USC §2475a (a).

Limitation (1) – As part of the evaluation of the costs and benefits of entering into an obligation to conduct a space mission in which a foreign entity will participate as a supplier of the spacecraft, spacecraft system, or launch system, the Administrator shall solicit comment on the potential impact of such participation through notice published in Commerce Business Daily at least 45 days before entering into such an obligation.

60 15 USC §5807 and 49 USC §70111.

61 49 USC §70102(4–5).

62 49 USC §70111(b)(1): "For the purposes of this paragraph, the term 'direct costs' means the actual costs that can be unambiguously associated with such use, and would not be borne by the United States Government in the absence of such use."

63 15 USC §5807.

64 "In my capacity as founder and chairman of Beal Aerospace, I previously testified to a congressional subcommittee that government subsidies to competing launch providers constituted the private sector's biggest business risk." Statement from Andrew Beal announcing Beal Aerospace was ceasing all business. Available at: www.bealaerospace.com/welcome.htm; Marcia S. Smith (2006) "CRS Issue Brief for Congress, Space Launch Vehicles: Government Activities, Commercial Competition and Satellite Exports," CRS brief, pp. 5 (domestic), 7 and 9 (international).

65 42 USC §2451. Congressional declaration of policy and purpose:

(a) Devotion of space activities to peaceful purposes for benefit of all mankind. The Congress declares that it is the policy of the United States that activities in space should be devoted to peaceful purposes for the benefit of all mankind.

66 Note 15 USC §5801(4) states

a timely extension of the excess third party claims payment provisions of chapter 701 of title 49 is appropriate and necessary to enable the private

sector to continue covering maximum probable liability risks while protecting the private sector from uninsurable levels of liability which could
hinder international competitiveness.

   This statement appears to have been made because the original 49 USC
§70113 had a sunset provision which was nearing.

67 42 USC §2458b, Insurance and Indemnification.
68 49 USC §70112 and 42 USC §2458b.
69 42 USC §2458(b) Indemnification:

   Under such regulations in conformity with this section as the Administrator shall prescribe taking into account the availability, cost and terms of
   liability insurance, any agreement between the Administration and a user
   of a space vehicle may provide that the United States will indemnify the
   user against claims (including reasonable expenses of litigation or settlement) by third parties for death, bodily injury, or loss of or damage to
   property resulting from activities carried on in connection with the
   launch, operations or recovery of the space vehicle, but only to the extent
   that such claims are not compensated by liability insurance of the user:
   Provided, That such indemnification may be limited to claims resulting
   from other than the actual negligence or willful misconduct of the user.

70 49 USC §70104.
71 The company is owned by Boeing of Seattle, Washington (40 percent); RSC-
   Energia of Moscow, Russia (25 percent); Kvaerner ASA of Oslo, Norway (20
   percent); and SDO Yuzhnoye/PO Yuzhmash of Dnepropetrovsk, Ukraine (15
   percent). See www.sea-launch.com/organization.htm.
72 International Launch Services was established in 1995 by Lockheed Martin and
   Khrunichev State Research and Production Space Center. See www.ilslaunch.
   com/whoweare.
73 The regulations implementing the CSLA can be found in Code 14 of Federal
   Regulations Chapter III, available at: http://ecfr.gpoaccess.gov.
74 Alaska Aerospace Development Corporation/Kodiak Launch Complex is the
   only licensed commercial launch site that is not co-located with a federal
   range.
75 49 USC §70117(g):

   Nonapplication. – This chapter does not apply to – (1) a launch, reentry,
   operation of a launch vehicle or reentry vehicle, operation of a launch site
   or reentry site, or other space activity the Government carries out for the
   Government; or (2) planning or policies related to the launch, reentry,
   operation, or activity.

76 15 USC §5601.
77 15 CFR §960.11.
78 15 CFR §960.1.
79 49 USC §70101(a)(7).
80 Agreement Establishing the World Trade Organization, April 15, 1994. Available at: www.wto.org/English/docs_e/legal_e/04-wto.pdf, pp. 33–35, ILM
   1125.
81 Agreement Establishing the World Trade Organization, p. 9.
82 General Agreement on Trade in Services, 201991, GATT Doc. MTN
   TNC/W/FA (GATS); Domenico Giorgi, "WTO and Space Activities," paper
   presented to the Third ECSL Colloquium, Perugia, Italy, May 1999. On February 5, 1998, the WTO's Fourth Protocol to the GATS for Basic Telecommuni-

cations Services took effect, requiring signatories to open their telecommunications markets to foreign competition.

83 Anders Hansson and Steven McGuire (1999) "Commercial Space and International Trade Rules: An Assessment of the WTO's Influence on the Sector" *Space Policy*, 15: 199–201; General Agreement on Tariffs and Trade, October 30, 1947, 55 UNTS 187, TIAS 1700 (GATT).

84 GATS, Part II, General Obligations, Article II, Most Favoured Nation Treatment, available at: www.wto.org/english/docs_e/legal_e/26-gats.pdf. See also, Peter Malanczuk (1999) "The Relevance of International Economic Law and the World Trade Organization (WTO) for Commercial Outer Space Activities," discussion paper presented to the Third ECSL Colloquium, Perugia, Italy, May 6–7.

85 Maj. Elizabeth S. Waldrop (2004) "Integration of Military and Civilian space Assets: Legal and National Security Implications," *The Air Force Law Review*, 55: 186.

86 GATS, Part III, General Obligations, Article XVII, Market Access, available at: www.wto.org/english/docs_e/legal_e/26-gats.pdf.

87 GATS, Part III, General Obligations, Article XVII, National Treatment, available at: www.wto.org/english/docs_e/legal_e/26-gats.pdf.

88 Fourth Protocol to the General Agreement on Trade in Services (WTO 1997), 36 ILM 354 at 366 (1997).

89 Fourth Protocol to The General Agreement on Trade in Services (1997) available at: www.wto.org/english/docs_e/legal_e/4prote_sl20_e.pdf.

90 Agreement On Subsidies And Countervailing Measures, available at: www.wto.org/english/docs_e/legal_e/24-scm.pdf. Part IV, Article 8, Non-Actionable Subsidies, available at: www.wto.org/english/docs_e/legal_e/24-scm.pdf.

91 Marcia S. Smith (2006) "CRS Issue Brief for Congress, Space Launch Vehicles" p. 5.

92 GATT Article XXI, GATS Article XIV.

93 Ryan Goodman (2001) "International Human Rights Law in Practice: Norms and National Security: The WTO as a Catalyst for Inquiry" *Chicago Journal of International Law*, 2: 101.

94 Frank Sietzen, Jr. (2000) "U.S. Launch Industry 'Xenophobic' Say Europeans," *space.com*, posted: July 18, 07:09 pm ET.

95 Goodman, "International Human Rights Law in Practice," 102.

96 States base this argument on the fact that the national security exception is not listed with other general exceptions (see GATT, Article XX) that are subject to a limiting introductory clause, and that the use of the term "it considers necessary" gives states more latitude than the "necessary" terminology used elsewhere in the agreements. The WTO dispute resolution procedure has not yet resolved this issue.

97 Ibid.

98 See www.nasaspaceflight.com/content/?id=4380 reporting on SpaceX antitrust lawsuit against Lockheed Martin and Boeing.

99 Ram Jakhu and Joseph Wilson (2000) "The New United States Export Control Regime: Its Impact on the Communications Satellite Industry," *Annals of Air and Space Law*, 25: 157–163.

100 Wyn Q. Bowen (1997) "US Policy on Ballistic Missile Proliferation: The MTCR's First Decade (1987–1997)" *The Nonproliferation Review*, 51: 21–45. The US and France had conflict over France's proposed transfers of technology to Brazil and India, for example.

101 Jere W. Morehead and David A. Dismuke (1999) "Export Control Policies and National Security: Protecting U.S. Interests in the New Millennium," *Texas International Law Journal*, 34: 173.

102 22 CFR parts 120–130.
103 EAA of 1979, Pub. L. No. 96–72, 93 Stat. 503 (1979); AECA, 22 USC §2778.
104 Ronald J. Sievert (2002) "Urgent Message to Congress – Nuclear Triggers to Libya, Missile Guidance to China, Air Defense to Iraq, Arms Supplier to the World: Has the Time Finally Arrived to Overhaul the U.S. Export Control Regime? The Case for Immediate Reform of Our Outdated, Ineffective, and Self-Defeating Export Control System," *Texas International Law Journal*, 37: 89. Within the DoS, the Office of Defense Trade Controls (DTC) monitor and control the shipment of items on the USML. The USML contains 21 categories ranging from "those unambiguously confined to military use, like Category II – 'Artillery Projectors', to some that can encompass items with civil application, like Category XV – Spacecraft Systems and Associated Equipment."
105 Jakhu and Wilson, "The New United States Export Control Regime."
106 22 USC §2278–2994.
107 22 CFR §120.3.
108 22 CFR §121.16.
109 15 CFR §730–774.
110 50 USC §2401–2420.
111 Sievert, "Urgent Message to Congress" (also noting there are many other US statutes and implementing regulations that directly or indirectly impact exports and, notably, may conflict with the EAR and the AECA. For example, the Trading with the Enemy Act (TWEA), International Emergency Economic Powers Act (IEEPA), Anti-Terrorism and Effective Death Penalty Act (AEDPA), Nuclear Non-Proliferation Act (NNPA), and various US Treasury directives (e.g. the Office of Foreign Assets Control (OFAC))).
112 15 CFR §734.3.
113 50 USC §1701–1706.
114 Cecil Hunt of Harris, Wiltshire, & Grannis, LLP (2003) "Overview of U.S. Export Controls," updated by Thomas M. deButts of Pilsbury Winthrop Shaw Pittman, LLP, p. 3.
115 22 CFR §120.17 and 126.1.
116 House Report 105-851, "Report One of the Select Committee On U.S. National Security And Military/Commercial Concerns With The People's Republic Of China," submitted by Mr. Cox Of California, Chairman.
117 15 CFR §743.2.
118 22 CFR §120.9.
119 15 CFR §740.5.
120 22 CFR §120.3.
121 Cecil Hunt, "Overview of U.S. Export Controls," p. 19.
122 22 CFR §125.4.
123 William A. Reinsch, Under Secretary of Export Administration, Department of Commerce, "U.S. Export Control Policies on Satellites and U.S. Domestic Launch Capabilities," statement before the Senate Committee on Foreign Relations Subcommittee on International Economic Policy, Export and Trade Promotion, June 24, 1999.
124 Ryan Zelnio (2006) "The Effects of Export Control on the Space Industry," available at: www.thespacereview.com/article/533/1.
125 Ibid.
126 National Security Strategy of the United States of America (1996) "A National Security Strategy of Engagement and Enlargement," available at: www.fas.org/spp/military/docops/national/1996stra.htm.
127 Treaty Between the United States of America and the Union of Soviet Socialist Republics on the Limitation of Anti-Ballistic Missile Systems, 23 UST 3435.

128 Eric M. Javits (2002) "Statement to the Conference on Disarmament" Geneva, February, 7, available at: www.usmission.ch/press2002/0207javits.htm. See also National Space Policy, August 31, 2006, available at: www.ostp.gov/html/ US%20National%20Space%20Policy.pdf.

129 UN General Assembly, Prevention of an Arms Race in Outer Space, UN GAOR, 58th Session, UN Doc. A/RES/58/36 (December 8, 2003) (PAROS Resolution 2003); Existing International Legal Instruments and Prevention of the Weaponization of Outer Space (a non-paper by Chinese and Russian delegations to the Conference on Disarmament) (August 26, 2004) (Chinese and Russian CD Paper).

130 See e.g. United Nations Institute for Disarmament Research (2004) "Safeguarding Space for All: Security and Peaceful Uses," conference report, Geneva, March 25–26) (Safeguarding Space). For this example of a proposed code of conduct, see Henry L. Stimson Center, Model Code of Conduct for the Prevention of Incidents and Dangerous Military Practices in Outer Space, available at: www.stimson.org/pub.cfm?id=106 (accessed November 3, 2004).

131 These concepts are summarized by one scholar in the following way:

> Two closely connected terms have been used: "liability" and "responsibility." Neither of these terms has been defined in space law but the term "liability" has been used to set the launching state's liability for damage caused by space objects, whereas the word "responsibility" has been used to mandate international responsibility by the appropriate state party for national activities in outer space. ... [I]n connection with "liabilities" we are dealing with legal consequences (mostly in terms of damages) arising from a particular behavior. In contrast, it seems that when we speak of responsibilities, we are dealing primarily with obligations imposed on people and institutions who are supposed to carry out certain activities or are accountable in given situations though not necessarily in the form of compensation for damages.

> Stephen Gorove (1983) "Liability in Space Law: An Overview" *Annals of Air and Space Law*, 8: 373 (discussing the two terms under domestic law and international law through two treaties: (1) Outer Space Treaty; and (2) Liability Convention).

5 Cooperation in space

International institutions

Tom Graham and Darren Huskisson

Introduction

Fostering international cooperation is a key element in the national space policies of numerous spacefaring states. It is an underpinning of the Outer Space Treaty, being addressed multiple times in the text of the treaty in the context of cooperation on legal and scientific issues as well as space exploration, with the ultimate purpose of maintaining friendly relations among states and maintaining international peace and security.[1] While there is a long history of using a multitude of mechanisms to foster international cooperation in various areas of international relations, mechanisms to foster cooperation on issues relating to outer space are comparatively new.

There has, however, been considerable progress made over the last five or six decades toward multilateral international cooperation in support of space activities. These efforts have a direct bearing on US national security. The United States has long recognized that the promotion of international cooperation can "further U.S. domestic, national security, and foreign policies."[2] Broad and sustained international cooperation in outer space activities is a key to a more secure future for the United States.

Various international institutions support the international goals relating to cooperation in outer space. It is the purpose of this chapter to describe the infrastructure underlying these institutions. First, the more formal system of the United Nations and how it supports space cooperation will be addressed, followed by regional telecommunications organizations, and then a discussion of various other methods used for international cooperation. Finally, we will examine what the future may hold for international space organizations and the role that the United States might play in those organizations.

The UN system

While examples abound of less formal avenues of international cooperation, probably the best known forum for space cooperation is the UN system. The United Nations has been involved in outer space activities from the early days of space exploration. Shortly after the launch of *Sputnik*, an ad hoc Committee on the Peaceful Use of Outer Space came into existence, becoming a permanent committee the following year.[3] Since that time, the outer space activities of the United Nations have greatly expanded and cover numerous functional areas.

United Nations Security Council

The Security Council of the United Nations is important in addressing UN support of space activities. While the UN Charter does not specifically mention outer space in its enumeration of the powers of the Security Council, under Chapters V, VI, and VII of the Charter, the functions and powers of the Security Council include the authority to maintain international peace and security; to investigate any dispute or situation which might lead to international friction; to recommend methods of adjusting such disputes or the terms of settlement; to formulate plans for the establishment of a system to regulate armaments; and to determine the existence of a threat to the peace or an act of aggression and to recommend what action should be taken, including calling on members states to apply economic sanctions and other measures not involving the use of force to prevent or stop aggression and military actions.[4]

These powers almost certainly apply to the resolution of disputes involving outer space. The international community has chosen to explicitly apply the UN dispute resolution system, including the Security Council, to outer space activities. Article III of the Outer Space Treaty states:

> States Party to the Treaty shall carry on activities in the exploration and use of outer space, including the Moon and other celestial bodies, in accordance with international law, including the Charter of the United Nations, in the interest of maintaining international peace and security and promoting international cooperation and understanding.

Thus, the Security Council has broad authority to take action in the case of a dispute or an incident in space which could affect space if it involves a threat to the peace, breach of the peace, or an act of aggression. While these terms have never been applied to an outer space dispute, they have been interpreted very broadly to include a wide variety of situations and would certainly be broad enough to cover numerous potential situations relating to outer space.[5] The day may come when the Security Council seizes itself an international dispute involving outer space, since a major dispute or incident in space would appear to almost automatically involve at least a threat to peace.

United Nations General Assembly

The United Nations General Assembly (UNGA), created under Chapter IV of the UN Charter, has, among other powers, the power to make recommendations on the maintenance of international peace and security, including the principles governing disarmament and arms regulation; to discuss any question relating to international peace and security; and, except where a dispute or situation is being discussed by the Security Council, to make recommendations on it; to make recommendations in furtherance of the promotion of international political cooperation, the development and codification of international law, the realization of human rights and fundamental freedoms for all; and international collaboration in economic, social, cultural, educational and health fields; to make recommendations for the peaceful settlement of any situation, regardless of origin, which might impair friendly relations among nations.[6] The General Assembly has less enforcement power than the Security Council but has been an active forum for the discussion of outer space issues.[7]

Since 1958 the UNGA has taken action on nearly 100 resolutions relating to outer space, with nearly one-third of those resolutions relating to the prevention of an arms race in outer space and one-fifth of those resolutions relating to international cooperation and peaceful use of outer space.[8] Perhaps the most well known and most authoritative actions of the General Assembly relating to outer space have been the attempts at "law making" through the issuance of declarations and principles covering various space activities. These include the 1963 Declaration of Legal Principles Governing the Activities of States in the Exploration and Use of Outer Space,[9] the 1982 Principles Governing the Use by States of Artificial Earth Satellites for International Direct Television Broadcasting,[10] the 1986 Principles Relating to Remote Sensing of the Earth from Outer Space,[11] the 1992 Principles Relevant to the Use of Nuclear Power Sources in Outer Space,[12] and the 1996 Declaration on International Cooperation in the Exploration and Use of Outer Space for the Benefit and in the Interest of All States, Taking into Particular Account the Needs of Developing Countries.[13] These declarations and principles do not in and of themselves constitute law; however, some aspects of these documents are recognized as reflective of customary international law. They can also be seen as a starting point in the formation of customary international law or form the basis of future treaty law. With the international treaty process as it relates to outer space stalled since 1979, it is likely that the General Assembly will continue to act in the issuance of international declarations and principles.

United Nations Office of Outer Space Affairs

While the Security Council and the General Assembly have wide mandates, some organs of the United Nations focus specifically on space-related issues.

The United Nations Office for Outer Space Affairs (UNOOSA), located in Vienna, is the UN office tasked with promoting international cooperation in the peaceful uses of outer space. In its implementation of this mandate it serves multiple functions. One of its most prominent functions is serving as the secretariat for the Committee on Peaceful Use of Outer Space (COPUOS), discussed in greater detail below. Additionally, UNOOSA implements the UN Programme on Space Applications (PSA).[14] PSA, created in 1971, seeks to further the knowledge and experience of space applications for countries that lack space expertise, with a focus on developing countries. Under PSA, the UNOOSA provides training and other activities in space-related topics such as basic space science, space law, remote sensing, satellite communications, satellite meteorology, search and rescue, and global navigational satellite systems. In recent years it has conducted workshops and training programs in Brazil, Australia, Algeria, Austria, Argentina, Japan, the United Arab Emirates, Nigeria, and China. A third function UNOOSA performs is providing secretariat support for the United Nations Conference on the Exploration and Peaceful Uses of Outer Space (UNISPACE). The latest UNISPACE conference was held in 1999, and resulted in numerous recommendations.[15] UNOOSA supports and participates in the implementation of those recommendations.

UNOOSA also promotes the sharing of information related to outer space activities. Both in the spirit of international cooperation and in compliance with international treaties, states provide information on their space activities to the Secretary General of the United Nations. On his behalf, UNOOSA maintains the Register of Objects Launched into Outer Space and disseminates information through the Online Index of Objects Launched into Outer Space.[16]

Committee on the Peaceful Use of Outer Space (COPUOS)

The Committee on the Peaceful Use of Outer Space, also located in Vienna, is the only committee of the General Assembly whose focus is exclusively international cooperation in the peaceful uses of outer space. It was created by the General Assembly in 1958, shortly after the launch of the first artificial Earth satellite by the Soviet Union in 1957, with 18 member nations: Argentina, Australia, Belgium, Brazil, Canada, Czechoslovakia, Egypt, France, India, Iran, Italy, Japan, Mexico, Poland, Sweden, the USSR, the United Kingdom, and the United States. These founding states of COPUOS include most of the important space powers today. The original mandate of the Committee was to consider: the resources of the United Nations relating to the peaceful uses of space; international cooperation in space appropriate under the United Nations; UN organizational arrangements to facilitate international space cooperation; and legal issues related to space. In 1958, the General Assembly established COPUOS as a permanent body and reaffirmed its mandate. In 1961, the

UNGA requested COPUOS to: maintain close contact with government and non-governmental organizations concerned with space matters; provide exchanges of information among governments relating to space activities; and assist in measures to promote international cooperation in space. The same UNGA Resolution requested the UN Secretary General to keep a public registry of launchings, based on information supplied by states launching objects into orbit or beyond. These terms have guided COUPOS ever since. In 1959 the Committee was increased to 24 members and currently has 67 members. In addition a number of intergovernmental and non-governmental organizations have observer status with COPUOS and its subcommittees.

COPUOS has two standing subcommittees – the Scientific and Technical Subcommittee and the Legal Subcommittee. The Scientific and Technical Subcommittee holds annual meetings to discuss matters within its purview and issues an annual report on its work.[17] In its forty-third session it planned to discuss matters ranging from remote sensing, space debris, and the use of nuclear power sources in space to space-system-based telemedicine, space-system-based disaster management support, and the physical nature and technical attributes of the geostationary orbit.[18]

The Legal Subcommittee has traditionally played a major role in the evolution of international space policies. It too meets annually and issues a report on its activities.[19] Recent activities include examination of the status and application of the five UN treaties on outer space, the definition and delimitation of outer space, equitable use of the geostationary orbit, the practice of states and international organizations in registering space objects, and a review of the concept of the "launching state" under international law. Some recent proposals for action in the committee have included reviewing the appropriateness and desirability of a new universal comprehensive convention on international space law, a review of the 1982 Principles on Direct Television Broadcasting by Satellite, and a review of existing norms of international law applicable to space debris.[20]

Conference on Disarmament (CD)

The Conference on Disarmament was established in 1979 as the single multilateral disarmament-negotiating forum in the international community. It is based in Geneva and is the successor to other Geneva-based disarmament fora: the Ten-Nation Committee on Disarmament (1960), the Eighteen-Nation Committee on Disarmament (1962–1968), and the Conference of the Committee on Disarmament (1969–1978). Originally it was a forum designed to bring East and West together to discuss disarmament at the height of the Cold War. By the conclusion of the negotiation of the Nuclear Non-Proliferation Treaty (NPT) in 1968, it was recognized as the central forum where multilateral arms control measures are discussed. As originally constituted in 1979, the CD had 40 members. The membership now stands at 65 nations. There are also other states that attend as observers. In addition to the NPT, the Comprehensive Nuclear Test Ban Treaty, the Chemical Weapons Convention, the Biological Weapons Convention, the Seabed Arms Control Treaty, and the Environmental Modification Convention were all negotiated at the CD or its immediate predecessor, the Conference of the Committee on Disarmament.

The CD has a special relationship with the United Nations; it adopts its own rules of procedure and its own agenda, taking into account the recommendations of the General Assembly and the proposals of its members. The CD reports to the General Assembly annually, or more frequently, as appropriate. Its budget is included in that of the United Nations.

For many years there has been an effort to begin a negotiation on space arms control at the CD in Geneva. This agenda item is entitled Prevention of an Arms Race in Outer Space, or PAROS. This effort has been pressed forward for years by China and Russia, as well as some NATO states. Since the CD operates on the basis of consensus, and the United States has never been comfortable with this process, a working group or subcommittee with a negotiating mandate for PAROS has never been established at the CD. Each year the process starts in Geneva and results in UNGA Resolutions in the fall urging a negotiation on PAROS which passes by a large margin.

Increasingly, the international community of spacefaring nations recognizes the need for restraint and seeks to develop some legal regime to preserve outer space as a non-weaponized realm. Many nations consider the possibility of weapons being deployed in space as highly threatening to their security. The Canadian government has unequivocally stated that it "draws the line at weapons in space." In mid-2003 both Canada and Russia agreed in principle to cooperate with the United States on ballistic missile defense, on the condition that no ballistic missile defense weapons are placed in space. The Outer Space Treaty bans the deployment of nuclear weapons and any other weapons of mass destruction in space, but the treaty does not apply to conventional weapons. While space has long been militarized with the deployment of military reconnaissance and navigation satellites and similar technologies, no offensive or defensive weapons have ever been stationed in outer space. Thus, pursuant to the initiative of President Dwight D. Eisenhower, who at the time of the establishment of NASA made it clear that it was US policy to keep space weapon-free, space remains free of weapons of all kinds.

During their administrations, George H.W. Bush and Bill Clinton asserted that there was no need for limitations beyond the existing Outer Space Treaty as no arms race or threat of an arms race in space existed. The Eisenhower policy held in the United States and was supported elsewhere as well. Consistent with the Bush–Clinton position, over the years, the United States at the CD, virtually alone, routinely opposed the creation of a negotiating mandate for outer space. The US position has become more isolated in Geneva in recent years as Russia and China, supported at least in principle by virtually all other states at the CD, have every year pressed for negotiations on a new treaty prohibiting weapons in space. The United States regularly opposes these efforts and this impasse has now affected other work, so that for some years now there has been a stalemate at the CD.

As part of this evolution, a number of years ago, in addition to draft treaties that have been proposed by Russia and China at the CD, a procedural effort began, both in Geneva and at the United Nations in New York, called the Prevention of an Arms Race in Outer Space (PAROS). This took the form each year of a draft mandate for negotiations in Geneva and a draft resolution in New York. The United States did not support this initiative, opposing the mandate in Geneva by refusing consensus and abstaining from voting on the resolution in the UN General Assembly each year. The resolution always passed the General Assembly by an overwhelming vote. Beginning in 2005, virtually alone among UN member states, the United States voted no on the resolution. Thus, there currently exists a complete impasse in Geneva on all work as a result of the disagreement between the United States and the rest of the world as to whether there should be negotiations on a new legal regime prohibiting the deployment of weapons in outer space.

Other space technology-utilizing organizations of the UN system

There are numerous other organs and specialized agencies within the UN system that deal with issues related to space activities. These organizations include those more typically associated with terrestrial activities, such as the World Health Organization, the World Bank and the United Nations Educational, Scientific and Cultural Organization, and include those who regulate the international air service, telecommunications, and maritime shipping.

While all of these organizations with interests in space are not under one controlling space office within the United Nations, their activities are to some degree coordinated within the United Nations. To coordinate its space activities, the United Nations annually convenes the United Nations Inter-Agency Meeting on Outer Space Activities to discuss "current and future space activities, emergent technologies of interest and other related matters."[21] The meeting produces a report for COPUOS and reports on the coordinated space-related activities of the UN system.[22] Below is a brief description of some of the more prominent UN organizations that have some interest in space activities in furtherance of the goals of their respective organizations.

International Telecommunications Union

One of the most important organizations with a stake in space operations is one whose existence predates even the United Nations itself, the

International Telecommunication Union (ITU). The ITU was founded in the nineteenth century after the invention of the telegraph to standardize and promote interconnectivity among nations. This ultimately resulted in the International Telegraphic Union and international regulations to govern international communication by telegraph. This was later expanded with the evolution of technology to cover international telephone traffic, and even further with the advent of wireless telegraphy in 1896 – the first type of radio communication. In 1932 at the Madrid Conference, it was decided to combine the International Telegraph Convention of 1865 and the International Radiotelegraph Convention of 1906 into the International Telecommunications Convention and the original International Telegraphic Union became the ITU.

In 1947 the ITU became a UN specialized agency, and the headquarters moved in 1948 from Bern to Geneva. At the same time, the International Frequency Registration Board was established to coordinate the management of the radio frequency spectrum. The Table of Frequency Allocations, first introduced in 1912, was made mandatory.

In 1963 the first geostationary communications satellite (Syncom-1) was placed into orbit, and a special ITU conference was held in Geneva to begin the allocation and put in place regulations governing the use by satellites of the radio-frequency spectrum and associated orbital slots. In 1992 allocations were made for the first time for a new kind of space service, using non-geostationary satellites. In the same year spectrum was identified for the ITU-developed next-generation global standard for digital mobile telephone.

Over the decades the role of the ITU has evolved as technology has developed. In 1989 the ITU officially recognized the importance of placing technical assistance to developing countries on the same footing at its traditional activities of standardization and spectrum management. In 1992 the ITU was divided into three sections, corresponding to its three main areas of activity: Telecommunication Standardization, Radio Communication, and Telecommunication Development. The ITU's strategic plan seeks to bridge the international digital divide by working toward the development of fully interconnected and interoperable networks and the development of tools to safeguard the integrity and interoperability of networks.

International Civil Aviation Organization

Another UN specialized agency that should be mentioned is the International Civil Aviation Organization (ICAO). It was established pursuant to the Convention on International Civil Aviation of 1944, commonly known as the Chicago Convention.[23] The organization is made up of an Assembly, a Council with limited membership with subordinate bodies, and a Secretariat. The Assembly meets every three years and decides which states are going to be members of the Council for the next three years. Council members are chosen on the basis of importance to air transport, contribution of facilities for air navigation, and to ensure that all regions of the world are represented. The Council gives continuing direction to the ICAO; it is in the Council that standards and recommended practices are agreed and appended as annexes to the Convention. The Secretariat, the administrative part of the ICAO, is divided into five main divisions: Air Navigation, Air Transport, Technical Cooperation, Legal, and Administrative.[24]

ICAO's participation in outer space activities is currently limited to how these activities affect the safety of international civil aviation. Some topics of interest to ICAO include sub-orbital flights and global navigational satellite systems.

International Atomic Energy Agency

The International Atomic Energy Agency (IAEA) is also an important UN specialized agency. The Agency and its Director-General, Mohamed El-Baredi, shared the Nobel Peace Prize for 2005. The IAEA is the world center for nuclear cooperation. It was set up in 1957 as the world's "Atoms for Peace" agency and as part of the UN structure. It was inspired by the famous speech by President Dwight D. Eisenhower before the UNGA in December of 1953. This led to the adoption of the IAEA statute, which was approved by 81 countries in the fall of 1956.

The IAEA was designed to be the agency to promote the peaceful use of atomic energy. However, as more countries mastered nuclear technology, concerns began to arise that nuclear weapons might sweep the world. France and China brought the number of nuclear weapon states to five, and it was known at least to some governments that Israel and India were far along in their programs. There were predictions during the Kennedy administration of 25–30 nuclear weapon states by the end of the 1970s. There was growing international support for legally binding constraints along with safeguards to stop the proliferation of nuclear weapons. This led to the negotiation and signing of the Non-Proliferation Treaty (NPT) in 1968 and its entry into force in 1970. The NPT set up a system of safeguards to ensure that peaceful nuclear power programs are not diverted into weapons programs. The IAEA administers the safeguards, and as of today, the NPT has been effective in that only four additional nations have acquired nuclear weapons since 1970 and two of them (India and Israel) were well along in their programs prior to that date. Thus, only Pakistan and perhaps North Korea are truly new additions to the list of nuclear weapon states in the last 35 years.

However, after the first Gulf War it was learned that Iraq had a clandestine nuclear weapon program and was perhaps only a year or two away from having the capability to construct a nuclear weapon. Thus the IAEA safeguards clearly had been inadequate; this led to the negotiation of an additional protocol to the Safeguards Agreements – signed in 1997 –

which considerably increased the power of the IAEA to conduct more intrusive inspections in NPT non-nuclear weapon states. All of this is important to space, because, as we have seen, nuclear power sources have been in space, and under the UNGA "Principles" document are formally permitted to be there, with certain limitations.

The IAEA has, in recent years, become the principal debating forum for international disputes in the nuclear nonproliferation field. The battle in the early 1990s over the North Korean nuclear weapon program and the threatened North Korean withdrawal from the Nuclear Nonproliferation Treaty was all played out at the IAEA. The same thing has been happening currently with the Iranian nuclear program. It is the IAEA that has been conducting the inspections and it is at IAEA Board of Governors meetings in the past where the debates have been held. While the Board of Governors last year voted to send the Iranian case to the United Nations Security Council, the IAEA is still very much involved – witness the recent agreement between Iran and the IAEA for Iran to fully disclose its past nuclear activities. The IAEA does have a tangential relationship with space issues in that, first, the Outer Space Treaty prohibits the deployment of nuclear weapons in space and second, pursuant to a UN Resolution, the deployment of nuclear power sources in space must be regulated. Ultimately, the IAEA will have to play an important role should there be further efforts to utilize nuclear power sources in space.

International and regional telecommunications organizations

Intelsat, Ltd

In 1945, the famous science fiction author Arthur C. Clarke published his article "Extra-terrestrial Relays" in *Wireless World*, setting forth his theory that three satellites positioned over each ocean region in geostationary orbit would provide global communications. These are the fundamentals by which Intelsat and all other geostationary satellite systems were formed.

In 1964, Intelsat established the first commercial global satellite communications system. Intelsat services involve the delivery of information and entertainment worldwide. The system uses a combination of satellite and terrestrial connectivity technology to help enterprises, governments, and service providers deliver content around the world. Fundamental services include: capacity services for quality point-to-point or point-to-multipoint transmissions; managed services to meet broadband, trunking and media requirements; specialized services, including such things as disaster recovery, co-location services and consulting and technical management support. The global integrated Intelsat communications satellite includes 28 geostationary satellites in prime orbital locations and a world-wide ground-based network of teleports, fiber and Internet points of presence at strategic exchange points.

Intelsat has more than 700 customers around the world, including the world's largest Internet service providers, telecommunications companies, broadcasters, and corporate network service providers. The total revenue of the company in 2004 was $1.04 billion; it has a staff of 800 and has offices in Brazil, Hong Kong, France, Germany, India, Singapore, South Africa, the United Arab Emirates, and the United Kingdom. The past history of Intelsat includes the first live global television broadcast of the *Apollo 11* Moon landing in 1969 and compressing 40,000 hours of programming into two weeks for the Sydney Olympics in 2000. In 2001 Intelsat became a private company after 37 years as an intergovernmental organization. The International Telecommunications Satellite Organization (ITSO) serves as the legacy international organization.[25] It is tasked with serving as the supervisory authority of Intelsat, Ltd., to ensure the performance of core principles for the provision of international public telecommunications services, with high reliability and quality, and to promote international public telecommunications services to meet the needs of the information and communication society. The ITSO is head-quartered in Washington, DC, and currently has 148 members.

Inmarsat

Inmarsat came into being as an intergovernmental organization in 1979 to provide global safety and other communications for the maritime industry. Starting with a customer base of 900 ships in the early 1980s, it grew rapidly to offer similar services to other users on land and in the air. In 1999 Inmarsat became the first intergovernmental organization to be transformed into a private company. It now supports links for telephone, fax, and data communication to more than 287,000 ship, vehicle, aircraft, and other mobile uses. The company operates a fleet of geostationary satellites that extend mobile phone, fax, and data communication to every part of the world, except the North and South Poles.

The Inmarsat satellites are controlled from Inmarsat's headquarters in London. Inmarsat communications connect over 80 countries around the world. London is also the home of an intergovernmental organization, the International Mobile Satellite Organization, whose mission is to supervise the Company's public-service duties to support the Global Maritime Distress and Safety System and satellite-aided air traffic control for the aviation community.

Arab Satellite Communication Organization

The history of the Arab Satellite Communications Organization begins in 1967 when the Arab Ministers of Information and Culture developed the vision of a satellite communications network to integrate the activities of the League of Arab States. On April 14, 1976 the vision came into reality when the Arab Satellite Communication Organization (ARABSAT) was established by the Arab League member states with the mission to provide satellite services to serve the needs of telecommunication, information, cultural, and education sectors of the Arab world.

ARABSAT is governed by multiple organs. The General Assembly is made up of the communications ministers of the member states or their representatives. It meets once a year and is the main body of the organization. The Board of Directors is tasked to secure, invest, and maintain the space sector and implement the policies of the General Assembly. It holds meetings every three months to carry out these duties. The Board consists of nine members from member states who are selected annually by the General Assembly. The Executive Organ carries out the day-to-day duties of the organization, and is led by the president and chief executive officer. Other executive officers in the organization include three vice presidents who head the Finance and Administrative Affairs Department, the Technical Department and the Marketing & Customer Service Department.

ARABSAT offers various satellite services, including television, telephone, Internet, and other interactive services. It owns and operates two control stations to control the satellites during the launch phase and in orbit, one in Saudi Arabia and one in Tunisia. Currently, Saudi Arabia and Kuwait hold the greatest share in paid capital in the organization, with their combined shares totaling greater than 50 percent. Libya, at 11 percent, and Qatar, at 9 percent, are the next largest contributors, with 17 others contributing less than 5 percent of paid capital.

Intersputnik International Organization of Space Communications

The Intersputnik International Organization of Space Communication (Intersputnik), one of the world's first satellite operators, has been offering telecommunication services since 1971. Originally nine countries signed an agreement on the establishment of a global satellite communications system, supporting cooperation and coordination of efforts aimed at its design, manufacture, operation, and development in the interests of its member countries. Today Intersputnik has 25 member countries, including Afghanistan, Azerbaijan, Bulgaria, Belarus, Hungary, Vietnam, Georgia, India, Yemen, North Korea, Kazakhstan, Kyrgyzstan, Cuba, Laos, Mongolia, Nicaragua, Poland, Romania, Russia, Syria, Tajikistan, Turkmenistan, Ukraine, and the Czech Republic. The organization has a three-level management structure made up of its Board, the main governing body which meets annually, the Operations Committee, responsible for supervisory control which reports to its Board, and the Directorate, which manages the operations of the organization.

In the past, Intersputnik operated Russian-built spacecraft only, which it leased from Russian space organizations. In the early 1990s the changing political and economic conditions caused Intersputnik to radically change its business practices and strategies. As part of its new strategy, Intersputnik, together with Lockheed Martin, established in June 1997 the Lockheed Martin Intersputnik (LMI) joint venture. LMI was tasked with full-scale satellite communications services encompassing the manufacture and launch of satellites as well as their long-term in-orbit operation. In 1999 the joint venture placed in geostationary orbit an advanced satellite, the LMI-1. Since April 2000, based on a long-term cooperation arrangement with the Russian Satellite Communications Company, Intersputnik offers services on the new-generation Russian-built Express-A-Series satellites. As a result of agreements signed with Eutelsat and Rascom (see below) in 2001–2002, the coverage area and the capabilities of the Intersputnik systems have been greatly expanded.

Eutelsat S.A.

The European Organization for the Exploitation of Meteorological Satellites was originally set up in 1977 as an intergovernmental organization, the European Telecommunication Satellite Organization, to develop and operate a satellite-based telecommunications infrastructure for Europe. It began operations with the launch of its first satellite in 1983. Initially established to address satellite communications demand in Western Europe, Eutelsat rapidly developed its infrastructure to expand coverage to additional markets, such as Central and Eastern Europe in 1989, and much of the rest of the world in the 1990s. Eutelsat was the first satellite operator in Europe to broadcast television channels direct-to-home. By the mid-1990s, through its five HOTBIRD satellites, Eutelsat was able to attract hundreds of channels to the same orbital location, thereby appealing to widespread audiences for consumer satellite TV.

In the context of the general liberalization of the telecommunications sector in Europe, Eutelsat's operations were transferred to a private company called Eutelsat S.A. in July 2001. In April 2005 the principal shareholders of Eutelsat S.A. placed their investment in a new entity, Eutelsat Communications, which is now the holding company of the Eutelsat Group, owning 95.2 percent of Eutelsat S.A. The group operates a fleet of 23 satellites (18 of which are fully owned) supplying capacity to operators who in turn provide their customers with radio and TV broadcasting services, professional data network solutions, and broadband Internet access. While Eutelsat is a private company, its shareholders are predominately governments. Eutelsat holds 12.4 percent of the global market share of the fixed satellite services industry.[26]

European Organization for the Exploitation of Meteorological Satellites

The European Organization for the Exploitation of Meteorological Satellites (EUMETSAT), headquartered in Darmstadt, Germany is an intergovernmental organization which was formed in 1986 to establish, maintain, and exploit a European system of meteorological satellites that provide Earth observation for its member states and 11 cooperating states. In particular, its systems have been used in weather forecasting and monitoring global climate change. It is also involved in projects fostering international cooperation, such as the WMO Global Observing System and the Global Monitoring for Environment and Security (GMES) initiative.

The organization has a vast array of data collection including the Meteosat series of satellites, ground stations, and the future EUMETSAT Polar System. The total EUMETSAT infrastructure represents several billion euro in investments by its member states.

EUMETSAT has two organs that control its operations, the Council and the Director-General. The Council is composed of representatives of each member state and it meets in ordinary sessions at least once a year. The Council has the power to adopt all the measures necessary for the implementation of this EUMETSAT Convention, with specific functions outlined in Article 5 of the organization's Convention.[27]

The Director-General is responsible for the implementation of the decisions taken by the Council. He acts as the legal representative of EUMETSAT and has numerous powers enumerated in the Convention, including ensuring the proper functioning of EUMETSAT and budgetary powers. The Director-General is supported by a Secretariat.

The Regional African Satellite Communications Organization

The Regional African Satellite Communications Organization (RASCOM) is an intergovernmental, commercially run organization. RASCOM was formed in May 1992, when African states meeting in Abidjan, Cote d'Ivoire decided to create an organization that could provide economical satellite communication service to the entire continent of Africa. This decision was based on the lack of development in the area of telecommunications on the continent and the non-existence of direct communication links among African states. RASCOM's stated mission is

> to provide to all the regions of African countries, efficient and inexpensive telecommunications facilities and meet their radio and television broadcast requirements, by having recourse to every appropriate technology, including a regional satellite communication system well integrated into existing and/or planned national network with a view to facilitating the development of the countries of Africa.[28]

RASCOM offers a variety of services from those designed for public/private telecommunication operators to sound- and TV-broadcasting companies. Among its specific services offered, one will find integrated telephone, thin route trucking services, and transponder lease services. RASCOM provides persons on the African Continent access to everyday services such as telephone, fax, data transmission, Internet access, and sound/TV reception.

RASCOM is organized around three entities. The first is the Assembly of Parties. The Assembly is made up of all governments who have signed the RASCOM Convention. It serves as the strategic and policy-orientation organ and meets every two years. The Board of Directors has a different function, being more involved in the operational aspects of the organization. It meets every three months and is responsible for the design, development, construction, establishment, operation, and maintenance of the RASCOM space segment and other RASCOM activities. Finally, its Executive Organ is responsible for the day-to-day operation of the organization.

Other venues for international space cooperation

European Space Agency

An important regional multilateral intergovernmental organization, the European Space Agency (ESA), is the gateway to space for Europe. It has 17 member states – Austria, Belgium, Denmark, Finland, France, Germany, Greece, Ireland, Italy, Luxembourg, the Netherlands, Norway, Portugal, Spain, Sweden, Switzerland, and the United Kingdom. Not all of the European Union's 25 members are members of ESA and two members of ESA are not members of the EU. The mission of ESA is to shape the development of Europe's space capability and to ensure that investment in space continues to deliver benefits to the citizenry of the EU. ESA develops Europe's space program and carries it through. By coordinating the financial and intellectual resources of its member states, ESA can undertake programs and activities far beyond the capabilities of any single European country. ESA's projects are designed to find out more about the Earth, its immediate space environment, the solar system and the universe, as well as to promote satellite-based services and technologies for Europe.

ESA is an independent organization that nevertheless maintains close ties with the EU. Through a Framework Agreement, ESA and the EU share a joint European strategy for space and together develop space policy for Europe. The headquarters of ESA are in Paris, where policies and programs are determined. ESA also has centers in other European countries.

The European Space Research and Technology Center is the design hub for most ESA spacecraft and technology development, and is located in the Netherlands; the European Space Operations Center is responsible for controlling ESA satellites in orbit and is located in Germany; the European Astronauts Center trains astronauts for future missions and is also located in Germany; and the ESA Center for Earth Observation, which is responsible for collecting and distributing Earth-observation satellite data to the ESA partners is located in Italy. ESA also has liaison offices in

Belgium, the United States, and Russia, a launch site in French Guiana, and ground and tracking stations around the world.

The governing body of ESA is its Council, which sets basic policy guidelines within which the agency develops Europe's space program. Each member state has one vote on the Council regardless of its contribution. The agency is headed by a Director-General with a four-year term, it has around 1,900 employees, and its 2005 budget is €2,977 million or about $3.5 billion.

Among planned future observation projects is the construction and launch of Cryosat II, a copy of an ice-measuring satellite that was destroyed in a failed October launch. The final go-ahead decision was scheduled for a February 2007 meeting. The ESA Council met in Berlin in December 2005, to decide on the agency's 2006–2008 budget. The intent was to include funding for a future mission to search for life on Mars, and to develop a "space plane" jointly with the Russian Federal Space Agency.

International Space Station

The International Space Station (ISS) is a cooperative effort among the United States, the members of ESA, Russia, Canada, Japan, and Brazil. The ISS is an enormous orbiting laboratory and shining symbol of international cooperation in space exploration. Its stated purpose is to provide for a permanently inhabited civil space station to enhance the scientific, technological, and commercial uses of outer space.

The United States has the lead role in the overall management and coordination of the ISS, as well as overall system engineering, integration, and safety issues.[29] The United States is also the primary contributor to this international endeavor, and will develop and operate major elements and systems of the ISS.

> The U.S. elements include three connecting modules, or nodes; a laboratory module; truss segments; four solar arrays; a habitation module; three mating adapters; a cupola; an unpressurized logistics carrier and a centrifuge module. The various systems being developed by the U.S. include thermal control; life support; guidance, navigation and control; data handling; power systems; communications and tracking; ground operations facilities and launch-site processing facilities.[30]

Complementing the US contribution to the ISS, the international partners are providing key components of the station. Canada, for example, is constructing a huge robotic arm that will be used for various external tasks on the ISS. ESA and Japan will provide laboratories, and Russia will provide multiple complements, including two research modules, living quarters, a solar power source, and Soyuz spacecraft for voyages to and from the station.[31] Additionally, Brazil is participating in the ISS via a separate agreement that allows the country to use a fraction of the US portion of the ISS in exchange for Brazil's agreement to provide parts and services for the ISS.[32]

The station is governed by an intergovernmental agreement (IGA) signed by the United States, Russia, ESA, Japan, and Canada.[33] The agreement covers such issues as registry, jurisdiction, and ownership, in which each partner retains ownership and jurisdiction over its elements of the station and agrees to separately register its elements. It also governs operation, communications, and evolution of the station, as well as providing a cross waiver of liability among the partner states and rules governing intellectual property and criminal jurisdiction. The IGA is further implemented by four separate Memoranda of Understanding (MOU) between NASA and the international partners. In accordance with the IGA and the implementing MOUs, NASA developed a code of conduct for the ISS crew.[34]

Inter-Agency Space Debris Coordination Committee

Some methods of international cooperation are less formal than others. The Inter-Agency Space Debris Coordination Committee (IADC) is not an international organization, but rather an international governmental forum. Its membership includes the following national space entities: the Agenzia Spaziale Italiana, the British National Space Centre, the Centre National d'Etudes Spatiales, the China National Space Administration, the German Aerospace Center, ESA, the Indian Space Research Organisation, the Japan Aerospace Exploration Agency, the US National Aeronautics and Space Administration, the National Space Agency of Ukraine, and the Russian Federal Space Agency. The committee is governed by the Terms of Reference (ToR) for the Inter-Agency Space Debris Coordination Committee, which establishes the basic principles related to its function.

The IADC coordinates worldwide activities related to issues involving space debris, both man-made and natural. Primarily, the members of the IADC exchange information on space debris research and facilitate opportunities for cooperation in space debris research. The members also review ongoing activities of the committee and identify options for space debris mitigation. Among its other actions, the IADC makes an annual presentation to the COPUOS S&T Subcommittee.[35] The IADC is made up of a Steering Group and four specified Working Groups.

The Steering Group provides guidance and management of the IADC. Among other things, it coordinates the Working Groups, organizes overall IADC activities, defines new areas for work, and represents the IADC in other organizations. It also appoints the chairperson and deputy in each Working Group, approves action items recommended by the Working Groups, and promotes education on space debris matters. The four Working Groups are WG 1 (measurements), WG 2 (environment and database), WG 3 (protection), and WG 4 (mitigation).

WG 1 focuses on all measurement techniques, both those currently in operation and those under development, to gain information on man-made and natural objects in near-Earth space. This includes ground and space-based techniques, onboard detectors and collectors, and the analysis of spacecraft surfaces exposed to the space environment. WG 1 also works on exchanging information related to activities in the area of measurements of orbital debris.

WG 2 is responsible for the characterization and modeling of space debris and the storage and access of this data by electronic means. This includes collision prediction and risk assessment, uncontrolled reentry, the establishment of a joint database for debris and meteoroids, and the development of models which characterize explosions or collisions in space.

Protection is the focus of WG 3, particularly involving the exchange of information relating to on-orbit impacts and shielding design performance and optimization, space vehicle fragmentation events, and research in the area of protection. Additionally, it focuses on test facilities and procedures, hypervelocity impact test data, simulation software, and design and test commonality.

Finally, WG 4 focuses on the area of mitigation. It studies all measures used to mitigate the creation of space debris and to reduce the hazards associated with space debris. WG 4 seeks to identify the sources of space debris and the design and operation of space systems that will mitigate the production of space debris.

Global Earth Observation System of Systems

The Global Earth Observation System of Systems (GEOSS)[36] is a historic initiative that will use a number of systems to revolutionize the understanding of our planet and how it works. Earth observation systems consist of measurements of air, water, and land made on the ground, from the air, or from space. Historically observed in isolation, the current effort is to look at these elements together and to study their interactions. On February 16, 2005 61 countries agreed to a plan that, over the next ten years, will revolutionize the understanding of Earth and how it works.

The agreement came from a 2003 Earth Observation Summit of 33 nations and many international organizations that adopted a declaration setting a goal of a comprehensive, coordinated and sustained system of systems to collect and disseminate data on the Earth.

In April 2004, US EPA Administrator Michael Leavitt and other senior US Cabinet members met in Japan with environmental ministers from more than 50 nations. They adopted the Framework Document for a ten-year implementation plan for GEOSS.

GEOSS is envisioned as a large national and international cooperative effort to bring together existing and new hardware and software, making it all compatible in order to supply data and information at no cost. The

United States and other developed nations have a unique role in developing and maintaining the system, collecting data, enhancing data distribution, and providing models to help all of the world's nations.

The stated goals of GEOSS are disaster reduction, integrated water resource management, ocean and marine resource monitoring and management, weather and air quality monitoring, forecasting and advisories, biodiversity conservation, sustainable land use and management, public understanding of environmental factors affecting human health and well-being, better development of energy resources, and adaptation to climate variability and change.

International Space University

The International Space University (ISU) also merits mention. The ISU is situated in Strasbourg, France and is designed to develop the future leaders of the world space community. The ISU provides graduate-level training to future leaders through its two-month summer session and its one-year Masters program at the Strasbourg campus as well as at other locations around the world. Since its founding in 1987, the ISU has graduated more than 2,200 students from around the world. It offers its students a unique core curriculum which covers all disciplines relating to space programs and enterprises – space science, space engineering, systems engineering, space policy and law, business and management, and space and society. Both its summer program and its one-year Masters program provide international graduate students and young space professionals the opportunity to solve complex problems together in an intercultural environment. At the ISU all programs are interdisciplinary, international and intercultural.

The programs at the ISU promote international cooperation and understanding, help develop an interactive global network of students, teachers and alumni, and encourage the innovative development of space for peaceful purposes. The ISU campus also serves as a neutral international forum for the exchange of knowledge and ideas on challenging issues related to space and space applications.

What does the future hold for space-related international institutions?

We now will briefly look at what the future might hold for the evolution of current and creation of future international space institutions. There are a number of space-related technologies that are ripe for growth and whose growth could result in the need for greater international cooperation. For example, the future growth of global mobile personnel communications systems (GMPCS), with the growing ability of every individual to hand-carry a communication device capable of operating outside a state's regulated communication systems, will undoubtedly necessitate some form of international cooperation. Intelsat already appears to be pursuing this market, but others may as well. The same can be said of expanded entertainment opportunities, remote sensing, manned spaceflight, global navigational satellite services,[37] and commercial activities.

There is considerable sentiment around the world, except in the United States, to develop a new arms control regime for space, which would prohibit all weapons, offensive and defensive, and provide for considerably more transparency than currently exists – perhaps through a provision for some kind of launch payload inspection. There might also be certain explicitly allowed functions, some of which might relate to space support of terrestrial military operations such as GPS. Will the desire for such expanded arms control regimes lead to agreements without US participation, as happened in the Ottawa process concerning landmines and the Rome Statute for an International Criminal Court, or will the position of the United States change over time?

Notes

1 See Outer Space Treaty, Preamble, Articles I, III, IX, X, XI.
2 Fact Sheet on National Space Policy, available at: www.ostp.gov/NSTC/html/fs/fs-5.html. International cooperation has long been a part of the US National Space Policy. See www.hq.nasa.gov/office/pao/History/spdocs.html.
3 The ad hoc committee was created by UN General Assembly Resolution 1348, Question of the Peaceful Use of Outer Space, available at: www.oosa.unvienna.org/SpaceLaw/gares/html/gares_13_1348.html. The committee was made permanent by UNGA Resolution 1472, International Cooperation in the Peaceful Uses of Outer Space, available at: www.oosa.unvienna.org/SpaceLaw/gares/html/gares_14_1472.html.
4 The UN Charter is available at: www.un.org/aboutun/charter.
5 A complete list of Security Council Resolutions is available at: www.un.org/documents/scres.htm.
6 For a more complete list of the functions, see www.un.org/ga/60/ga_background.html.
7 Under the so-called Uniting for Peace Resolution, it has been argued that the General Assembly has enhanced powers in the event of an impasse in the Security Council. For a copy of the resolution, see www.un.org/Depts/dhl/landmark/pdf/ares377e.pdf.
8 A complete list of General Assembly Resolutions related to outer space is available at: www.oosa.unvienna.org/SpaceLaw/gares.
9 See www.unoosa.org/oosa/en/SpaceLaw/lpos.html.
10 See www.unoosa.org/oosa/en/SpaceLaw/dbs.html.
11 See www.unoosa.org/oosa/en/SpaceLaw/rs.html.
12 See www.unoosa.org/oosa/en/SpaceLaw/nps.html.
13 See www.unoosa.org/oosa/en/SpaceLaw/spben.html.
14 For more information on the Programme on Space Applications, see www.oosa.unvienna.org/sapidx.html.
15 UNISPACE III documents can be found at www.oosa.unvienna.org/unisp-3/index.html.
16 See www.oosa.unvienna.org/OSOIndex/index.html.

17 Annual reports of the committee dating back to 1998 are available at: www.oosa.unvienna.org/COPUOS/stsc/repidx.html.
18 See UN document A/AC.105/C.1/L.283, available at: www.oosa.unvienna.org/Reports/AC105_C1_L283E.pdf.
19 Reports of the Legal Subcommittee dating to 1993 are available at: www.oosa.unvienna.org/SpaceLaw/legrepidx.html.
20 See UN document A/AC.105/826, p. 20, available at: www.oosa.unvienna.org/Reports/AC105_826E.pdf.
21 See www.uncosa.unvienna.org.
22 For copies of these reports, see www.uncosa.unvienna.org/reports/index.html.
23 The full text of the Chicago Convention is available at: www.icao.int/cgi/goto_m.pl?/icaonet/dcs/7300.html.
24 For more information on ICAO, see the organization's website at www.icao.int.
25 Information regarding the International Telecommunications Satellite Organization is available at: www.itso.int.
26 For information on Eutelsat, see www.eutelsat.com/home/index.html.
27 The Convention to establish EUMETSAT is available at: www.eumetsat.int/groups/cps/documents/document/pdf_leg_convention.pdf.
28 RASCOM Mission Statement, available at: www.rascom.org/index2.html.
29 See Article 7of the Inter Governmental Agreement.
30 See "The International Space Station," available at: www.shuttlepresskit.com/ISS_OVR.
31 Ibid.
32 See "Brazilian International Space Station Program," available at: www.inpe.br/programas/iss/ingles/use/Default.htm.
33 A. Yakovenko (1999) "The Intergovernmental Agreement on the International Space Station" *Space Policy*, 15: 79–86.
34 See Code of Conduct for the International Space Station Crew, 14 CFR Part 1214.
35 See www.iadc-online.org/index.cgi?item=docs_pub.
36 See the GEOSS homepage at www.epa.gov/geoss.
37 Jiefang Huang, (1997) "Development of the Long-Term Legal Framework for the Global Navigational Satellite System," *Annals of Air and Space Law*, XXII-I: 585.

6   The policy process

Eligar Sadeh and Brenda Vallance

The evolution of civil, commercial, or military space policy in the United States is a critical part of public policy. Visions of a spacefaring future in which humanity settles the solar system, mines asteroids, or deploys solar arrays to meet future power requirements captivate the imagination, but space policy has to meet practical requirements in the present at acceptable cost. Unfortunately, this seldom happens. Instead, the process of policy-making frequently interferes with rational outcomes. The purpose of this chapter is to examine those processes, and to discuss a defense policy process model useful in examining and explaining space and defense policy-making.

In contrast to the rational decision-making associated with the creation of grand strategy, this model assumes that the actors associated with space and defense policy-making – like their counterparts in other areas of strategic planning and implementation – frequently serve individual and organizational interests and goals, rather than being guided by an objective, rational standard. Alternatively, the actors identify agency or personal interests with objective, national interest. Space policy also has some unique aspects, such as the broad range of political demands and interest groups that span from practical issues of national defense, commerce, science, and technology, to less quantifiable characteristics like the contributions of human space exploration. The resulting decision-making model can be illustrated in Figure 6.1.

In this chapter, the defense policy process model is explained as it applies to space policy-making. This entails an examination of two recent and significant events that impacted space policy: (1) the 2001 Space Commission Report; and (2) the September 11, 2001 (9/11) terrorist attacks on the United States. The model is then applied to assess commercial satellite export controls. Finally, the usefulness of the model to space policy-making is evaluated.

The defense policy process and policy-making on space

Theoretically, there are three stages in the making of space policy; these take place in the following order: (1) the setting of goals by the national

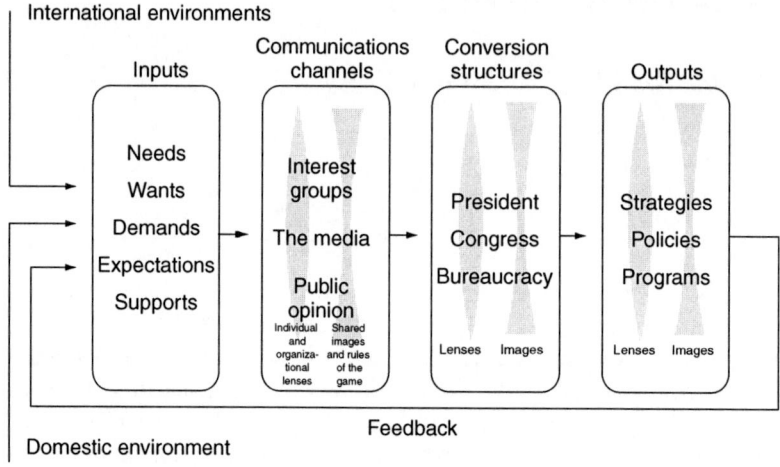

Figure 6.1 Defense policy process model (source: Hays, Peter, Brenda J. Vallance, and Alan R. Van Tassel, eds., With a foreword by Brent Scowcroft. *American Defense Policy*, Seventh Edition, fig. 4 p. 13. Copyright 1996 The Johns Hopkins University Press. Reprinted with permission of the Johns Hopkins University Press).

leadership; (2) the establishment of appropriate means by administrative agencies to achieve those goals; and (3) the allocation of resources by the administration and congress. In reality, the initial impulse for space policy will often come not from the administration leadership, but rather from advocacy coalitions that include political leaders, individuals from administrative agencies, private interest groups, academia, and industry. Illustratively, the policy embodied in President Reagan's Strategic Defense Initiative of 1983 can be traced to an initiative that originated in an advocacy coalition composed of academics, former military officers, and individuals like Edward Teller representing the Lawrence Livermore National Laboratory.

Aside from these advocacy coalitions, there are institutional players who are regular sources of policy initiatives. Most notable are the civil and military space programs administered by the National Aeronautics and Space Administration (NASA) and the United States Air Force (USAF). In recent years, this list has expanded to include non-governmental actors, especially commercial enterprises, both established industrial organizations like the major space contractors, and entrepreneurial enterprises utilizing private capital. Examples of the influence of these actors are found in the areas of telecommunications; global positioning, navigation, and timing (global positioning system (GPS)); remote sensing; and space launch vehicles to support civil, military, and commercial activities. The major institutional players are listed in Box 6.1.

Public policy reflects an aggregation of interests, both public and private. The process of aggregation involves discovering players whose perceived interests coincide with those of the policy initiator – modifying the policy proposal as necessary to garner the widest possible support. For example, an initiative to use commercial assets to perform services for the military in space might call on the support of the private companies who own those assets, congressional delegations who represent the districts in which those companies do business, interest groups lobbying for efficiency in government, and ideological supporters of the free market. The breadth of support depends as well on the feasibility of the proposal, and on the coincidence between the proposal and existing national goals.

The space shuttle was originally proposed as a follow-on to the Apollo program to sustain human spaceflight, and therefore justify the continued existence of NASA. The shuttle concept evolved to meet the requirements of an expanded coalition, involving the military and commercial sectors. In this process, claims for the program gradually expanded, so that eventually it was portrayed as a "single launch

---

Box 6.1 US government actors and space policy

Central Intelligence Agency (CIA)
Congress
Department of Commerce (DoC), Office of Space Commercialization
Department of Defense (DoD)
Department of Transportation (DoT)
    Federal Aviation Administration, Administrator for Space
    Transportation (FAA-AST)
Federal Communication Commission (FCC)
National Aeronautics and Space Administration (NASA)
National Geospatial Intelligence Agency (NGA)
National Oceanic and Atmospheric Administration (NOAA)
National Reconnaissance Office (NRO)
National Security Council (NSC)
Office of Management and Budget (OMB)
Office of Science and Technology Policy (OSTP)
President
State Department (DoS)
United States Air Force (USAF)
United States Geological Survey (USGS)

Source: Eligar Sadeh.

vehicle" for all civil, military, and commercial purposes. In the end, the shuttle coalition became so powerful that then Defense Secretary Weinberger became both a supporter and an advocate; it was Weinberger's intervention that convinced President Nixon to adopt the shuttle as the centerpiece of his space policy. In short, an expanding coalition of interest groups succeeded in both establishing the feasibility of this program and setting the Nixon Administration's agenda on space policy. The result of this coalition building was an over-optimization of the shuttle system as functions were added to the shuttle's mission that made it less capable of optimizing any single goal, whether civil, military, or commercial. It proved to be ideal for none of its stakeholders. By the time this result became apparent, the strength of the shuttle coalition was so pervasive that attempts to coalesce support behind alternative systems failed. Only the *Columbia* disaster enabled policymakers to adopt a replacement system, the Crew Exploration Vehicle and Project Orion of today. The shuttle example illustrates how the interplay of differing agendas among the actors in space can make rational policy-making more difficult. Some of those agendas are summarized in Table 6.1 below.

In order for political agendas to move from conception to policy, by way of presidential decision, congressional legislation, and administrative rule-making, the actors need to enlist the support of advocacy coalitions. These coalitions can be motivated by ideology, political conviction, or private interest, and can include engineers, scientists, military officers, policy-makers, space business leaders, and space enthusiasts in grassroots organizations. They fall into four general categories: (1) human destiny; (2) space science; (3) national defense; and (4) space business.[1] The characteristics of these groups are defined in Box 6.2.

Table 6.1 Space and defense policy agendas

Activity	Initial agenda	Agenda before 9/11	Future agenda
Military	Military space Freedom of space	Force enhancement Force support	Space control Force application
Scientific	Space science	Earth science Astronomy	Planets Environment
Civil	Human spaceflight	Space shuttle Space Station	Human exploration Human habitation
Commercial	Technology Development	Commercial applications	Economic competition

Source: Eligar Sadeh.

Box 6.2 Advocacy coalitions and space policy

Human Destiny

The Human Destiny coalition envisions humanity as a spacefaring species and calls for the settlement of the Moon and Mars. It is focused primarily upon purposive benefits to society and is made up of the so-called NASA (and NASA associated) "old hands" of the Apollo era (Wernher von Braun and Carl Sagan), and a number of space-based professional and grassroot organizations, such as the American Institute of Aeronautics and Astronautics, American Astronautical Society, International Astronautical Federation, National Space Society, Planetary Society, and Mars Society. Support for this coalition's view is also scattered throughout US federal agencies, universities, and parts of the public.

Space Science

The Space Science coalition includes research scientists in industry, academia, government, and their allies in congress and the federal agencies. Professional associations representing scientific disciplines are politically involved. The relationship between this coalition and others, like Human Destiny, is often tense. Even though federal funding for all areas of science and engineering increased after *Sputnik*, many in the US scientific community came to view NASA and the human spaceflight program with suspicion. This coalition seeks a mix of material and purposive benefits. In some cases, e.g. the Hubble Space Telescope, these groups persuaded congress to fund large-scale space science programs. They also, with less policy success, publicly criticized other NASA projects (e.g. the International Space Station).

National Defense

The National Defense coalition is driven by national security concerns. This coalition constitutes the relevant parts of the DoD (Ballistic Missile Defense Organization and the USAF Space Command), private defense contractors with whom the DoD works (e.g. Boeing and Lockheed Martin), and the US State Department (DoS). It receives additional support from the House and Senate Armed Services Committees in congress, and from prominent conservative and other pro-military organizations and individuals.

Space Business

The Space Business coalition is focused on material benefits and is made up of those interests that regard the space environment primarily as a venue for making profit. It includes the commercial space sector and their political supporters in the US Departments of Transportation and Commerce (Office of Space Commercialization), and parts of NASA. This group is also closely aligned with the Republicans on the House Space and Aeronautics Subcommittee and the Senate Commerce, Space, and Transportation Committee. Parts of this coalition are committed to free market capitalism and regard government involvement in space as a constraint on the commercial development of space.

Source: Eligar Sadeh.

The shuttle campaign expanded to include all of these groups, to the ultimate disadvantage of the program. The initiative that led to the International Space Station (ISS) also sought to tap into support from all these groups, and has struggled to maintain that support through difficulties in implementation. As a result, the ISS lost support of the science, defense, and business sectors. Despite this development, the program was sustained when President Clinton linked it to post-Cold War foreign policy goals, and thus brought into the calculation a powerful new constituency, the international collaborators on the ISS, including Russia.

Policy formulation, by its very nature, reflects many views, perspectives, and interests, and involves political accommodation leading to goal modification. The political process of formulation represents a compromise between what space organizations may want and might regard as most effective, efficient, or feasible, and what organizations perceive as the appropriate response to political forces. The political forces are shaped by historical conditions; and policy formulation is framed by the extent to which the chosen policies are congruent with the prevailing rationales and goals (see Table 6.1).

The defense policy process model mentioned in the introduction to this chapter elucidates the dynamics of agenda setting and formulation. As shown in Figure 6.1, the process entails the basic inputs, communications channels, conversion structures, and outputs, and the overall international and domestic environments that characterize agenda setting and formulation of policy. This process deals with day-to-day decision-making rather than crisis decision-making. In the daily decision-making process, associated with strategic policy, systems acquisition, and budgeting among other long-term actions, the domestic environment influences the space policy process through a number of mechanisms, including US strategic culture, the state of technology, and national security objectives. International regimes such as the International Telecommunications Union (ITU), which manages radio frequency spectrum and geostationary orbital slot allocations, the Missile Technology Control Regime (MTCR), the US International Traffic in Arms Regulations (ITAR) overseeing the transfer of technology associated with space launch vehicles and satellites, and international space law represent influences in the international environment in which policy-making occurs. It is important to recognize that feedback from the actions undertaken, and the strategies and policies created and resulting from the process, influence future decisions and actions.

The process accounts for the complicated and multidimensional inputs into decision-making that lead to policy formulation. In doing so, the model emerges from the literature on bureaucratic politics and political psychology. As an example, even at the inputs stage, the earliest stage of the process that equates with agenda setting, popular demands for a particular policy or approach (demands) may be moderated by the budget or service interests (supports). Similarly, expectations and demands from the international environment may significantly contrast with needs as defined by domestic events. This complicated situation is compounded as inputs are communicated through a number of channels and acted on by the executive and legislative branches of government and the bureaucracy. These actors serve as conversion structures that receive the varied, and frequently conflicting, system inputs and convert them into the formulated decisions of government (outputs).[2]

While this conversion sounds straightforward, the actual process is far more difficult and is characterized by coalition building, bargaining, compromise, and goal accommodation through formal and informal mechanisms. For example, governmental actors (see Box 6.1) and advocacy coalitions (see Box 6.2) often hold competing and conflicting views of the situation. Advocacy coalitions will seek to influence different congressional members; the DoD and DoS may seek different resolutions to the situation based on organizational interests and goals; and members of the presidential staff, such as the OSTP, may experience increased scrutiny and criticism by the media when deciding on appropriate actions to pursue. Further complicating this process are the images and lenses brought to the process by each of the participants, something else that must be considered when seeking to understand the policy process. The agendas discussed earlier shape the context of the images and lenses. These images and lenses represent the ways in which individual perspectives and organizational cultures, interests, and goals shape and influence the political outputs.

As inputs enter the system from multiple sources, they are acted on through both formal and informal processes. The interagency, budget, and acquisition processes represent some of the formal approaches to resolving policy issues. Organization and agency cultures and values may converge around "widely shared values and images of international reality" and "rules of the game" as framed by the constitutional provisions, laws, regulations, and bureaucratic structures through which decisions are made.[3] Conversion structures frame the decision-making process and outputs of policy formulation. Although these outputs appear rational, the process leading to them definitely does not conform to the rational decision-making model. The pulling and hauling that characterizes bureaucratic policy-making often results in formulated outputs that are in fact far from rational and more closely resemble incremental decision-making.[4]

The outputs in the form of policies, strategies, and programs must be acted upon through implementation. Policy implementation involves the development of the enabling space technologies and their application to the actual building of the hardware and systems to support space-related programs and projects.[5] The technical skills of the implementers, i.e. national space agencies and commercial industries, are influenced by political forces. To clarify, complex space technology often presents a twofold challenge for implementers: (1) to manage technology to avoid failure; and (2) to maintain capacity for meeting external political challenges. The space shuttle *Challenger* and *Columbia* accidents were in part a result of political pressures for an operational state for the technological system. Political and organizational factors eclipsed technical concerns in the *Challenger* launch decision,[6] and in the management decisions that precipitated the *Columbia* accident.[7] These unsuccessful outcomes are an example of how management emphasized political issues more than technical concerns.

The fact is that the implementation organizations, like NASA or the DoD, are obligated to seek political support and be accountable to political sponsors, namely Congress and the president. The implementation process is influenced by a trade-off relationship between political pressures for accountability, and the need for organizational autonomy and methods to realize successful programmatic and project outcomes. To add to this, administrative organizations are left to grapple with imprecise language and the interpretation of congressional intent, further complicating implementation. The case of ITAR and the export control of commercial satellites discussed in this chapter illustrate this issue. The consequences of the implementation difficulties highlighted previously frequently lead to reformulation of the policy and then reimplementation as efforts are made to correct these problems. The ISS program, in particular, went through such a process involving at least six major redesigns and efforts in congress to terminate the program (albeit, unsuccessful).

2001 Space Commission Report

Increasing budgets, the problem of coordinating the management of activities, and the tension created by the competing needs of air and space requirements experienced by the USAF led the congressional House and Senate Armed Services Committees to direct the formation of the Space Commission, chaired by Donald Rumsfeld prior to his selection by President Bush as the Secretary of Defense. The resulting Space Commission report noted that the United States is highly vulnerable "to surprises in space" which were equated to a "Space Pearl Harbor."[8] According to the Space Commission, this vulnerability results from the US dependency on space assets, the limited attention given to analyzing and assessing threats to these assets, and the lack of responsibility and accountability resulting from the limited coordination and oversight of the disparate space activities and agencies.

The reasons for these concerns are due in part to the many government organizations involved in space-related activities, and therefore those involved in policy decision-making. The decision process is further complicated by additional actors, some of which were noted already as making up the international environment affecting decision-making, and others in the domestic environment, such as the various advocacy coalitions and industrial contractors included in Box 6.2 who are involved in and influence the decision process.

The Commission proceeded to make a number of recommendations intended to "achieve greater responsibility and accountability" in space activities.[9] Given the Commission's charter of reviewing space policy management and organization, these recommendations focused on the need for national leadership supported by a presidential space advisory group and a senior interagency group, and a closer working relationship between the Secretary of Defense and the Director of Central Intelligence. It recommended designation of an Under Secretary of Defense for Space, Intelligence, and Information, and made several recommendations at the service level. Perhaps, the most significant of these was the recommendation to amend Title 10 of the US Code (USC), assigning the USAF "responsibility to organize, train, and equip for prompt and sustained offensive and defensive air and space operation. In addition, the Secretary of Defense should designate the Air Force as Executive Agent for Space within the Department of Defense."[10]

Results of the Space Commissions recommendations

Many of the Space Commission's recommendations were not acted on immediately because of the change in administration, resulting in the appointment of Rumsfeld as Secretary of Defense, and the change in security focus resulting from the events of 9/11. Since then a number of the recommendations were put into action, and in some cases, decisions and actions taken as a result of the Commission's findings have been reversed. Following publication of the Space Commission's findings, Secretary Rumsfeld designated the USAF as the DoD Executive Agent for Space in a memorandum dated October 2001 and a draft directive was written to that effect.[11] The actual directive formalizing this designation was completed in 2003, delaying any attempts to develop an implementation plan.[12] Despite this, the USAF did undertake organizational changes as suggested by the Space Commission and the Rumsfeld memorandum, to include realignment of headquarters and formation of a National Security Space Integration Office that combined with the National Security Space Architect in May 2004 to form the National Security Space Office.[13] Figure 6.2 depicts these changes.

One of the most significant actions taken in response to the Space Commission was the designation of an acquisition executive, an important action given the complicated and complex challenge of coordinating space needs. This challenge is discussed further in Chapter 13. In addition to this action, several other changes were initiated in response to the Commission. The new Under Secretary of the Air Force was appointed Director of the National Reconnaissance Office in December 2001. Following designation of the Under Secretary of the Air Force as the USAF Acquisition Executive for space in 2002, the Under Secretary of

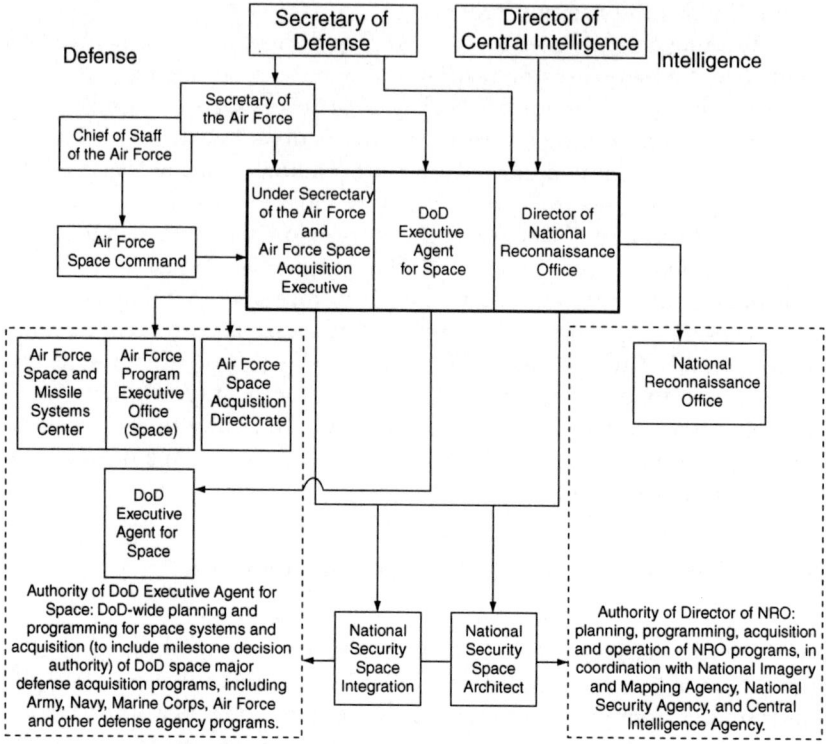

Figure 6.2 US DoD's and the USAF's Organization for National Security Space
(February 2003) (source: USAF).

Defense, Acquisition, Technology, and Logistics delegated the Secretary
of the Air Force (with authority to delegate authority to the Under Secretary) for milestone decision authority of major DoD space programs. The
former Acting Secretary of the Air Force, Peter Teets, held the three jobs
of NRO Director, Air Force Under Secretary, and Pentagon Space Acquisition Chief prior to his retirement in March 2005.

   In related military service and command changes, the Air Force formed
the USAF Space Command separate from the Space Command and North
American Aerospace Defense Command in 2002, and Space Command
was combined with Strategic Command. The Air Force Space and Missile
Systems Center separated from Air Force Materiel Command and was
realigned under USAF Space Command. The 14th Air Force at Vandenberg Air Force Base opened the Joint Space Operations Center in 2005,
serving as the space operational component of US Strategic Command.[14]

   One of the significant recommendations emerging from the Space
Commission was the requirement to identify, train, and employ a cadre of

space professionals. This is an area on which much recent attention has been placed, including in President Bush's national space policy of October 2006. As a result of criticism in this area, USAF Space Command initiated the Space Professional Development Program. The National Security Space Institute was created in October 2004 with the goal of developing credentialed space professionals. This will be accomplished by offering courses grouped under the Space Tactics School that provides Advanced Space Training courses on specific mission areas, the Space Operations School for warfighter preparation and space familiarization training, and the Space Professional School for continuing education at appropriate career points.[15]

Despite the changes discussed above, the many challenges associated with the organization and management of defense policy are not completely resolved. Many space analysts believe that Secretary Teets personified the rare individual who could balance the three positions he held and make things work, while others argued that the three jobs stretched one person beyond what was reasonable, and that Teets found himself focusing on NRO business at the expense of other duties. With Secretary Teets' retirement, astronaut and USAF Reserve Maj. Gen. Ronald Sega was appointed Under Secretary of USAF and Executive Agent for Space, and Donald M. Kerr was appointed Director of the National Reconnaissance Office, thus again separating the two responsibilities. Currently, there is discussion of reversing some of the reorganization actions resulting from the Space Commission recommendations, and there is purportedly an initiative within the Air Force to reduce the rank of the Commander of USAF Space Command from a four-star to a three-star officer, effectively downgrading the significance of the position.[16]

Post 9/11 changes, the policy process model, and space policy

The defense policy process model presented in Figure 6.1 and discussed in the section above set the tone for a study intended to address space and defense policy in the evolving post-Cold War world. The ending of the Cold War was seen by many, and certainly by the authors here, as the most significant change in the international security environment to occur in a generation. Yet, another momentous event took place – the tragedies of 9/11. It would seem, intuitively, that such a tragedy would result in a massive review of space and defense policy-making that would potentially effect changes in the model.

In fact, a review and assessment was conducted, although the results of the largest and most significant effort were not published until 2004. *The 9/11 Commission Report* reviewed the historical materials leading up to the terrorist attacks, provided a detailed account of the decision-making and actions taken prior and in response to the attacks, and made recommendations on how to prevent further devastating attacks. As noted by many analysts, the changes in government organizations and processes undertaken in the years between the attacks and the completion of the commission report were not incorporated into the report, meaning that some of the recommendations were obsolete when made.[17] Nonetheless, based on the four areas of failure identified by the commission (imagination, policy, capabilities, and management),[18] the commission did propose a number of changes with a potential impact on the space arena, particularly in the intelligence community.[19]

The question is whether the changes that resulted from the terrorist attacks, either prior to the commission report or because of the recommendations, effectively alter the space policy process captured by the model in Figure 6.1. As an example, the passage of the Intelligence Reform and Terrorism Prevention Act of 2004 resulted in designation of a Director of National Intelligence (DNI) among other reforms,[20] potentially altering the inputs and conversion channels illustrated in the model.

Multiple inputs continue to flow into the policy process depicting individual and organizational perspectives, world views, standard operating procedures, and rules of the game, among other factors. The legislation to reform the intelligence community experienced the compromise and consensus-building characteristic of politics, with the end result appearing less like radical reform and more like incrementalism and satisficing. While the legislation may lead to a more consolidated targeting of intelligence assets against specific threats, it changes little within the model.

Commercial satellite export controls

The case of export controls on US commercial satellites exemplifies the defense policy process model shown in Figure 6.1. This case is characterized by bureaucratic politics leading to policy outcomes that are not rational, i.e. the desired outcome of national security is not met and commerce in the satellite sector is harmed. The constraints on rational policy-making are a result of competition, conflict, and protectionism, the bureaucratic "pulling and hauling," among the relevant actors, including the president and Congress, the DoS, the DoC, and the DoD. It is this pulling and hauling that results in policies for licensing the export of commercial satellites that are far from orderly, stable, or predictable.[21]

The crux of the political issue revolves around bureaucratic control and jurisdiction over the licensing process for export of commercial satellites. Since commercial satellites represent a dual-use space technology,[22] bureaucratic pulling and hauling exists between the framing of export controls as a matter of national security as opposed to a matter of business and commerce. The national security advocates, among them the president, Congress, the DoS, and the DoD, view commercial satellites and the related technologies as items to be controlled for export within the same legal regime that controls export and trafficking of arms. The DoS, through the Office of Defense Trade Controls Policy, is the bureaucratic entity that governs this regime, known as ITAR and the associated Munitions Control List (MCL). The DoD, through the Defense Threat Reduction Agency, assists the DoS in implementing its regulatory authority. The commercial space advocates, among them the president and Congress, especially from 1988 to 1998, the DoC, and the aerospace and defense industries, view commercial satellites as an indicator of US leadership with a strong market share in the global commercial satellite sector. Logically, the way to regulate export of these satellites is through the legal regime that governs dual-use technologies used commercially. This is the Export Administration Regulations (EAR) administered by the Commerce Bureau of Industry and Security. Commerce governs exports through the Commerce Control List (CCL), and from 1992 to 1999 this regime applied directly to the export of commercial satellites.

This commercially oriented approach enabled China to compete within the US market for the launch of commercial satellites. From 1992 to 1996, the Chinese Long March rocket failed in launching commercial satellites manufactured by US companies Hughes Space and Communications (purchased by Boeing in 2000) and Space Systems Loral. As required by the insurance companies covering these companies' assets, investigations into the launch failures were concluded and submitted to the DoC for approval. The DoC then authorized Hughes and Loral to communicate the technical reports to the Chinese launch officials. The transfer of the reports sparked political controversy over the statutory authority of the DoC to allow such a transfer without the proper review and oversight by the DoS. Specifically, the controversy focused on the export of knowledge dealing with the reliability of space launch vehicle technology, and more generally, was linked to the issue of ballistic missiles and US–Chinese relations. Congress investigated this issue of transfer through the *Report of the Select Committee on U.S. National Security and Military/Commercial Concerns with the Peoples' Republic of China* (known as the *Cox Report*), and determined that Hughes and Loral transferred to China, in violation of US export control laws – the Arms Export Control Act of 1976 and the ITAR regime – missile design information and knowledge that improved the reliability of the Chinese Long March rocket useful for civil and military purposes.[23] The congressional response led to the National Defense Authorization Act for Fiscal Year 1999 that directed sole export control responsibility to the DoS using the ITAR/MCL regime for commercial satellites. The DoS' jurisdiction began in March of 1999, and continues through this writing in 2007. According to many space leaders, the application of ITAR to commercial space technologies is a misapplication of the regime and is one of the top space policy issues requiring congressional redress.[24]

International and domestic environments

In applying the defense policy process model to the case of export controls and commercial satellites, it is important to explain how national security and commercial space advocates' respective policy preferences, needs, wants, demands, and expectations are influenced by the international and domestic environments. The international and domestic environments date back to the Cold War and the issue of how to control dual-use technologies. The concern, then and now, is that such technologies can be used for the development of arms that, in turn, can lead to proliferation of ballistic missiles, and nuclear, biological, and chemical weaponry. Dual-use technologies with these potential applications are viewed by national security advocates as sensitive items to be controlled.

One aspect of control lies with the statutory authority within the United States for dual-use technologies. This authority lies with the Export Administration Act (EAA) of 1979 in which congress delegated to the executive branch the legal authority to regulate foreign commence by controlling and licensing exports. EAA is the domestic environment from which the DoC's EAR regime (discussed previously) emerged. Of note, the EAA expired in September 1990; reauthorization of EAA took place for short periods with the last incremental extension expiring in August of 2001. Since then, no new congressional legislation has been passed to either reauthorize or rewrite EAA, and the regime functions on the basis of presidential authority under the International Emergency Economics Powers Act.

Within the context of the post-9/11 environment and the resulting emphasis on national security, at times to the detriment of commercial interests, the congressional failure to act on the EAA further strengthens and maintains the DoS-led ITAR regime for control of commercial satellites. Furthermore, the origins of the EAA are Cold War-related and originate from the Export Control Act of 1949. Even though the EAA of 1979 represents a lessening of restrictive export control in comparison to the Export Control Act and subsequent amendments to that Act, the legal regime is a relic of Cold War-era international politics and national security rivalries.[25] As a result, the EAA has not been sufficiently adapted as an export control regime for the post-Cold War international environment of non-traditional security concerns, developments in space technologies, capabilities and applications, and the emergence of global commercial space activities.

A second aspect of control deals with the Arms Export Control Act of 1976, the basis for the ITAR export control regime. This regime also was established during the Cold War environment and has not undergone any statutory changes. Further, neither the DoS nor the DoD made any changes to the implementation modalities of any of these Cold War regimes.[26] During 1999–2000, both the President and Congress noted the need to review the arms export control regime to streamline the processing of export license applications. Neither the DoS or DoD acted on these recommendations. The issue of delays and the cost of bureaucratic compliance in the granting of export licenses is one of the key concerns of commercial space advocates; these concerns translate into an economic issue for the commercial satellite sector. The economic issue also posits a barrier to entry for new space commercial companies, often referred to as alternative space (alt.space), that are attempting to enter into existing markets such as space launch services, or to develop new markets, such as space tourism.

A third aspect of the control issue exists at the international level. In 1949, a multilateral export control regime called the Coordinating Committee for Multilateral Export Controls (CoCoM), involving North Atlantic Treaty Organization (NATO) allies, was established. This regime mirrored US domestic controls as established with the Export Control Act of 1949. CoCoM advanced restrictive export controls on sensitive dual-use technologies at the multilateral level. The regime was dissolved in 1994 and replaced in 1996 by the Wassenaar Arrangement on Export Controls for Conventional Arms and Dual-Use Goods and Technologies. The Wassenaar Arrangement, as compared to CoCoM, lessened export controls of dual-use technologies at the international level and is more loosely organized with more limited institutional structures. It relies on consensus by state members, frequently resulting in a lowest common denominator approach for multilateral export control, minimal reporting requirements preventing pre-export consultations among state members, and a lack of authority among state members to block transactions of other state members.[27] In addition, the liberal multilateral regime that emerged with Wassenaar no longer sought multilateral control over commercial satellite technology or expertise. This development influenced the US environment and raised national security concerns when dealing with the export of dual-use technologies. In the end, the liberalization of the international legal regime is a factor that favors the national security space advocates' position and their preference for ITAR as the regime to control and license exports of commercial satellites.

Communications channels

In this case of export controls, the communications channels in the defense policy process model are a function of the relevant bureaucratic strategic cultures. The strategic cultures of the national security advocates versus the space commerce advocates frame the political debates and arguments. This framing represents the organizational lenses, images, and rules of the game regarding export controls of commercial satellites.

Commercial space advocates frame the export control issue through the lens of foreign availability of technology. The contention is that the proliferation of technology cannot be effectively controlled, and US dominance of space technology cannot be assumed. The globalization of space commerce points to the fact that unilateral controls will not stop foreign states from acquiring the technologies. Thus, US dominance in space commerce is diminished, while foreign businesses win new markets and gain incentives to enter into new markets.[28] This is complicated by the fact that as space commerce is increasingly global, many components in the commercial satellite sector are manufactured worldwide and considered commercial commodities. ITAR is not designed to deal with the global nature of the industry, and the outcome is one where foreign commercial satellite developers seek to reduce dependence on US satellite components due to delays associated with the US export licensing process. The emerging trend is one where US satellite manufacturing companies, which must adhere to ITAR restrictions, are at a growing disadvantage as inventory of "ITAR-free" (no US-manufactured components) satellites expands abroad.[29]

In addition to the economic argument, space commerce advocates see a link between national security and robust export control industries, and hence, favor an export control regime that is streamlined, less complex, and therefore not an impediment to exports. As an example, the DoC presumes that the issuing of an export license is routine unless good cause can be shown otherwise.[30] Space commerce advocates argue that national security is undermined when exports are impeded, resulting in the loss of US market share, and the limitation of US satellite components through export controls leads to greater foreign research and development (R&D) investments in this area. In turn, these foreign R&D investments can be leveraged to achieve parity and even surpass the US technological lead. In conclusion, space commerce advocates frame commercial satellite technology as possessing no inherent strategic or military relevance, a view shared with the state members of the Wassenaar Arrangement with the exception of the United States.[31]

In contrast, the national security advocates maintain that there is a need to control commercial space exports as sensitive military technologies. This control prevents the proliferation of technologies that could be used by hostile, rogue states against the United States or its allies, secures the DoD's reliance on the commercial sector for R&D as a result of declining defense budgets in the 1990s, and sustains the US military use of commercial space assets for operations, including commercial satellites for telecommunications and remote sensing purposes. National security is framed in ideological and warfighting terms – limiting the diffusion of technology advances US foreign policy interests and enhances national security. The framing of export control as a national security issue compelled congress to place commercial satellites and related technologies within the authority of the ITAR/MCL regime.[32] The Chinese Long March "satellite scandal" and the events of 9/11 served to strengthen this worldview and weaken political attempts to reform the export control regime.

Conversion structures and outputs

The model's conversion structures – how the president, congress, and bureaucracies interact – exemplify bureaucratic politics. It is the nature of bureaucratic politics that results in policy outputs that are not rational. A rational policy-making process suggests outputs that serve the desired communications channels of at least one group of advocates. In this case, the policy outputs, albeit unintended, do not ideally realize the policy preferences of either the national security or commercial space advocates. On one hand, ITAR can damage national security by placing legal and bureaucratic restrictions on the US military use of commercial space assets that rely on a robust satellite industry.[33] This includes risks to the military use of: commercial satellites for operational support; advanced satellite technologies developed in the commercial sector; and foreign suppliers for satellite components and services needed for operations. On the other hand, export control of commercial satellites vis-à-vis ITAR has made the US space and satellite component industry less competitive internationally and contributed to a weakening of US market position.[34]

How did the issue of export controls of commercial satellites result in policy outputs that are not desired? The answer to this question lies in the nature of how the relevant political actors serve as conversion structures. Prior to 1992 export control of commercial satellites fell within the purview of the ITAR regime, but beginning in 1988 President Reagan began to loosen export restrictions on commercial satellites to keep US industry competitive in global markets, and to advance national space policy for the development of the commercial space sector. The following Bush and Clinton administrations shared these policy preferences and acted to these ends. Bush and Clinton used presidential legal authority to waive trade sanctions with China, put in place through congressional legislation following the Tiananmen Square massacre. The sanctions waived included commercial satellites for export to launch on the Chinese Long March. The policy conflict between the President and Congress set the stage for the Chinese satellite scandal and the resulting 1999 congressional legislation that reversed the loosening of export controls initiated by Reagan.

The theme of policy conflict persisted as Bush made use of presidential authority to extend the EAA and pocket-vetoed a congressional bill that would have amended and extended the full EAA on a permanent basis.[35] In this bill, congress took more of a national security position on the export of dual-use items, in conflict with Bush's post-Cold War commercial view for the increased role of economics in national security. Bush sustained this view by removing all items from MCL that were on the CoCoM dual-use list. This, in turn, led to split jurisdiction, in the period 1992–1996, between the DoS and DoC for export controls. An interagency review process initiated by Bush determined which of the dual-use items listed on MCL could be transferred to CCL. Under the DoC's business-friendly licensing process, these transfers made it easier to export some commercial satellites for foreign launches. Less advanced commercial satellites were exported as commercial goods under the EAR regime. Throughout the story of commercial satellite export controls, the DoS and DoC have both sought influence and authority, and split jurisdiction was viewed by the actors as a compromise to resolve this dispute.[36] Nevertheless, the differences in strategic cultures of each bureaucracy sustained the struggle for political influence over export controls.

As a result of split jurisdiction, the technical parameters for determining whether commercial satellites should be treated as munitions or dual-use commercial goods became unworkable by 1995.[37] One of the issues that emerged was that the export regulatory bureaucracies at the DoC, DoS, and DoD lacked the requisite technical expertise to determine which technologies to control as munitions versus which could be exported as commercial commodities.[38] This was further exacerbated by the fact that regulatory monitors were asked to implement near impossible tasks – apply overlapping, self-contradictory rigid sets of rules, and track all hardware for export without explicit guidance on what to protect for reasons of national security and what was a commercial commodity. Consequently, split jurisdiction was abandoned as a policy preference by the actors. In October 1996, and until March 1999, congress assigned primary jurisdiction to the DoC. Since then, commercial satellites and related technologies are listed on MCL and regulated for export by the DoS.

The moves undertaken by the political actors to transfer jurisdiction to the DoC were met with countermoves by DoS export officials determined to exert their full authority to the extent permissible by law. The political process underlying the transfer to DoC jurisdiction was characterized by bureaucratic politics and conflicts. Both export control bureaucracies sought regulatory authority, and their self-interest to do so became a goal in-and-of-itself. The bureaucratic politics concept that "where you sit defines who you are" applies directly in this case; DoS and DoC regulators were explicitly tied to the strategic cultural perspectives of their organizations. As policy preferences for DoC jurisdiction moved to fruition by 1996, the DoS pursued enforcement regulations that made it increasingly difficult and costly for satellite companies to export if even a single component remained subject to DoS control through MCL.

Congressional reaction to the Chinese affair and the sustained efforts of national security advocates advancing their case for export controls led to congressional legislation that resulted in sole DoS jurisdiction in 1999. In many ways this action was reactive rather than rational. One indication of this is that the export violations committed by Hughes, Loral, and Boeing did not damage US national security in any material way; the expertise transferred to China only marginally benefited Chinese missile programs by improving launch reliability.[39] In fact, many of the breaches were little more than technical violations of DoS export control regulations dealing with services that could "in theory" be applied for national security purposes.[40]

In this situation, the policy output of DoS jurisdiction is suboptimal; rather than seeking a compromise, the DoS countered the preferred policy preferences of the commercial space advocates. Given the drive for bureaucratic self-preservation, the DoS took the congressional mandate for sole jurisdiction and unilaterally implemented its approach, through administrative rule-making, to realize its national security perspective. This raises a number of issues. First is the issue of what was intended by the *Cox Report* recommendations, which had prompted congress to give the DoS commercial satellite licensing authority. It is not clear whether the recommendations intended to control the export diffusion of technology from solely a national security standpoint, or to control the technology diffusion in a way to satisfy both national security and commercial advocates' preferences. This ambiguity provided the DoS with the opportunity to advance their national security perspective. At the same time, officials at the DoS expressed their desire to work with space commercial businesses by facilitating and approving ITAR applications, and viewed the political problem as rooted in the congressional mandate for the department's sole jurisdiction and enforcement of the export control law.[41] In fact, the DoS does approve the vast majority of export license applications.[42] The issue with the export control of commercial satellites within the ITAR regime is not one of denial of licenses, but rather in how the DoS enforces the law. Enforcement leads to excessive delays and bureaucratic compliance with the export regulations that are a cost to the commercial satellite sector.

What is also clear is that the DoS is enforcing the law in ways that are not necessarily in line with what Congress intended, yet Congress itself fails to act on this problem. To illustrate, the *Cox Report* called for: congressional reauthorization of EAA; continuous updating of the export control regime; and streamlining the licensing procedures to provide greater transparency, predictability, and certainty. In all these areas, neither the DoS or Congress took any substantive actions. Not only did the DoS act unilaterally to do other than what was recommended by the *Cox Report*, but Congress also failed in its basic oversight role to hold the DoS accountable to congressional policy intentions and preferences. This dynamic, together with the events of 9/11, stalled reform advocates' efforts. Although there is pending legislation in Congress to follow through on the *Cox Report* recommendations, the advocates are in the minority. The proposed congressional Satellite Trade and Security Act of 2001 went as far as to restore DoC jurisdiction, but the measure failed to advance, and through the 109th Congress there have been no serious attempts to introduce subsequent legislation or to put the issue on the agenda.[43] Other barriers to reform include export risks and organizational constraints on expediting DoS processes for exporting commercial satellites.[44] These barriers stem from the fact that technical expertise at the DoS and DoD is lacking. Even though some incremental advances in addressing these barriers have taken place, as recommended by the *Cox Report*, the policy lesson of split jurisdiction is that determining risk is in many ways unworkable and the control of satellite exports through the national security lens does not readily lend itself to streamlining the licensing process.[45]

The policy dynamic discussed earlier, the DoS countering the DoC, resulted in sole DoS jurisdiction. The DoS unilaterally reversed the DoC approach that exempted many items from requiring licenses,[46] extended ITAR controls to US allies for commercial satellites,[47] and advanced regulations that required the return of hardware to its state of origin for repair.[48] The DoS also issued retroactive regulations for the Technology Assistance Agreements governing technology transfers for satellites that had been licensed by the DoC. These agreements require marketing discussions and the exchange of basic technical information with insurance companies and launch service providers for satellites exported and launched under DoC jurisdiction. The retroactive approach adopted by the DoS created a situation where new technology transfer licenses had to be issued for satellites already operating in orbit. The DoS even acted to reverse Reagan's decision that exempted fundamental research information from an export license.[49] Export directives to control such information affect NASA, universities, and industry R&D as they require licenses for any collaboration with foreign nationals on fundamental research. In addition, the DoS and DoD practice intrusive monitoring, allowing monitors' access to proprietary knowledge. Despite this, industry has not objected due to a fear of congressional reaction and their dependence on governmental contracts.

The end of the commercial satellite export story is the resulting damaging consequences for US technology and business leadership in space. The political process began with the incremental political liberalization of export controls in response to the changing international post-Cold War environment and the rapid increase in space commerce globally. The process then transitioned to congressional action to overturn the then existing satellite export control regime in favor of DoC jurisdiction.[50] All the while, the process was driven by bureaucratic politics between the DoC and DoS. In the context of the post-9/11 world and the security concerns the attack generated, the general sense was that US business and commercial interests should never trump national security interests. The DoS succeeded in advancing their national security worldview as the US national interest, a costly situation for commercial space and their advocates.

Does the model hold?

The remaining question for this chapter is whether space policy is in some way different, unique, or special, placing it outside the scope of the analytical model discussed earlier (Figure 6.1). Some argue that the use of space assets for commercial, intelligence, science, and defense purposes means that there is a unique convergence (or divergence) of interests and uses in this one medium. Others add that this medium is particularly expensive, with budgets controlled by multiple organizations and agencies such as the DoD, individual military services, and NASA. Finally, there are those who note the "black" versus "white" nature of space, referring to the fact that it is both a highly restricted intelligence environment as well as an open, commercial, and scientific environment.

While all of these arguments are true, they do not negate the applicability of the model presented and discussed in this chapter. In many ways these arguments further illustrate the complex environment, interests, and policy preferences that the model seeks to illustrate. Space policy-making, whether in the civilian or defense sector, involves a multitude of advocacy coalitions, governmental actors and agencies (space bureaucracies), and commercial corporations, competing over resources and objectives, and control of space programs and projects. If there is a shortcoming in the model, it is that it omits the significant area of implementation. As the case study above indicates, implementation is often clouded by the varying interpretations of organizations and agencies (again through their organizational lenses) of congressional intent.

Conclusions

Space policy decision-making has come full circle in many ways. Attempts to streamline management and organization associated with this area, the goal of the 2001 Space Commission report, produced a spotty record of success. While at this writing it is too early to argue that no progress, whether through streamlining or consolidation, was made as a result of the Commission's recommendations and the subsequent changes introduced, it is also too early to conclude that space policy decision-making falls outside of the model discussed in this chapter.

This chapter proposed the model as a tool to examine and better understand policy-making related to space. The defense policy process model illustrated in Figure 6.1 through its linear presentation shows a less dynamic process as compared to the reality. Having said this, it does allow the reader and policy analyst to look inside the individual model steps identified with agenda-setting and policy formulation, to better capture the lenses and images influencing the decision process. Using the model presented when analyzing space and defense policy debates and issues helps to better understand this complicated policy arena.

Notes

1 Christopher J. Bosso and W.D. Kay, "Advocacy Coalitions and Space Policy," in Eligar Sadeh (ed.), *Space Politics and Policy: An Evolutionary Perspective* (The Netherlands: Kluwer Academic Publishers, 2002).

2 Schuyler Foerster and Edward N. Wright, "The Twin Faces of Defense Policy: International and Domestic," in Schuyler Foerster and Edward N. Wright (eds.), *American Defense Policy*, 6th edition (Baltimore: The Johns Hopkins University Press, 1990).
3 Morton H. Halperin and Arnold Kanter, "The Bureaucratic Perspective: A Preliminary Framework," in Morton H. Halperin and Arnold Kanter (eds.), *Readings in American Foreign Policy: A Bureaucratic Perspective* (Boston, MA: Little, Brown, 1973).
4 Incremental policy making does not solve an identified national problem and is more apt to represent a continuing government commitment. See Charles E. Lindblom's classic essays, "The Science of Muddling Through," *Public Administration Behavior* 19(2) (1959): 79–88; and "Still Muddling ... Not Yet Through," *Public Administration Review*, November/December (1979): 517–526.
5 The task of implementation is "rocket science." It is terribly complicated. Launchers explode and spacecraft disappear. No one wants failure. Good enough is not good enough for technology within which thousands of components must work in tandem for a mission to succeed. Howard E. McCurdy, "Bureaucracy and the Space Program," in Eligar Sadeh (ed.), *Space Politics and Policy: An Evolutionary Perspective* (the Netherlands: Kluwer Academic Publishers, 2002).
6 *Report of the Presidential Commission on the Space Shuttle Challenger Accident, Volume 1* (Washington, DC: Government Printing Office, 1986); and Diane Vaughan, *The Challenger Launch Decision: Risky Technology, Culture, and Deviance at NASA* (Chicago, IL: University of Chicago Press, 1995).
7 *Columbia Accident Investigation Board, Volume 1* (Washington, DC: Government Printing Office, 2003).
8 *Report of the Commission to Assess United States National Security Space Management and Organization*, January 11, 2001, p. xiii (available at: www.dod.gov/pubs/space20010111.html. Accessed December 14, 2005).
9 *Report of the Commission to Assess United States National Security Space Management and Organization*, p. ix.
10 *Report of the Commission to Assess United States National Security Space Management and Organization*, p. xxxiv.
11 *Inside the Pentagon*, March 7, 2002, pp. 15–18; cited in Benjamin S. Lambeth, *Mastering the Ultimate High Ground: Next Steps in the Military Uses of Space* (RAND: Project Air Force, 2003) Appendix (available at: www.rand.org/pubs/monograph_reports/MR1649. Accessed December 12, 2005).
12 United States General Accounting Office, *Report to Congressional Committees; Defense Space Activities: Organizational Changes Initiated, but Further Management Actions Needed*, GAO-03-379, April 2003, pp. 8–9 (available at: www.gao.gov/new.items/d03379.pdf. Accessed January 4, 2006).
13 Several of the Commission's recommendations were not fulfilled. As an example, the Secretary of Defense chose not to seek amendment of Title 10 USC, choosing to delegate authority to the Air Force instead.
14 "14th Air Force History" (available at: www.vandenberg.af.mil. Accessed March 13, 2006).
15 Lt. Col. Frank Gallagher, USAF, "HQ National Security Space Institute: Answering the Space Commission's Call to Create Credentialed Space Professionals," *High Frontier* 2(1): 40–43 (available at: www.petersonaf.mil. Accessed March 13, 2006).
16 Press release, United States Senator Wayne Allard, March 1, 2006 (available at: http://allard.senate.gove/public/index.cfm. Accessed 17 March 2006).
17 Richard A. Falkenrath, "The 9/11 Commission Report: A Review Essay," *International Security* 29(3) (2004/5): 182. Falkenrath notes that:

This period was of course a time of exceptional governmental activity and change in the United States and indeed the world: the war on terror had been launched; many new laws and regulations had been enacted; security spending had increased dramatically; the institution of the executive branch had been extensively reorganized and redirected; and major new efforts were under way to secure the homeland – all for the express purpose of preventing another terrorist attack on the scale of 9/11.

18 *The 9/11 Commission Report: Final Report of the National Commission on Terrorist Attacks upon the United States* (New York: W.W. Norton, 2004), p. 339.
19 See the final chapter of *The 9/11 Commission Report* titled "How To Do It? A Different Way of Organizing the Government," which lists five major recommendations:

> unifying strategic intelligence and operational planning against Islamist terrorists across the foreign–domestic divide with a National Counterterrorism Center; unifying the intelligence community with a new National Intelligence Director; unifying the many participants in the counterterrorism effort and their knowledge in a network-based information-sharing system that transcends traditional governmental boundaries; unifying and strengthening congressional oversight to improve quality and accountability; and strengthening the FBI and homeland defenders.
>
> (pp. 399–400)

20 See Title I, also called the National Security Intelligence Reform Act of 2004, US Congress.
21 Joan Johnson-Freese, "Alice in Licenseland: U.S. Satellite Export Controls since 1990," *Space Policy* 16(3) (2000): 195–204.
22 Commercial satellites are clearly intended for commercial use and applications, but do represent applications and technologies that could be used for military purposes and military satellite development.
23 *Report of the Select Committee on U.S. National Security and Military/Commercial Concerns with the Peoples' Republic of China* (United States House of Representatives, 1999). See www.house.gov/coxreport (accessed December 11, 2006). Both Boeing and Loral were fined by the US federal government for the export violations; both companies paid the fines in 2002. Boeing was also charged with similar export violations concerning Sea Launch – a joint venture with Russian, Ukrainian, and Norwegian companies – during this same period.
24 *The Space Report: The Guide to Global Space Activities* (Colorado Springs: Space Foundation, 2006); and *Space 2030: Exploring the Future of Space Applications* (Paris: OECD, 2004).
25 Ian F. Fergusson, *The Export Administration Act: Evolution, Provisions, and Debate* (United States Congressional Research Service, the Library of Congress, updated May 5, 2005).
26 *Defense Trade, Arms Export Control System in the Post-9/11 Environment* (United States Government Accountability Report, February 2005).
27 Ian F. Fergusson, *The Export Administration Act*.
28 Export controls on space commerce create risk through uncertainties, result in losses of markets because of impacts on the space industry's ability to serve international markets, and prevent efficient industry restructuring to the forces of globalization (OECD, *Space 2030*).
29 In Europe, Alcatel Alenia Space and the European Aeronautic Defense and Space Company have both made it company policy to build ITAR-free commercial satellites.

148   E. Sadeh and B. Vallance

30 Johnson-Freese, "Alice in Licenseland."

31 Wassenaar Arrangement state members, in addition to the United States, include: Argentina, Australia, Austria, Belgium, Bulgaria, Canada, Croatia, the Czech Republic, Denmark, Estonia, Finland, France, Germany, Greece, Hungary, Ireland, Italy, Japan, Latvia, Lithuania, Luxembourg, Malta, the Netherlands, New Zealand, Norway, Poland, Portugal, Republic of Korea, Romania, Russian Federation, Slovakia, Slovenia, South Africa, Spain, Sweden, Switzerland, Turkey, Ukraine, and the United Kingdom.

32 It is the sense of the US Congress that business interests must not be placed above national security interests. Strom Thurmond National Defense Authorization Act for Fiscal Year 1999.

33 Thomas Moorman, *U.S. Space Industrial Base Study* (McLean: Booz-Allen & Hamilton, 2000).

34 Since the application of the ITAR regime for export control of commercial satellites in March 1999, US global share of commercial satellite manufacturing revenues fell to 41 percent in 2005 from 51 percent in 2000; US commercial satellite component suppliers captured 90 percent of the global market in 1995, whereas by 2000 they retained only 56 percent; US satellite firms lost approximately $5 billion between March 1999 and the end of 2001; and, from 1999 to 2004, it was estimated that US share of the lucrative geostationary satellite market declined by 16 percent. See *State of the Satellite Industry Report* (Washington, DC: Futron Corporation, June 2006); Robert D. Lamb, *Satellites, Security, and Scandal: Understanding the Politics of Export Controls* (University of Maryland, College Park, Center for International and Security Studies at Maryland, 2005); OECD, *Space 2030*; and *State of the Space Industry* (Washington, DC: International Space Business Council, 2000).

35 The congressional bill pocket-vetoed by President Bush was the Omnibus Export Amendments Act of 1990.

36 Marcia S. Smith, *Space Launch Vehicles: Government Activities, Commercial Competition, and Satellite Exports* (United States Congressional Research Service, the Library of Congress, updated January 1, 2006).

37 George Abbey and Neal Lane, *United States Space Policy, Challenges and Opportunities* (American Academy of Arts and Sciences, 2005).

38 *Preserving America's Strength in Satellite Technology, a Report of the CSIS Satellite Commission* (Washington, DC: Center for Strategic and International Studies, 2002).

39 *Report of the Select Committee on U.S. National Security and Military/Commercial Concerns with the Peoples' Republic of China.*

40 Robert D. Lamb, *Satellites, Security, and Scandal.*

41 Interview, Ann Ganzer, Director of the Office of Defense Trade Controls Policy, State Department, the Space Show, February 12, 2006.

42 Since the listing of commercial satellites within the ITAR export control regime in 1999, only 1–2 percent of all export license requests are denied.

43 In addition to the Satellite Trade and Security Act of 2001, congressional sponsors have proposed amendments to the EAAv and other separate bills that would return export licensing authority for commercial satellites to the DoC.

44 The inability to accurately measure risk to national security is one of the most serious problems for the system of export controls.

45 A congressional bill to expedite the DoS process for exporting commercial satellites, particularly to states considered friendly to the United States, such as NATO allies and other major non-NATO allies, was signed into law in 2004. With this bill, every effort was made to allay national security concerns, while attempting to find ways to not only sell commercial satellites abroad, but to allow the transfer of information necessary to bid on new projects as well as

respond to business requests for information on existing systems. Of note is that in 2000, following the *Cox Report* recommendations, Congress allocated additional funds to the DoS to allow for addressing the issues of technical expertise and expediting the licensing process. At that time, the DoS unilaterally acted to shift these funds within the bureaucracy away from the congressional intent. The 2004 mandate by Congress is more closely monitored, and the DoS is working to deal with the expertise and delay barriers. One significant effort is the development of an electronic filing system for export licenses at the DoS.

46 The DoC exempted basic items, like screws and knobs for example, from export control.
47 The Strom Thurmond National Defense Authorization Act for Fiscal Year 1999 included language stating that MCL shall not necessarily apply to the "export of a satellite or related items for launch in, or by nationals of, a state that is a member of NATO, or that is a major non-NATO ally of the United States." In implementing ITAR, the DoS interpreted this exception to apply only to the mandated monitoring activities. Further, the expanded definitions of satellite-related components, and the additions of defense technical services and space insurance business meetings as new areas needing export licenses, led to the bureaucratic microregulation of the US commercial satellite industry in response to accusations initially related to China.
48 The ITAR regime as applied can also have extraterritorial elements. A Technology Assistance Agreement may specify the states of which a foreign national may be a dual citizen and still have access to the transferred data or hardware. Eric Choi and Sorin Niculescu, "The Impact of U.S. Export Controls on the Canadian Space Industry," *Space Policy* 22(1) (2006): 29–34.
49 In 1985, President Regan issued an ITAR exemption for fundamental research conducted at US universities. National Security Decision Directive 189, September 21, 1985.
50 Lewis R. Franklin, "A Critique of the Cox Report Allegations of PRC Acquisition of Sensitive U.S. Missile and Space Technology," in M.M. May, (ed.), *The Cox Committee Report: An Assessment* (Stanford University, Center for International Security and Cooperation, December 1999).

7 Space and the military

Peter L. Hays

Space capabilities are essential at all levels of military planning and operations.

Honorable Ronald M. Sega, former Under Secretary of the Air Force

US and coalition forces achieved rapid and decisive military successes during operations Desert Storm in the Persian Gulf in 1991, Allied Force in Serbia in 1999, Enduring Freedom in Afghanistan in 2001, and Iraqi Freedom in 2003 that illustrate a new American way of war empowered by a space-enabled global reconnaissance, precision-strike complex. A primary goal of Secretary of Defense Donald Rumsfeld and the Department of Defense (DoD) is to continue and accelerate this military transformation by developing even lighter and more easily deployed forces that are able to strike more precisely from greater distances and leverage network-enabled operations to empower users at the edge of the network and achieve decision dominance over adversaries. Space capabilities often provide the best and sometimes the only way to pursue these ambitious transformational goals. There are also, however, many longstanding, difficult, and fundamental issues related to space and defense policy including: the place of space in dealing with the novel security challenges of the post-Cold War and post-9/11 environment, the ability of space capabilities to dissuade and deter competition from potential military peers, the role of space in enabling the information revolution and the new American way of war, overlaps and changing roles in traditional space missions caused by growth in commercial space activity, and the current state of flux in many military space organizations and acquisition programs. Cumulatively, these factors make military space issues more indeterminate, complex, multidimensional, and controversial than ever before.

Moreover, the United States faces enormous challenges as it seeks to modernize and improve its space systems. Almost all major military space systems are currently being improved or replaced by new systems and most of these acquisition programs have encountered significant cost overruns and deployment delays. It is unclear if Congress and the Air

Force will be able to sustain the political will needed to continue funding these nearly simultaneous modernizations, if these new and improved space capabilities will deliver on their promise of accelerating transformational effects, and if the technology required for these future systems can be developed and integrated on cost and on time.

The national security space (NSS) sector encompasses DoD activities, conducted primarily by the Air Force, to enhance national security and National Reconnaissance Office (NRO) programs to collect intelligence data from space. The NSS sector is sometimes also divided into separate sectors known as the military or defense space sector and the intelligence space sector. Following implementation of one of the recommendations of the January 2001 Commission to Assess National Security Space Management and Organization (Space Commission) Report, the DoD now uses an accounting procedure known as the Virtual Major Force Program (vMFP) to track spending on national security space programs.[1] According to the Congressional Research Service, the total DoD request for space spending amounted to $22.12 billion in fiscal year 2005 and is $22.66 billion for fiscal year 2006.[2] Using these classifications and discussing the nearly simultaneous modernization of most major military space systems, a September 2005 Congressional Budget Office (CBO) study found that unclassified military space acquisition spending is poised to grow from $4.9 billion to $6.9 billion, or more than 40 percent, between 2005 and 2006.[3] Perhaps the most significant trend in all planned major military space acquisition through 2024 is in risk of cost growth. Space acquisition expenditures are expected to peak at $14.4 billion in 2010, or nearly triple from their present levels if current programs follow the historic trend that space research, development, engineering, and testing costs on average have risen 69 percent, while space procurement costs on average have grown by 19 percent.[4] Clearly, the United States faces a difficult if not unsustainable path toward deploying its currently planned military space improvements and modernizations. Many of the most daunting military space challenges are discussed at the end of this chapter after introduction of conceptual frameworks for analysis, an overview of the most important military space organizations, and discussion of major acquisition programs.

Sea power and air power analogies

One of the most important and obvious set of factors shaping perceptions is the oft-invoked analogy between space power and sea power or air power. There is, of course, a rich literature on sea power and air power theory. Seminal theorists who developed enduring perspectives on military operations in these two mediums include Alfred Thayer Mahan, Julian Corbett, Giulio Douhet, William "Billy" Mitchell, John Boyd, and John Warden.[5] Some of the key concepts that these theorists developed or applied to the air and sea mediums are command of the sea, command of the air, sea lines of communication, common routes, choke points, harbor access, concentration and dispersal, and parallel attack. Several of these concepts have been appropriated directly into various strands of embryonic space theory; others have been modified slightly and then applied. For example, the ideas of Mahan and Corbett about lines of communications, common routes, and choke points have been applied quite directly to the space medium. Sea power and air power concepts that have been modified to help provide starting points for thinking about space power include harbor access and access to space, and command of the sea or air and space control.[6] But, of course, to date no comprehensive space power theory has yet emerged that is worthy of claiming a place alongside the foundational theories listed above.[7]

There are also many fundamental questions concerning the basic attributes of the space medium and how appropriate it is to analogize directly from sea power or air power theory when attempting to build space power theory. Few concepts from sea power theory translate directly into air power theory – why should we expect either sea power or air power theory to apply directly for the distinct medium of space? Questions concerning the attributes of space and the proper way to build space doctrine have sometimes also been at the heart of the disagreements between the Air Force and rest of the DoD over whether air and space should be treated as a seamless operational medium (previously defined as aerospace by the Air Force) or regarded as distinct air and space mediums (as seen by the rest of DoD).[8]

Many of the problems with the aerospace concept and the development of space power theory and doctrine have been repeatedly addressed within the Air Force. Dennis Drew, Charles Friedenstein, and Kenneth Myers and John Tockston published three of the best analyses during the 1980s.[9] These interrelated articles build on Drew's doctrine-tree model – the idea that doctrine should grow out of the soil of history, develop a sturdy trunk of fundamental doctrine, branch out into doctrine for specific environments, and only then attempt to sprout the organizational doctrine analogous to "leaves." This approach provides a comprehensive way to examine the aerospace concept and the Air Force's first official space doctrine, Air Force Manual (AFM) 1–6, *Military Space Doctrine*, released in 1982.[10] Friedenstein finds that "there is no doctrinal foundation for the term *aerospace*" (emphasis in original) and critiques the Air Force for attempting to produce "leaves on a nonexistent branch" because it had not developed environmental doctrine before issuing the organizational doctrine in AFM 1–6.[11] Myers and Tockston strongly critiqued the Air Force's tendency to "force-fit" space doctrine into the mold of air doctrine and argued that the three major characteristics of space forces are not air power's speed, range, and flexibility but rather emplacement, pervasiveness, and timeliness.[12]

Space power and the frontier analogy

The image of a frontier to be tamed evokes powerful images, particularly for Americans, and it is therefore not surprising that it has become one of the most popular ways to describe space. Frederick Jackson Turner first advanced his frontier thesis in 1893 as a way to describe and explain what he perceived to be distinctive characteristics of American history and American political thought.[13] For Turner, numerous American cultural traits could all be attributed to the influence of the frontier – "that coarseness and strength combined with acuteness and acquisitiveness; that practical inventive turn of mind, quick to find expedients; that masterful grasp of material things ... that restless, nervous energy; that dominant individualism."[14] In short, he argued that the frontier represented "the line of most rapid Americanization."[15]

A very short list of important specific references to space as a frontier would include the beginning of Captain James Tiberius Kirk's opening monologue on the 1960s *Star Trek* series; the title of Space Studies Institute founder Gerard K. O'Neill's 1977 book, *The High Frontier*, the report of the 1986 National Commission of Space, *Pioneering the Space Frontier*, and former Senator Bob Smith's (R-N.H.) numerous references to space as the "permanent frontier."[16] As with most other concepts associated with space power, there is much more agreement on describing space as a frontier than on the national security implications of this association. The US military obviously played a very important role in opening the American frontier. It took on exploration missions such as Lewis and Clark's expedition, surveys for railroad routes by the topographical engineers, construction of navigable waterways by the Corps of Engineers, and protection for pioneers. Clearly, the military helped to explore, survey, and pacify the American frontier – are these activities analogous to what will be conducted in space and is the military the proper organization to carry them out?

Spaceflight, the overview effect, and religious implications for space power

A final set of perspectives on space power may shape our views in the most subtle and pervasive ways. At their core, these perspectives link space to humankind's purpose and destiny. Humankind has pondered its relationship with the cosmos for millennia, and perceptions about space form foundational components of many religious beliefs. In the modern era, the visions of spaceflight produced by Jules Verne and H.G. Wells helped to lay the foundation for the new genre of science fiction and were echoed in the quasi-religious zeal of spaceflight pioneers such as Konstantin Tsiolkovsky and Wernher von Braun as they laid the conceptual framework for spaceflight and began to create some of the tools needed to "leave the cradle." Later science fiction authors such as Arthur C. Clarke, Robert Heinlein, and Isaac Asimov, combined with the increasing popularity of this genre for television and films, has pervaded the human psyche with the boundless possibilities of space and rendered our actual achievements in space mundane by comparison. Yet, as humans entered space, many people and groups believed that the rationale and import-ance of spaceflight took on increased significance. Mainstream views on spaceflight cover a broad range. Individuals such as Gerard K. O'Neill build on Turner's frontier thesis and emphasize exploration as a cathartic and defining human characteristic. Carl Sagan speaks for those who view spaceflight in scientific and ecological terms and see it as essential to the survival of the human species. Visions about spaceflight undoubtedly culminate in what Frank White labels "the overview effect" – nothing less than space opening the door to the next phase of human evolution.[17]

Likewise, the links between space and religious beliefs are still very important in the modern era. The first Soviet cosmonauts, for example, went to great pains to emphasize that they had not seen God during their travel through the heavens, and this prompted Western retorts question-ing whether they were pure of heart. The reading of the first ten verses from *Genesis* by the crew of *Apollo 8* as they became the first humans to view an Earthrise from lunar orbit on Christmas Eve 1968 evoked strong religious feelings. As McDougall tells us, humankind has never "been able to separate our thinking about technology from teleology or eschatol-ogy."[18] The very framework of his book warns that technocracy in general and spaceflight in particular cannot serve as humankind's Guarantor of Destiny; instead, his instinct tells us that

> our science and technology, feeble as they are in controlling Nature, are so acute in studying it that they will soon reveal their limits. It is then that man must confess the mortality of his works, without turning on them or himself with contumely. It is then that the ortho-dox message is a sure guide: God made us, is disappointed in us, but loves us anyway, by which we are redeemed. Technology is our subcre-ation. We made it, we will be disappointed in it, but we must love it anyway, or it cannot be redeemed.[19]

The message for analysts attempting to understand space power is simple: the medium is the message. To a greater degree than any other physical domain, space is shaped in fundamental ways by our very broad-ranging perceptions about it. Any comprehensive analysis of the strategic utility of space power must attempt to take these factors into consideration.

Post-Cold War perceptions about space power and the new American way of war

Recently, two less controversial concepts, force enhancement and the growing commercial space sector, have emerged as the most important dimensions of US space policy. Force enhancement has long been an important military mission but has seldom captured the military's imagination in the same way as force application or space control missions. In the post-Cold War era, however, force enhancement or space support to the warfighter is clearly the dominant military space mission. This development follows decades of incremental improvements that have created increasingly capable space systems and the widespread recognition of their significant contributions to the stunning coalition victory in the Gulf War.[20] Perhaps just as significantly, within the past few years, commercial space activity has grown to make it a large and important space sector – this development will have increasingly important implications for the military, intelligence, and civil space sectors.

The flag follows trace: USSPACECOM perceptions of space

The growing commercial importance of space and claims that it had become an economic center of gravity (COG) was the central theme in US Space Command's (USSPACECOM) public discourse during the latter half of the 1990s. The increased use of the term COG to describe the commercial space sector coincided with rapid actual growth in commercial space activities in this period, but it was predicated even more directly on projections of exponential growth. Forecasts during 1997 and 1998 called for growth at a "blistering rate of 20 percent a year" to support a "gold rush in space."[21]

> 550 satellites today are in Earth orbit, performing numerous critical defense and civil functions. Nearly half of them belong to the US, and half of those are commercial. US space investment now exceeds $100 billion, and the stakes are about to go higher.
>
> Expectations are that the US and the world's other spacefaring nations, over the next five years, will pump another $500 billion into space. They will launch at least 1,000, and possibly 1,500, new satellites. Most will be commercial systems. Many will have military significance.
>
> "We'll see commercial use of space go out of sight," said USAF's Chief of Staff, Gen. Michael E. Ryan.[22]

Gen. Howell M. Estes III developed and articulated one of the most powerful visions for space of any USSPACECOM Commander. Early in his tenure (August 1996–August 1998) he began emphasizing the emergence of space as an economic COG at virtually every opportunity. In one of his earliest and most sweeping speeches, delivered at the US Space Foundation's annual symposium in April 1997, he introduced several major themes that he reiterated during the remainder of his term:

> Our investment in space is rapidly growing and soon will be of such magnitude that it will be considered a vital interest – on par with how

we value oil today…. Now while it might seem appropriate that I should be more concerned with military space, I must tell you that it is not the future of military space that is critical to the United States – it is the continued commercial development of space that will provide continued strength critical for our great country in the decades ahead. Military space, while important, will follow.

Commercial space … will become an economic center of gravity … and the U.S. military … will be expected to protect this new source of economic strength.[23]

Gen. Richard B. Myers, Gen. Estes' successor, was confirmed as Chairman of the Joint Chiefs of Staff in 2001. He served as USSPACECOM Commander from July 1998 until February 2000 when he became vice chairman of the Joint Chiefs of Staff. Gen. Myers generally reiterated Gen. Estes's emphasis on space as an economic COG but added three important changes: first, that space was already a COG; second, that space was a *military* and economic COG; and third, that United States' reliance on commercial space had created vulnerabilities easily exploited by potential adversaries. One of his first pronouncements along these lines came in Los Angeles at the Air Force Association Space Symposium in November 1998: "space has become a military and economic center of gravity. So much of the world's standard of living, so much of its commercial wealth depends on space."[24] Later in his tenure, Gen. Myers put more emphasis on how US reliance on commercial space was creating new vulnerabilities: "Clearly, our reliance on commercial space has created a new center of gravity that can easily be exploited by our adversaries."[25] Just before leaving his USSPACECOM tour, Gen. Myers summarized his position and emphasized the importance of space control in an editorial for *Aviation Week & Space Technology*:

Space is a military and economic center of gravity. We can't afford to take it for granted. Only through a robust space control and modernization vision can we thwart military or terrorist attacks, and manage the space "gold rush," while continuing to reap tremendous benefit, both in economic and national security terms.[26]

Gen. Ralph E. Eberhart became the last Commander of USSPACECOM in February 2000. In October 2002, US Space Command was disbanded and its functions moved under US Strategic Command (STRATCOM). At the same time a new command responsible for homeland defense, US Northern Command (NORTHCOM), was created and Gen. Eberhart was named its first Commander. As USSPACECOM Commander, Gen. Eberhart usually avoided using the term COG to describe the economic and military importance of space and, in general, did not place as much emphasis on the growth and importance of the commercial space sector as did his predecessors. His approach reflected the slowdown in commercial space and was in line with the major recommendations in the *Space Commission Report.* Instead, Gen. Eberhart stressed personnel issues such as retention problems; the USSPACECOM's efforts to come to grips with its newest missions, computer network defense (CND) and computer network attack (CNA); and, especially, the need for space control.[27] He also recommended the formation of a space tactical school to "develop space warfare concepts" and has created the "Space Aggressor Squadron, whose job it is to play against the Air Force and other services in war games such as Red Flag and to heighten both military and civilian awareness of the threat."[28] One of the best illustrations of these subtle changes in emphasis came in Gen. Eberhart's November 2000 interview in *Aviation Week & Space Technology:*

> Integration [of air and space capabilities] has been exactly the right thing to concentrate on these last 5–10 years, as we tried to harness the national systems post-[Operation] Desert Storm…. The fact that we heard so much about [the need for integration] after Desert Storm, and didn't after Kosovo, tells me that we're on the right track. Now, we need to make sure we can protect the capabilities that resulted from that integration…. I don't think we would be good stewards of space if we only thought about "integration." We also need to be spending resources and intellectual capital on space control and space superiority…. The importance of space control and space superiority will continue to grow as our economy becomes more reliant on space…. If we only look at space in terms of "integration," in my view, we'll fall into the same trap we fell into with the airplane…. We [initially] thought of it in terms of intelligence, surveillance, reconnaissance, communication and weather [support]. If we only think of space in these ways, [it's just] a "higher hill" as opposed to a center-of-gravity. We [also] have to be able to surveil, protect and negate under this space control mission.[29]

Is space an economic center of gravity?

More than most, commercial space is a volatile industry that has been through several boom and bust cycles and has often delivered less than promised. It is also highly complex because it is closely tied not only to economic cycles but also to many other factors such as technological developments, international politics, and domestic regulation. USSPACECOM's assertions during 1997–1999 that space is an economic COG were made based on projections drawn from the commercial space sector's strongest ever growth cycle. The "gold rush" mentality of firms seeking competitive niches in the communications spectrum or in specific markets reinforced perceptions that commercial space would remain in a cycle of continuing upward acceleration. The resulting projections too often relied on best-case scenarios rather than more somber economic analysis, and they also suffered from the lack of an objective and timely overall market survey.

Analysts currently have far better insight into these issues due to the slower actual development of the markets over time, and the Futron Corporation's new annual *Satellite Industry Guide* helps to address the later problem.[30]

Space activities clearly enhance and enable many economic activities; space should undoubtedly be considered a strategic sector of the global information infrastructure and the world economy. However, when comparing actual development with projected space activity, one overarching conclusion immediately jumps out: the commercial space sector simply has not come close to developing in the directions and magnitude projected in the late 1990s. Despite the significant growth of the commercial space sector in the second half of the 1990s, the trajectory of actual developments fell well short of the projected vector ($500 billion investment and 1,000–1,500 launches by 2003) that had been touted in forecasts as late as the end of 1998. The period of most rapid growth in this sector has been followed by its most rapid decline. Instead of being perceived as a gold rush, commercial space is now often seen as a burst bubble.

Moreover, it is questionable whether it is useful to characterize commercial space activity as an economic COG, even in its most vibrant state. Aerospace corporations form an important part of the economy, but in pure dollar terms they – like any other single industry – are simply not a dominant sector or an economic COG in terms of overall value, revenues, or market capitalization. The main reason for this is the huge size of the US gross domestic product (GDP). The Department of Commerce (DoC) estimated the 2000 US GDP at $9.873 trillion, a value that dwarfs the value of any individual sector.[31] Anyone watching the financial markets during the past several years knows that revenues and market valuations are highly volatile; but space-only revenues and valuations have never been that big of a part of the US economy at any time or under any classification scheme. Consider revenues: the 2001 *Fortune Magazine* list of the top 500 US corporations by revenue does show a scattering of aerospace companies among the top 100 firms – Boeing at number 15, Motorola at 34, United Technologies at 64, Lockheed Martin at 69, Honeywell at 71, and the AMR Corporation at 98.[32] But this listing reflects *all* revenues for these firms rather than their space-only revenues. When the space-only revenues are examined, the picture becomes quite different. According to the *Space Commission Report*, global commercial space activities generated a total of $80 billion in revenues in 2000, and while this is clearly a lot of money in absolute terms, it represents only 8.9 percent of the revenues of just the top five US corporations (Exxon Mobil, Wal-Mart, General Motors, Ford Motor, and General Electric) from the *Fortune* 500 list for 2001.[33]

Should we consider commercial space "on par with how we value oil today"? Space is not there yet in dollar terms: the total revenues of energy corporations from the *Fortune* 500 list for 2001 was more than three times the value of the revenues from aerospace corporations.[34] What about the market valuation of space corporations? At the end of 1999 the combined market valuation for all major US aerospace firms (Boeing, Honeywell, United Technologies, General Dynamics, Textron, Lockheed Martin, Raytheon, TRW, Northrop Grumman, and Litton Industries) amounted to approximately $150 billion but was still less than the market valuation of Home Depot Corporation.[35] The intent of these comparisons is not to depreciate the importance of commercial space activities; rather, they are designed to show that commercial space activities are not a COG for the economies of the United States or the world. The comparisons also help to illuminate the true strategic utility of commercial space activities and highlight that these activities should be thought about and valued in a variety of ways other than just in terms of economics.

Major military space programs and budgets

The sections below examine major national security space programs within each of the four military space mission areas: space support, force enhancement, space control, and force application. As noted above, the vast majority of current US national security space efforts fall into the force enhancement mission area.

Space support

The space support mission area includes two main areas: satellite control programs that provide global communications systems for telemetry, tracking, and control (TT&C) of satellites (the Air Force's Satellite Control Network is one example), and launch and range programs that maintain and improve the infrastructure at the DoD's two launch sites, the Eastern Range at Cape Canaveral Air Force Station and the Western Range at Vandenberg Air Force Base. Under the 2006 Future Years Defense Program (FYDP) and CBO's long-term projection, annual investment funding for those activities would average about $180 million through 2024.[36] Part of those resources would go toward modernizing the Launch and Test Range System, which provides tracking, telemetry, flight safety, and other support for space launches and ballistic missile tests.

Space launch

Space launch systems include the Evolved Expendable Launch Vehicle (EELV) the DoD now uses to put most of its satellites into orbit and the new Operationally Responsive Spacelift (ORS) program, designed to develop launchers capable of rapidly placing payloads into orbit. The EELV program currently uses two types of launchers: the Boeing Delta IV and the Lockheed Martin Atlas V rockets. Both types can carry medium-sized payloads (about 10–15 metric tons), but only the Delta IV family now includes an operational heavy lift variant for putting larger payloads (up to about 25 metric tons) into low-Earth orbit. As of March 2006, there have been 11 successful EELV launches, three government and eight commercial. Due to several factors, including a decline in commercial launch demand, in 2004 the EELV program experienced a cost increase of more than 25 percent, triggering a Nunn–McCurdy certification breach.[37] The 2006 FYDP implies total funding needs of about $28 billion for approximately six or seven EELV launches each year through 2024.[38] A 2002 ORS mission-needs statement established the requirement for responsive, on-demand access to, through, and from space and a Defense Advanced Research Projects Agency (DARPA)–Air Force joint program office was established in December 2002. The Falcon joint technology development program is the product of this effort and focuses on the development and transition of more mature technologies into a future weapon system capable of delivering and deploying conventional payloads worldwide from and through space such as Joint Warfighting Space satellites. Given the lack of significant funding and well-defined goals for the program, however, it is difficult to envision how it might evolve beyond anything but a research and development (R&D) program.

Force enhancement: military satellite communications systems

As the nuclear command and control system, the glue that binds together commanders and forces globally, and the enabler of netcentric operations, military satellite communications systems probably provide the single most important military space capability. It is logical, therefore, that this force enhancement area has the largest number of programs and receives large investments. What may be surprising, however, is that even with this large number of systems and significant investments, the DoD remains heavily reliant on commercial satellite communications systems to carry the bulk of its communications traffic. Estimates are that during operations Enduring Freedom and Iraqi Freedom, over 60 percent and over 80 percent, respectively, of wideband military communications were carried on commercial systems.[39] The benefits of using commercial systems include lower acquisition and operations costs as well as greater flexibility, but these must be balanced against drawbacks such as high cost of buying commercial services on the spot market, questionable availability of services, and less secure and protected systems. In addition, the DoD has recently benefited from the widespread availability and relatively low costs associated with the overbuilt commercial satellite communications created during the dot-com bubble of the late 1990s. However, this market oversupply appears to be correcting, and the DoD should not rely on these conditions in the future.

The three main types of military satellite communications each primarily use different parts of the radio spectrum: wideband on super high frequency (SHF), protected on extremely high frequency (EHF), and narrowband on ultra high frequency (UHF). CBO estimates for military satellite communications system spending through 2024 total $27 billion in Air

Force funding for wideband and protected capabilities and $5 billion in Navy funding for narrowband capabilities.[40] Wideband systems now in operation include the Defense Satellite Communications System (DSCS), a constellation of five primary satellites plus a number of older residual-capability satellites, and the Global Broadcast System (GBS), which consists of payloads on three Navy UHF Follow-On (UFO) satellites. DSCS satellites were originally expected to have a service life of ten years, however, as in the case of many space systems, they are lasting longer than anticipated. The DSCS constellation is supposed to remain operational through 2015 and the GBS through 2010 or slightly beyond. The Air Force plans to replace both systems with the Wideband Global System (WGS). The first of the WGS constellation of six satellites was launched in October 2007 and the system is scheduled to remain in service until 2024.

Protected satellite communications is currently provided by five Milstar satellites that are expected to be operational at least until 2014. Like DSCS satellites, Milstar satellites are exceeding their design lifetime and some may be available beyond that time. Beginning in 2009, a constellation of three new Advanced extremely high frequency (AEHF) satellites will be launched to begin replacing Milstar. In addition to these satellites designed to provide coverage of the globe up to about 65° of latitude, the Air Force is also pursuing programs to improve protected communications over the northern Polar Regions: two Interim Polar payloads are scheduled to be operational by 2006 and then be replaced by two Enhanced Polar payloads around 2013. Finally, the Transformational Communications System (TSAT) constellation is currently being designed in two blocks (two satellites in block one and three in block two); that will provide high data-rate wideband and protected services. The first TSAT launch is now scheduled for 2009.

Nine Navy UFO satellites now in orbit (out of 11 launched) provide the current DoD narrowband-communications capability. Beginning in 2010, the UFO constellation is due to be replaced by five Mobile User Objective System (MUOS) satellites that the Navy is developing. Three of the satellites in the UFO constellation may fail by the end of 2008, leaving little margin for slippage in the initial launch date for the MUOS constellation. The major military satellite communications systems that are under development or being upgraded are discussed individually below.

Global Broadcast System

The GBS provides high-rate transmission of data such as imagery, logistics and weather information, maps, operational orders, and video directly to dispersed warfighters via small, inexpensive user terminals. Requirements for the system were established by the Joint Requirements Oversight Council (JROC) in April 1997 and updated in May 2001 and January 2005. The space segment is hosted on Navy UFO satellites 8, 9, and 10 and is augmented with leased Ku-band transponders. During fiscal year (FY) 2006, the system underwent an analysis of alternatives to determine the best way to transition between Internet Protocol version (IPv)4 and IPv6. This upgrade will eliminate obsolete hardware and software and improve the performance of the system and its ability to support netcentric operations.

Wideband Global System

The WGS is a joint Air Force–Army program that will augment the DoD's current DSCS X-band and GBS Ka-band capabilities; in addition, it will establish a new two-way Ka-band service. The requirements for the WGS were established by the Joint Space Management Board-approved military satellite communications architecture in August 1996, the military satellite communications capstone requirements document approved by the JROC in October 1997, and the JROC-approved WGS operational requirements document in May 2000. First launch had been delayed for over three years before it took place in October 2007. In October 2007 the Australians entered into a $927 million partnership on WGS, agreeing to buy the sixth satellite scheduled for launch in 2013.[41] In December 2002, the Office of the Secretary of Defense (OSD) directed the addition of two more WGS satellites as part of the transformational communications architecture; satellites four and five are based on the Boeing 702 satellite and will support increased bandwidth requirements for Airborne ISR and other missions.

Advanced Extremely High Frequency Military Satellite Communications System

The AEHF system includes the satellites, mission control element, and cryptography necessary to provide worldwide, survivable, and anti-jam protected communications for strategic and tactical warfighting. The system will provide much higher data rates than the Milstar system it is replacing. Lockheed Martin is the prime contractor on the system and Northrop-Grumman is developing the satellite payload. AEHF is a cooperative program with Canada, the United Kingdom, and the Netherlands and forms part of the DoD bid to provide the North Atlantic Treaty Organization (NATO) with protected satellite communications. Development of new, complex information assurance products by the National Security Agency has been a technically challenging and high-risk area for the AEHF program; delays in the delivery of these products and other factors resulted in a cost overrun of approximately $1 billion and a Nunn–McCurdy breach at the 15 percent threshold in December 2004, as well as a one-year launch delay to each of the three satellites, from April of 2007–2009 to April of 2008–2010.[42] Plans to expand the program beyond the three satellites being developed are under consideration, but the program is currently structured to launch just three satellites and not achieve what had been planned originally as the fully operational capability of the AEHF system until the launch of the first TSAT.

Polar Military Satellite Communications System

This program provides protected communications (anti-jam, anti-scintillation, and low probability of intercept) for users in the North Polar Region. The first portion of this system has funded three low data-rate Milstar packages for three classified host satellites as an interim solution to providing protected communications capabilities in the region. Two enhanced packages that take advantage of AEHF technology will replace the previously planned Advanced Polar System and should be available for launch in FY 2014 and FY 2016.

Transformational Communications System

TSAT is now scheduled for first launch in FY 2019 and will provide the DoD with high data-rate wideband and protected communications as the space segment of the global information grid (GIG) and a key component of the transformational communications architecture. TSAT may use both radio frequency and laser cross-links to provide an essential foundation for military transformation and will support global network-enabled operations. As the result of Quadrennial Defense Review (QDR) deliberations, the program was restructured and is now following an incremental approach with two blocks of satellites planned (two in the first block, three in the second, and a spare). The first block of two satellites will have reduced requirements for laser communications links and Internet-like processing routers.[43] The five satellites may establish a laser cross-link ring in GEO. By employing advanced communications technologies including Internet protocol (IP) routing, packet switching and encryption/decryption, dynamic bandwidth and resource allocation techniques, and efficient protected bandwidth modulation, TSAT will connect thousands of users simultaneously through networks rather than using limited point-to-point circuits and will enable communications- and networking-on-the-move. TSAT capabilities will be particularly important for future high data-rate connections to space and airborne intelligence, surveillance, and reconnaissance (ISR) platforms. Technology demonstration and validation efforts are underway and are on schedule, but if a technology fails to mature, less-capable technology off-ramps exist and can be used to preserve schedule. Even the technology off-ramps will significantly enhance warfighter capabilities and the advanced technology can be spiral developed into the second block of spacecraft. At a September 2004 Senior Warfighters Forum, the Combatant Commanders gave unanimous support for TSAT as a critical warfighting enabler and an interim program review the next month directed continuation of the plan to achieve AEHF full operational capability-equivalency following on-orbit checkout of the first TSAT satellite.

A contract to acquire operational satellites is scheduled to be awarded during 2010. Congress, however, repeatedly has expressed concerns about the direction and technical maturity of the program; it approved only $436 million in funding for TSAT for FY 2006, about half the administration's requested $836 million.[44] Congress also recommended the Air Force focus on developing and integrating the technologies needed for the program, rather than pursuing it as a formal acquisition effort, and noted that the Government Accountability Office (GAO) had found that only one of its seven critical technologies is mature, although no new fundamental discoveries or breakthroughs are required for TSAT. In addition, in the FY 2006 National Defense Authorization Act (NDAA), Congress directed the Secretary of the Air Force to prepare a contingency plan for delivering transformational communications capabilities via WGS and AEHF if TSAT is not developed on schedule and for the National Security Space Office to deliver to Congress by April 15, 2006 an analysis of alternatives on the trade-offs required for this contingency plan.[45]

Mobile User Objective System

The Navy is managing and Lockheed Martin is the prime contractor for MUOS, the next generation advanced narrowband system for the DoD that will replace the UFO system. The MUOS program is building on state-of-the-art technologies and best commercial practices to create a highly responsive system for the joint warfighter. First launch for MUOS is scheduled for 2010, but contingency plans are being developed to accelerate this date due to potentially serious difficulties with the UFO constellation beginning in 2008. The MUOS program is being developed using advanced concepts and technologies that will allow the system to address significant growth in narrowband demand and meet all requirements specified in the joint operational requirements document.

Force enhancement: position, navigation, and time (PNT)

A National Security Presidential Directive (NSPD) issued on December 8, 2004 that replaced National Science and Technology Council (NSTC)-6, "U.S. Global Positioning System Policy," from March 1996, updated the US policy framework for PNT issues. The fundamental goals of the NSPD are

> to ensure that the United States maintains space-based positioning, navigation, and timing services, augmentation, back-up, and service denial capabilities that: (1) provide uninterrupted availability of positioning, navigation, and timing services; (2) meet growing national, homeland, economic security, and civil requirements, and scientific and commercial demands; (3) remain the pre-eminent military space-based positioning, navigation, and timing service; (4) continue to provide civil services that exceed or are competitive with foreign civil space-based positioning, navigation, and timing services and augmen-

tation systems; (5) remain essential components of internationally accepted positioning, navigation, and timing services; and (6) promote U.S. technological leadership in applications involving space-based positioning, navigation, and timing services.[46]

To manage the system, the NSPD established the National Space-Based PNT Executive Committee, chaired jointly by the Deputy Secretaries from the DoD and the Department of Transportation (DoT), and an Advisory Board and Coordination Office to support the Executive Committee. In a reflection of how important and widespread GPS applications have become, the NSPD also assigned a large number of specific responsibilities to the Secretaries of Defense, Transportation, Commerce, and State, along with the NASA Administrator and the Director of National Intelligence. One of the most difficult of these responsibilities calls for the Secretary of Defense to develop navigation warfare capabilities to operate GPS effectively despite adversary jamming; deny GPS use to adversaries; not unduly disrupt civil, commercial, or scientific uses of GPS outside an area of military operations; and identify, locate, and mitigate on a global basis any interference that adversely affects GPS use for military operations.[47] The policy also seeks to ensure that foreign systems are compatible with GPS and re-emphasized that the US government will continue to provide PNT services "on a continuous, worldwide basis" that are "free of direct user fees" and the intention of the US government to keep the selective availability feature off (SA was turned off on May 2, 2000).[48]

The Air Force acquires and operates the GPS constellation that currently contains 32 satellites developed through a series of block upgrades. The baseline configuration consists of four satellites in each of the six semi-synchronous orbital planes, but the current system provides greater accuracy by including additional satellites that have exceeded their design life. In September 2005, the Air Force began launching Lockheed Martin block IIR-M satellites, which incorporate two new military signals and a second civilian signal. It plans to start launching Boeing block IIF satellites, which will broadcast a third signal for civilian use, in FY 2009. The first block III satellites are projected for launch in FY 2014 and are being designed to include improvements such as better anti-jam capability and satellite cross-links for more-accurate signals. Based on the FY 2006 President's Budget, CBO projected that the total investment spending on the GPS would be $12.5 billion through 2024.[49]

Force enhancement: intelligence, surveillance, and reconnaissance

Many components of the US ISR network are classified, but at least portions of the Space Radar (SR) and Future Imagery Architecture (FIA) programs, two of the most important future ISR systems, are public knowledge.

Space Radar

The name of the Space-Based Radar program was changed to Space Radar in 2005 and a new Integrated Program Office was established in Chantilly, Virginia to highlight the fundamental restructuring of this joint DoD and intelligence community (IC) program. DoD and IC leadership have agreed on a path to converge to a single common space radar system that affordably meets the nation's needs by delivering a capabilities-driven system as part of a horizontally integrated system-of-systems designed to meet both national intelligence and joint warfighter requirements. The SR system is being designed to provide a global capability to detect and track mobile targets in all weather conditions and would be launched beginning about 2015. The plan, approved by the JROC in February 2006, calls for a constellation of approximately nine satellites with somewhat reduced capabilities; earlier plans had envisioned using at least 24 satellites in order to provide near-continuous tracking capability. The system is being designed to use an active electronically scanned array capable of providing high-volume, readily taskable synthetic aperture radar (SAR) imaging; surface moving target indications (SMTI); and high-resolution terrain information (HRTI). It is also being designed from the ground up to enhance horizontal integration through agile, responsive collections using near-real time tasking and data dissemination. The SR system will peer into denied areas of interest in all weather, day or night, with minimal risk to personnel or equipment. The program will leverage NRO, National Geospatial-Intelligence Agency, DARPA, and Air Force Research Laboratory (AFRL) activities to ensure both DoD and IC requirements are satisfied in the baseline SR effort. Previous plans for a small-scale on orbit demonstration to validate SR costs and technology maturity are currently on hold. Cancellation of the SR program of record was announced in March 2008 and the restructured program received no funding in the FY 2009 appropriation.

The Air Force had requested $226 million from Congress for the SR program in FY 2006, but that figure was slashed to $100 million in the 2006 Defense Appropriations Act, and both the House and Senate Armed Services Committees commented extensively on the program, emphasizing the need to integrate the SR into a broader architecture of radar capabilities, including airborne radars. Based on the 2006 FYDP, CBO projected costs of $19 billion through 2024 for the space segment of the SR program. Recent reports indicate that the DoD estimates total life-cycle costs for a nine-satellite SR constellation (including the ground segment) would be $34 billion.[50]

Future Imagery Architecture

In September 1999, the NRO rocked the aerospace industry by choosing Boeing to build its next generation of secret imaging satellites, bypassing Lockheed Martin, the decades-long incumbent on the program. The original design for the FIA electro-optical (EO) and radar-imaging satellites called for a constellation that split collection functions among smaller, simpler, and more numerous satellites. FIA, however, failed to escape the dysfunctions plaguing most government satellite programs. The EO satellites, the largest problem area in the FIA program, were five years behind schedule, with first launch delayed until at least 2009. Moreover, the price tag for FIA had grown from $6 billion to at least $15 billion when Director of National Intelligence (DNI) John Negroponte began his first review of technical programs in the summer of 2005.[51] As the result of that review there is a new way ahead for imagery satellites: Boeing will keep developing the radar-imaging satellite portion of FIA and remain the lead contractor for the overall program, but, at the DNI's direction, Lockheed Martin will develop the more complicated satellite that would take highly precise pictures with visible-light or infrared cameras. As a final reflection of FIA difficulties, it is also noteworthy that Roger Roberts, the senior Boeing official overseeing FIA, was recently forced to retire and the NRO's two most recent FIA program managers were both passed over for military promotions after their stints in the program.

Commercial imagery systems

The Land Remote Sensing Policy Act of 1992 and the March 1994 Presidential Decision Directive (PDD)-23 were designed to create conditions and incentives for the development of a dominant US commercial high-resolution remote sensing industry. These incentives, however, were also balanced by factors such as the ability of the US government to exercise shutter control over these imaging systems if warranted by national security considerations, and the necessity for new systems to obtain operating licenses from the National Oceanic and Atmospheric Administration (NOAA). When the United States initiated Operation Enduring Freedom in Afghanistan in October 2001, the National Imagery and Mapping Agency (NIMA, now the National Geospatial-Intelligence Agency, or NGA) bought exclusive rights to Ikonos imagery of that area so that no one else could use the data without NIMA's approval, a practice dubbed "checkbook shutter control" in the media. The government apparently did not limit access to commercial satellite imagery during Operation Iraqi Freedom in 2003. Partially in response to concerns with the commercial viability of the industry, President Bush signed a new commercial remote sensing policy in April 2003, intended to sustain and enhance the US remote sensing industry. The 2003 Bush policy and Congressional direction strongly encourage the NGA to purchase commercial imagery to augment classified imagery and for all parts of the US government to use US commercial remote sensing space capabilities to the maximum extent practicable. Foreign commercial remote sensing space capabilities may be used consistent with national security and foreign policy objectives.

There are currently two US companies that operate high-resolution commercial remote sensing systems, DigitalGlobe and GeoEye (formed in December 2005 by the merger of Space Imaging and Orbimage). The merger that formed GeoEye is a reflection that the market for high-resolution commercial remote sensing has not grown as large or as quickly as had been hoped. Though not as precise as military reconnaissance satellites, the three operating US private sector satellites, Ikonos 2 (GeoEye), QuickBird (DigitalGlobe), and Orbview 3 (GeoEye) produce imagery with resolution (the ability to "see" an object or feature of a certain size) of 1 meter or less. Competitors include French, Russian, Indian, and Israeli companies that offer imagery with 2.5-meter, 1-meter, 1-meter, and 1.8-meter resolution respectively. Use of commercial imagery has several advantages for the DoD and IC, including lower costs for acquiring imagery commercially instead of building their own imagery systems, availability of wider views of more areas, and easier sharing of data with allies and coalition partners. Potential drawbacks include uncertainties about the availability and protection of the systems and their data. The NGA has ClearView and NextView contracts with Digital Globe and GeoEye, each worth up to $500 million, for delivery of high-resolution imagery into the next decade.

Force enhancement: integrated tactical warning and attack assessment

Currently, the Air Force maintains a constellation of GEO satellites, called the Defense Support Program (DSP), to provide warning of ballistic missile launches and some data on the type of attack and the missile's intended target. The last DSP satellite (DSP-23) was launched in November 2007.

Space-Based Infrared Systems (SBIRS)

The successor to the DSP is the SBIRS program, a system designed to satisfy operational military and technical intelligence overhead non-imaging infrared (ONIR) requirements, provide improved detection, and supply foundational assessment capabilities for ballistic missile defense. Lockheed Martin is the prime contractor and the operational SBIRS constellation was originally envisioned to include four GEO satellites, two highly elliptical orbit (HEO) payloads riding on classified host satellites, one spare GEO satellite (procured to protect against launch or early on-orbit failure), and both fixed and mobile ground elements. In addition, the Missile Defense Agency (MDA) is considering a system being developed by Northrop-Grumman, formerly known as SBIRS-Low and now named the Space Tracking and Surveillance System (STSS). The MDA plans to launch two demonstration satellites in 2009. If the STSS demonstrators work well in tracking missile launches and warheads as part of a missile defense system, an operational system consisting of six to nine satellites could follow, with a first launch in about 2016–17. The first SBIRS HEO payload was delivered in August 2004, and the first GEO satellite is expected to launch in late 2009.

Unfortunately, however, SBIRS is probably the single most troubled NSS acquisition effort. The Defense Science Board led by Tom Young said the program "could be considered a case study for how not to execute a space program." Total cost estimates have jumped to nearly five times the original estimates, and the program has triggered four required reports to Congress for Nunn–McCurdy Act breaches.[52] In December 2005, Under Secretary of Defense for Acquisition, Technology, and Logistics Kenneth Krieg and Under Secretary of the Air Force Ronald Sega significantly restructured the program. There will be no more than three GEO SBIRS spacecraft in the restructured program and purchase of the third satellite will be contingent on performance of the first satellite. In addition, the restructuring calls for Dr. Sega, in his role as DoD Executive Agent for Space, to conduct a 120-day study on alternative infrared satellite systems that could be launched around 2015 and substitute for SBIRS, and for Mr. Krieg to retain milestone decision authority for the SBIRS program. Under the 2006 FYDP and CBO's projection of its implications, investment spending for DSP and SBIRS would total about $11 billion through 2024.[53]

Force enhancement: environmental monitoring

The DoD presently uses data from five environmental monitoring (weather) satellites that are part of the Defense Meteorological Satellite Program (DMSP), plus the data from two NOAA Polar-Orbiting Operational Environmental Satellites (POES). By about 2015 those systems are to be replaced by three satellites of the National Polar-Orbiting Operational Environmental Satellite System (NPOESS) and one European Meteorological Operational Satellite. CBO projects that carrying out the plans in the 2006 FYDP would require a total of $3.4 billion in investment funding through 2024 for the DMSP and NPOESS programs.[54] Under current plans, four DMSP satellites remain to be launched, with the last launch planned for April 2012, and the last POES launch is scheduled for February 2009.[55]

National Polar-Orbiting Operational Environmental Satellite System

Presidential Decision Directive/National Science and Technology Council-2 (PDD/NSTC-2) in May 1994 directed the DoD, NASA, and the DoC to save money by establishing a single, converged national polar-orbiting weather satellite program. The Air Force (DoD) and NOAA (DoC) provide equal funding for the program. A tri-agency Integrated Program Office was established in October 1994 to manage the acquisition and operations of the converged system. Northrop-Grumman is the prime contractor for NPOESS. The program will buy four satellites, operate three satellites at a time, and be the nation's primary source of global weather and environmental data for operational military and civil use (polar satellites currently provide 90 percent of the data used in DoD and DoC weather prediction models). Original plans called for the launch of the NPOESS Preparatory Project (NPP) satellite in October 2006 and the first NPOESS launch in November 2009, but development of the NPOESS sensors has run into several significant snags in the past few years. Cost overruns of 15 percent in September 2005 and 25 percent in January 2006 triggered required reports to Congress under the Nunn–McCurdy Act; the program is now more than $3 billion over its total original budget of $6.5 billion and at least three years behind schedule.[56] Under restructuring announced in June 2006, the government now plans to: spend $12.5 billion for four slightly less capable satellites, launch the NPP by 2010, delay the first NPOESS launch by about four years until 2013, and operate the system until 2026. Cancellation, restructuring, or further delay in the NPOESS program is not likely to present many attractive options, however, as it is very doubtful other systems can be developed in time to avoid gaps in coverage, especially if there is a failure with any of the DMSP or POES satellites on orbit or awaiting launch.

Space control

Space control programs focus on developing ground- and space-based sensors to enhance space situational awareness (SSA) (knowledge of activity and events – in or that could affect – circumterrestrial space), improve capabilities to protect friendly space assets from enemy attack, and develop capabilities to negate enemy space capabilities. SSA programs include Spacetrack, which is developing radar and optical sensors, and the Space-Based Surveillance System and other ground systems that are designed to track objects of interest in space. Other space control programs – such as the Rapid Attack Identification, Detection, and Reporting System (RAIDRS) and the Counter Communications System (CCS) – focus on developing technology to protect friendly systems or to disrupt, deny, degrade, or destroy enemy space capabilities. Joint Publication 3–14, *Joint Doctrine for Space Operations*, discusses ways to gain or maintain space control by providing freedom of action in space for friendly forces through protection and surveillance or to deny freedom of action in space to enemy forces through prevention and negation.[57] Air Force doctrine, by contrast, aligns space control doctrine like air doctrine, as offensive counterspace (OCS) and defensive counterspace (DCS). OCS missions call for the ability to disrupt, deny, degrade, or destroy space systems, or the information they provide, if used for purposes hostile to US national security interests. DCS missions include both active and passive measures to protect US and friendly space-related capabilities (satellites, communications links, and supporting ground systems) from enemy attack or interference. This includes development efforts to prevent adversarial ability to use US space systems and services for purposes hostile to US national security interests.[58] Funding for the Orbital Deep Space Imager, a space-based system designed to track objects in GEO, was eliminated from the President's FY 2007 budget request. Under the 2006 FYDP, research, development, testing, and evaluation funding for space control programs would increase from $195 million in 2006 to $768 million in 2011.[59] CBO's long-term projection assumes a constant level of funding for those activities through 2024. SSA and space control are areas of particular concern in Congress as reflected by the tasking to the Secretary of Defense in section 911 of the FY 2006 NDAA to report to Congress about these topics in April and June 2006, respectively.[60]

Force application

Development of systems with the potential to apply force to, in, or especially from, space is an area of even greater congressional, public, and international concern. These concerns are exacerbated by significant difficulties in distinguishing between concepts and technologies being developed for the ballistic missile defense, protection, space control, and force application mission areas as well as by the fact that some of the systems are being developed in classified programs. Groups opposed to space weaponization, such as the Center for Defense Information (CDI) and the Stimson Center, argue that the momentum created by a number of experiments testing space control and force application concepts in space will create "facts in orbit," driving US policy toward space weapons without debate by either Congress or the public.[61] The programs of greatest concern to these groups include the MDA's Space-Based Interceptor Test Bed, Micro Satellite, and Near Field Infrared Experiment (NFIRE); as well as the Air Force's Experimental Satellite System and Autonomous Nanosatellite Guardian for Evaluating Local Space (ANGELS).[62] It is difficult, however, to see how the United States could effectively continue improving its space protection and ballistic missile defense capabilities without conducting these relatively small-scale experiments, how the experiments could appreciably change any facts in orbit, or how they might lead to full-scale space weaponization without triggering significant public debates, especially given all the woes of acquiring new force enhancement systems detailed above. Indeed, the cumulative effect of all the current NSS acquisition problems has contributed to a small but perceptible shift in priorities away from space control and force application. Comparison of the most recent Space Posture statements to Congress by the Under Secretary of the Air Force (who is dual-hatted as the DoD Executive Agent for Space) shows the emphasis that Peter Teets placed on assured access to and freedom of action in space while his successor, Ronald Sega, did not focus on this area but consistently emphasized a "back-to-basics" approach to acquisition.[63]

Common Aero Vehicle

The prompt global strike mission-needs statement established the requirement for rapid conventional strike worldwide to counter the proliferation
of weapons of mass destruction and provision of a forward presence
without forward deployment. In December 2002, the Deputy Secretary of
Defense directed the Air Force and DARPA to establish a joint program
office to accelerate the Common Aero Vehicle (CAV) effort to meet this
requirement. The CAV program was originally envisioned as a conventional warhead that would be launched from an intercontinental ballistic
missile (ICBM) or potentially from an orbiting space platform and was
part of the Force Application and Launch from Continental United States
(FALCON) program. Because of FY 2005 congressional language, the
FALCON portion of the CAV program was restructured by DARPA and
the Air Force to ensure it meets the intent of Congress. Now known as
Falcon (lower case), the program is focused on the development and
transition of more mature technologies into a future weapon system
capable of delivering and deploying conventional payloads worldwide
from and through space. Within the Falcon program, CAV has been
redesignated the Hypersonic Technology Vehicle and all weaponization
activities have been excluded from Falcon. The FY 2008 Defense Appropriations Bill shifts funding from CAV and the Navy's conventional
Trident Modification to provide $100 million for research into promising
prompt global strike technologies. CBO's projection assumes the limited
deployment of 40 CAV-equipped ICBMs in about 2015, at which point the
demand for investment resources would peak at $600 million.[64]

Major space organizations

The major organizational stakeholders in military space have included the
DoD, the Navy, the Army, the NRO and the intelligence community, the Air
Force, AFSPC, and USSPACECOM.[65] These organizations and their cultures
form the bureaucratic environment in which NSS decisions will be made.

Generally speaking, only a few major military space policy inputs have
been generated at the DoD level, and most of these have been designed to
adjudicate roles and mission disputes between the Services rather than to
provide overall guidance. For example, in November 1956 Secretary of
Defense Charles E. Wilson issued a memorandum that amplified on the Key
West Agreement on roles and missions by reiterating the Army's exclusion
from developing or employing ballistic missiles with ranges beyond 200 miles.
This made it very difficult for Wernher von Braun and his "rocket team" at
the Army Ballistic Missile Agency (ABMA) to pursue their dream of developing space launch vehicles.[66] In September 1959 the DoD reinforced Air Force
primacy in space when Secretary of Defense Neil H. McElroy ruled against
Army and Navy efforts to create a unified (multi-Service) space command

and formally assigned to the Air Force the responsibility for development, production, and launching of space launch vehicles.[67] The DoD further consolidated Air Force control over space and established what remains the current basic structure for military space by making the Air Force the DoD's "executive agent" for space under Directive 5160.32 issued in March 1961.[68] This directive gave the Air Force control over nearly all DoD space programs from inception through launch and TT&C. It is important to emphasize, however, that the DoD civilian decision-makers behind these policies were not trying to expand military space or the Air Force's turf. In fact, just the opposite was true – they were eager to consolidate and streamline DoD space activities in order to rein in the scope of military space and save money.[69] The DoD's March 1987 Space Policy statement officially laid out the four military space missions discussed in the section above and, although short on many specifics, remains the DoD's single most complete public policy statement.[70] Another piece of DoD-level military space organizational structure was put into place in 1995 when Congress called for the creation of two new offices: the Deputy Under Secretary of Defense for Space – DUSD (Space) – and the Space Architect. These offices were intended to provide top-level policy review and an architectural mechanism to integrate systems and capabilities across space sectors. Unfortunately, however, the DUSD (Space) part of this new organizational structure has already been dismantled as the result of the Defense Reform Initiative announced by Secretary of Defense William S. Cohen in November 1997.[71]

The Navy's military space efforts have centered on missions that are crucial for maritime forces such as surveillance, communication, and navigation, but of course it is also quick to point out its long tradition of operating exploratory missions on new seas.[72] The Naval Research Laboratory developed the Vanguard booster for the IGY program, the Galactic Radiation and Background (GRAB) experiment (the world's first electronic intelligence satellite), the Transit system (the world's first space-based navigational aid), and throughout the space age the Navy has emphasized developing improved communication capabilities such as those currently deployed in the UFO communications satellite system. In the late 1950s, the Navy led the unsuccessful attempt to create a unified space command and, in 1983, created its own Naval Space Command.

At the end of World War II, the Army was arguably the organization in the best position to become America's space Service. It had captured the lion's share of von Braun's team, along with their blueprints, files, and parts for 100 V-2s.[73] Moreover, the Army believed it was well suited to open the space frontier due to its rich tradition in civil exploration, such as the Lewis and Clark Expedition, and its civil engineering efforts led by the Topographical Engineers and the Corps of Engineers. Of course, America's first satellite did eventually ride into space atop a modified Jupiter missile assembled by von Braun's team and derived from their V-2, but this proved to be the high watermark of the Army's space efforts. Following this achievement, the Army was largely stripped of its space capabilities by a series of DoD decisions to award the military space launch mission to the Air Force and national policy-level decisions to create NASA and emphasize the civil space mission. The Army's Jet Propulsion Laboratory became NASA's first center, and von Braun's team at ABMA became NASA's Marshall Space Flight Center. By the 1980s, the Army had regained some of its enthusiasm for space activities and, following the Navy's lead, created Army Space Command in 1988. Like the Navy, the Army prides itself on its ability to effectively use the force enhancement capabilities provided by current systems.

As indicated in the sections above, throughout most of the space age the sanctuary doctrine and intelligence sector dominated US space policy. The NRO perpetuated this emphasis in US space policy and was in many ways the most powerful military space bureaucracy during the Cold War, even though it was a black organization from its creation in 1960 until its existence was declassified in 1992.[74] On August 25, 1960 President Eisenhower decided that spy satellites required a new national-level organization to channel their strategic intelligence feed directly to top-level decision-makers, to hide and streamline the process of developing and operating spysats, and because he was leery of military control of spysats.[75] As the NRO's technical prowess grew, it became increasingly tied to arms control and National Technical Means of Verification. There was a subtle but important link between the NRO's improving capabilities and the units of limitation in arms control agreements. When the first Strategic Arms Limitation Treaty (SALT I) was signed in 1972, NTMV was responsible for counting large, fixed facilities such as missile silos and phased array radars. By the time of SALT II in 1979, NTMV was expected to track mobile launchers, distinguish between different types of missiles, and even monitor the number of warheads per missile.[76] Evidence of the NRO's secret but heavy hand in guiding space policy can be seen in the 1969 cancellation of the Manned Orbiting Laboratory (MOL) in favor of unmanned reconnaissance systems, in the way the Space Transportation System (STS or space shuttle) cargo bay design specifications evolved to accommodate future generation spysats, and by the fact that prior to the 1986 *Challenger* disaster the NRO was the only US government organization that was allowed to develop an alternative space launch capability to back up the STS (the complementary expendable launch vehicle (CELV)).[77] From its new headquarters in Chantilly, Virginia, the NRO must now adapt to its changed role as an unclassified organization in a post-Cold War environment where collection on a wide and difficult array of intelligence targets currently takes priority over its traditional NTMV function.[78]

The Air Force is the most powerful and important military organization in the military space sector, but its position as the DoD's executive agent for space also presents the Air Force with very difficult conceptual and organizational challenges. The Air Force often must walk a difficult tightrope on space issues, and the stakes involved may include its very existence. It must be proactive enough to satisfy those groups both inside and outside the Air Force that believe the military should be doing more to pursue force application and space control or that it is not working hard enough on force enhancement. At the same time, it cannot be so proactive on space that it alienates the pilots and air power mission at its institutional core. If the Air Force is not perceived to be moving quickly enough on space, it may inflame those who believe that only a separate space Service can properly advocate for military space; if it moves too quickly, it may undercut its responsibility to develop air power capabilities in pursuit of a chimera. Of course, in an ironic twist, the Air Force's organizational dilemmas in pursuing space power are exacerbated by the institutional baggage surrounding its own creation as a separate Service as well as by the continuing doctrinal ferment over the efficacy of air power.

Given these organizational sensitivities and the sanctuary doctrine mindset of the Cold War, the Air Force's sometimes hesitant and halting approach toward space is not surprising. In the late 1950s, Air Force Chief of Staff Gen. Thomas D. White advocated the "aerospace" concept – a concept that has helped to define the importance of space for the Air Force and the Air Force's role in space ever since.[79] The aerospace concept sees air and space as one indivisible medium and strongly implies that the Air Force is the Service best prepared to conduct aerospace military operations in order to control and project power from this medium.[80] In keeping with the aerospace concept and anticipating approval of significant space control and force application missions, the Air Force pushed hard for manned space systems such as Dyna-Soar or MOL. When these programs were cancelled, however, the Air Force recognized the full strength of national policy support for the sanctuary doctrine and the civil space sector.[81] For the remainder of the 1960s until the late 1970s the Air Force was resigned to the sanctuary doctrine and did not give much consideration to expanding military space operations.

Several factors, such as the breakdown of détente, the first STS flights, and President Ronald Reagan's SDI speech of March 23, 1983, helped to reignite Air Force thinking about military space. When the USAF stood up AFSPC on September 1, 1982 it created its first new major command in 32 years and signaled that space had become a primary operational focus for the Air Force. AFSPC centralized and consolidated a number of functions that had been performed primarily by Air Force Systems Command. Transforming the USAF space community's organizational culture from its roots in the Systems Command R&D mindset to the operational focus desired by AFSPC proved to be a long and difficult process, however. This transformation – and not the innovative space doctrine leadership desired and expected by most of those who had championed the creation of AFSPC – would consume much of AFSPC's time and attention for the remainder of the Cold War. A very significant change in NSS management and organization came in October 2002 when USSPACECOM was merged into United States Strategic Command (USSTRATCOM). Although this was originally described as a merger of equals, in practice this major organizational shift quickly amounted to the absorption of USSPACECOM into USSTRATCOM and left very few vestiges of the original USSPACECOM. Instead of space being the sole focus of one of just nine unified commands, under the new structure space now competes for attention among a very wide array of disparate mission areas that include global strike, homeland defense, information operations, and missile defense.

The establishment of USSPACECOM in September 1985 created the DoD's most effective military space advocate, but also added an additional bureaucratic layer that sometimes complicates organizational loyalties and military thinking about space. In terms of bureaucratic politics, the creation of the unified command was the quid pro quo demanded from the USAF by the DoD and the other Services as the "price" that it had to pay in order to create AFSPC.[82] It is also interesting that this outcome was the opposite of the situation in the late 1950s when the Air Force had been successful in blocking the attempts of the other Services to create a unified command for space. Finally, it is also worth noting that to at least some observers USSPACECOM suffers from "Rodney Dangerfield Syndrome" because it does not believe it receives the respect it deserves as a unified command. It has sometimes been marginalized both by AFSPC (which controls about 90 percent of military space personnel and funding) and by the regionally organized unified commands that have much better defined warfighting roles and areas of responsibility than USSPACECOM.

Near-term space and defense issues: space weaponization

At a fundamental level, virtually all issues of space strategy turn on broad questions related to weaponizing space such as whether space will be weaponized, how and when that might happen, which states and other actors might be most interested in leading or opposing weaponization, and how any of these space weaponization issues might best be controlled. Space weaponization will be shaped by technical, fiscal, and political considerations; political factors are probably most important of all. At the political level, there is, of course, a broad spectrum of opinion but most of the major tenets in mainstream views on weaponizing space can usefully be grouped into four major camps: space hawks, inevitable weaponizers, militarization realists, and space doves.[83]

Space hawks

Adherents to the space hawk camp believe that space already is or holds the potential to become the dominant source of military power. Accordingly, they advocate that the United States move quickly and directly to develop and deploy space weapons in order to control and project power from this dominant theater of combat operations. According to former Senator Bob Smith, for example, the concerted development of American space weapons "will buy generations of security that all the ships, tanks, and air-planes in the world will not provide..... Without it, we will become vulnerable beyond our worst fears."[84] In addition, space hawks often point to space-based BMD as a potentially decisive weapon capable of fundamentally reordering the strategic balance. Space hawks tend to oppose virtually all space-related arms control or regulation because of its potential to slow or derail rapid and direct space weaponization by the United States.

Inevitable weaponizers

This group believes that space, like all other environments man has encountered, will eventually be weaponized. Yet they differ from space hawks in two important ways: they are not convinced that space weaponization would be beneficial for US or global security, and they are unsure that space will prove to be a decisive theater of combat operations. The *Space Commission Report* is a good example of this camp:

> we know from history that every medium – air, land and sea – has seen conflict. Reality indicates that space will be no different. Given this virtual certainty, the United States must develop the means both to deter and to defend against hostile acts in and from space.[85]

Inevitable weaponizers take a nuanced view of space arms control and regulation. They generally support transparency- and confidence-building measures (TCBMs) and other mechanisms designed to slow military competition and channel it in predictable ways. But they are less supportive of broad efforts to ban space weapons because they see them as futile or even dangerous due to their potential to lull the United States into complacency or otherwise cause it to be outmaneuvered by states that successfully circumvent space weaponization accords.

Militarization realists

Members of this camp oppose space weaponization because they believe US security interests are best served by the status quo in space. They believe that the United States has little to gain but much to lose by weaponizing space because it is both the leading user of space and, enabled by this space use, the dominant terrestrial military power. Militarization realists also believe that if the United States takes the lead in weaponizing space, it would become easier for other states to follow due to lower political and technological barriers. Militarization realists support space-related arms control and regulation that precludes other states from weaponizing or even militarizing space. Most of them believe, however, that this support must be balanced against the increased attention that formalized arms control efforts could draw to the United States' already formidable space-enabled force enhancement capabilities and the political, military, and arms control fallout this increased scrutiny might cause.

Space doves

A wide range of organizations and viewpoints can be grouped together in the space dove camp because they all oppose space weaponization for a variety of reasons including moral, arms control, conflict resolution, stability, and ideology arguments. Most space doves also oppose any militarization of space beyond the limited missions they see as stabilizing – NTM, early warning, and hotline communications – because they see any military missions beyond these as the "slippery slope" to space weaponization. Most space doves emphasize how destabilizing most space militarization and all space weaponization would be:

> Unlike the strategy for nuclear weapons, there exists no obvious strategy for employing space weapons that will enhance global stability. If the precedent of evading destabilizing situations is to continue – and that is compatible with a long history of US foreign policy – one ought to avoid space-based weapons.[86]

They also highlight the deep roots of President Eisenhower's "space for peaceful purposes" policy and argue that, especially in the post-Cold War era, there is no rationale for space weaponization that is strong enough to overturn the basic strategic logic America developed at the opening of the space age. Space doves support space arms control and regulation more strongly than any other camp. Since they do not believe the United States (or other states) would reap strategic benefits from weaponizing space, they are not overly concerned about the numerous arms control challenges identified by the other camps.

Near-term space and defense issues: high-altitude nuclear detonations

The threat caused by high altitude nuclear detonations (HAND) is sufficiently different and potentially damaging that it warrants discussion and analysis separate from the broad space weaponization issues raised above. Just one such detonation holds the potential to disable *all* non-hardened LEO satellites. Today, these assets are worth tens of billions of dollars – this class of assets is likely to be worth far more in the future – and this threat poses daunting detection, deterrence, and defense challenges, not least of which is the fact that such an attack would take place outside the sovereign jurisdiction of any state and not directly kill a single person.[87] As such, HAND is a unique asymmetric threat that is the single "most potentially disruptive and dangerous possibility."[88]

HAND can destroy or disrupt LEO satellites in two primary ways. First, prompt X-rays can upset or burnout the electronics for the 5–10 percent of each LEO constellation within line of sight of the explosion.[89] Second, in weeks to months, potentially all non-hardened LEO satellites can fail due to the cumulative effects of phenomena such as transient-radiation effects on electronics (TREE) and system-generated electromagnetic pulse (SGEMP) as the satellites orbit through the greatly increased radiation belts the explosions cause in LEO.[90]

One of the largest problems, however, in assessing the specific level of threat posed by HAND is a lack of experimental data on the effects of HAND on satellites (especially on modern satellite systems) and this contributes to a range of assessments concerning the severity of the threat.[91] The United States conducted two high-altitude nuclear test series before such testing was banned by the LTBT; the tests were conducted in August and September of 1958 and again during the summer and fall of 1962.[92] The Argus series was designed to test and did confirm the theory of Nicholas Christofilos of the University of California's Radiation Laboratory that the high-energy electrons produced in a high-altitude explosion would become trapped in Earth's magnetic field.[93] As predicted, these trapped particles do "pump up" the radiation belts through which LEO satellites pass during each orbit and slowly build a potentially fatal radiation dose for the satellites' electronics.

What are the best technical and political options for the United States to mitigate the risks associated with HAND? Watts is surely correct is his assessment that "for the next 15–20 years, the most sensible stratagem for preventing an exo-atmospheric nuclear detonation is a combination of deterrence and hardening the satellites themselves."[94] However, as with the other most difficult security challenges such as counter proliferation, a comprehensive, layered, and synergistic approach to this threat would seem to offer the best prospects for success. For these cases the United States should pursue a range of policies designed to move up the escalatory ladder from denial, to reassurance and dissuasion, cooperative and involuntary reversal, deterrence, passive and active defenses, through counterforce operations including preemptive strikes.[95] For HAND more specifically, the United States should begin by continuing its arms control efforts such as the Non-Proliferation Treaty and Missile Technology Control Regime designed to deny potential adversaries the tools necessary to carry out a HAND. It should reassure and dissuade these actors by attempting to embed them in the global information infrastructure by sharing the fruits of LEO architectures (perhaps by cross subsidies, as required) and by positive and negative security assurances.[96]

Moving up the ladder, the United States may, unfortunately, face very difficult challenges in the areas of deterrence, passive and active defense, and counterforce. Deterrence may be strengthened because it would be very difficult to launch a HAND without attribution in almost all scenarios; but deterrence could be undermined because the attack takes place outside US territory and does not directly kill anyone, so it might not generate support for a swift and sure response. Unambiguous declaratory policy statements that the United States will respond forcefully to any purposeful interference with its space systems might strengthen deterrence, but there are a number of difficult cross-cutting issues associated with this problem, such as how best to deter attacks on multinational and commercial systems.[97] The potential for active defense against HAND is extremely limited due to the nature of the threat and the defensive technologies available. Active defenses would have just a fleeting moment in which to operate because of the very limited flight time of a HAND booster (a couple of minutes at most), would somehow have to discriminate between HAND-weapon launches and all other launches within this fleeting window, and, to be effective, would need to be 100 percent successful since leakage of even one can spell the end for all non-hardened LEO satellites. Likewise, the prospects for the United States successfully carrying out a preemptive strike on a HAND weapon prior to launch are minimal. Not only would the preparations for such a strike require breathtakingly prescient intelligence data, but also, and perhaps more significantly, carrying out a preemptive strike would require a level of political will the United States has seldom displayed in the past.

Hardening of LEO satellites to withstand HAND-induced radiation stands out as one of the least costly and potentially effective means of addressing this threat. It is not a panacea but, according to Wilson, "[h]ardening of a space system's elements is the single most effective survivability measure."[98] The national security space protection strategy framework signed by Air Force Under Secretary Peter Teets in March 2005 requires future LEO national security space systems to be hardened against total radiation dose failures. Finding the resources to implement this framework may prove difficult given the national security space acquisition bow wave described above, but it is an important first step toward requiring better balance between new space capabilities and more robust protection.

Near-term space and defense issues: high-resolution commercial remote sensing

High-resolution commercial remote sensing is an evolving, complex issue area that requires carefully considered approaches to balance several interdependent goals.[99] High-resolution commercial imagery creates opportunities and risks across a wide range of diplomatic, military, economic, and political considerations. Among the largest are: how these systems contribute to global transparency and the implications of a more transparent world; economic competition and the viability of the high-resolution commercial remote sensing industry worldwide; competition from other remote sensing providers and the quality, timeliness, and types of products offered by space-based systems; the optimal mix between commercial and government systems; and mechanisms for controlling and regulating this industry.

US commercial remote sensing policy

Following the end of the Cold War, the United States completely reoriented its remote sensing policy away from the secret spysat regime crafted by the Eisenhower administration at the opening of the space age. Instead of a secret government monopoly, under the new policy

> U.S. companies are encouraged to build and operate commercial remote sensing space systems whose operational capabilities, products, and services are superior to any current or planned foreign commercial systems. However, because of the potential value of its products to an adversary, the operation of U.S. commercial remote sensing space systems requires appropriate security measures to address U.S. national security and foreign policy concerns. In such cases, the United States Government may restrict operations of the commercial systems in order to limit collection and/or dissemination of certain data and products, e.g. best resolution, most timely delivery, to the United States Government, or United States Government approved recipients.[100]

By attempting to dominate this market worldwide, the United States hopes to preserve its defense industrial base and workers trained in this sector, use commercial systems for many government needs, and shape global standards for such systems. Building on the Land Remote Sensing Policy Act of 1992 and Presidential Decision Directive 23 of March 1994, the Bush Administration's April 2003 policy indicates that the US government will:

- rely to the maximum practical extent on US commercial remote sensing space capabilities for filling imagery and geospatial needs;
- focus US government remote sensing space systems on meeting needs that cannot be effectively, reliably, or affordably satisfied by commercial providers;
- develop a long-term, sustainable relationship between the US government and the US commercial remote sensing industry;
- provide a timely and responsive environment for licensing and exports;
- enable US industry to compete successfully as a provider for foreign governments and foreign commercial users while ensuring appropriate measures are implemented to protect national security and foreign policy.[101]

Strategic implications of high-resolution commercial imagery

Like most dual-use technologies, high-resolution commercial remote sensing holds both beneficial and threatening potential. By increasing transparency, these systems should help to dampen the security dilemma by illuminating the actual force levels of states and they should also help to increase stability by revealing preparations for an attack. Conversely, by pinpointing potential targets, such systems may create incentives for preemption – especially by states that possess highly accurate, long-range weapons. The utility of actually attacking following a planned exercise or during crisis demobilization also would seem to be increased. Paradoxically, the amount of data available from proliferated commercial imagery systems might actually place even greater value on the type of information usually only available from human intelligence sources – the intentions of potential adversaries. In addition, this technology can empower non-state actors and provide them with information to support a wide variety of goals.

Two congressionally mandated studies reemphasize the importance of high-resolution commercial remote sensing to the IC and its tasking, processing, exploitation, and dissemination (TPED) processes. According to the NRO Commission report, for example, the US government: "could satisfy a substantial portion of its national security-related imagery requirements by purchasing services from" US firms; it "must" develop a "clear national strategy that takes full advantage of the capabilities of the US commercial satellite imagery industry"; and it should create a system similar to the DoD's industrially funded airlift account to help efficiently focus government systems "on targets where their unique capabilities in resolution and revisit times are important, while commercial systems would be used to provide processed 'commodity' images."[102] The NIMA Commission report goes even further. It found the IC to be "collection centric," "that NIMA was not a good, dependable business partner," and recommended creating a "central commercial imagery fund" to help mitigate problems resulting from the fact "that national technical means (NTM) imagery appears to be 'free' to government agencies, while use of commercial imagery generally requires a distressingly large expenditure of (largely unplanned, unprogrammed) O&M [operation and maintenance] funds."[103] The commission recommended that the central commercial imagery fund start at about $350 million annually for "raw imagery and vendor's value-added offerings."[104] The NIMA Commission saved its harshest critique for NIMA's TPED shortcomings. These problems "increasingly strain at the fabric of the NIMA organization as a whole" and undermine confidence "that NIMA currently has the system engineering experience, acquisition experience, appropriate business practices, and performance measures" to acquire a cutting-edge TPED system.[105] The commission concluded that NIMA's TPED efforts simply cannot "get there from here" and recommended:

> creation of an Extraordinary Program Office (EPO) armed with special authorities ... to be constituted within NIMA from the best national talent ... charged with and resourced for all preacquisition, systems engineering, and acquisition of imagery TPED – from end to end, from "national" to "tactical." The first milestone shall be completion of, understandable, modern-day "architecture" for imagery TPED.[106]

High-resolution commercial imagery and deception

Digitized data streams designed to produce imagery are ideally suited for deception. This is because digitized data must always be mathematically processed to create images, and this processing is subject to manipulation in a variety of ways – many of which are not available for manipulating film images. As Steven Livingston explains:

> Mathematically altering the value of the pixels alters *seamlessly* the representation. "Since it is purely a mathematical process, the source images can be altered fundamentally and undetectably before and/or during their production." Elements can be added or subtracted, changed in color, brightness, or contrast. Changes are made not by altering the computer code that produces the image, and not in the image itself as in analog manipulation. In fact, it is more accurate perhaps to say that no image exists beyond the mathematical equations that create a particular array of pixels. The equations are the image. Therefore as computer processors become faster and more powerful, so too does the ability to alter digital information.[107]

The phrase "altered fundamentally and undetectably" is absolutely loaded with implications. For starters, it means that virtually *anything* can be added, subtracted, or changed in digital imagery (or to any digital information) and that even experts cannot necessarily detect these changes. The possibilities for deception through manipulating digital imagery are literally unlimited. Perhaps even more alarmingly, all of this can happen in real time as the data stream is converted into manipulated imagery. This also offers the potential for novel means of control, exploitation, and deception by altering data streams rather than limiting dissemination. It is no wonder that the digital age creates a number of legal conundrums and that the veracity of digitized information is increasingly being questioned in courtrooms.[108] At the very least, as *No More Secrets* summarizes, "[c]ommercially available high-resolution satellite imagery will trigger the development of more robust denial and deception and antisatellite countermeasures."[109] Given this potential for deception, the US government and the news media should adopt a "dual phenomenology" requirement as a way to attempt to confirm the veracity of digitized imagery before it is broadcast or acted upon.

Control of high-resolution commercial imagery

During Operation Enduring Freedom in Afghanistan, NIMA limited data dissemination by signing a $1.9 million "agreement of assured access" with the Space Imaging Corporation on October 7, 2001. Under the terms of the agreement, Space Imaging could not sell or share its Afghanistan theater imagery with anyone except the US government until after December 5, 2001.[110] This agreement opens many interesting issues related to the utility of limiting information dissemination for public diplomacy, the media, and exploitation of enemy information channels. It also raises the issue of whether it has set a precedent of using market mechanisms (checkbook shutter control) that might well make it more difficult in the future for the United States to limit data dissemination without first purchasing the data.

Near-term space and defense issues: global utilities

Because of the growth in space systems and the services they provide, some analysts believe they should now be considered in a new way – as global utilities that provide an essential foundation that enables the global information infrastructure. In some ways, the concept of global utilities is just another recognition of how much the commercial space sector has grown and how important it has become; but it is also clear that the global information infrastructure as it currently exists simply could not function without space systems. Global utilities have been defined as: "Civil, military, or commercial systems – some or all of which are based in space – that provide communication, environmental, position, image, location, timing, or other vital technical services or data to global users."[111] To date, all space-based global utilities provide information services, but they are analogous to Earth-bound utility services that provide a foundation for modern life such as water and electricity. And like these Earth-bound utility services, space-based global utilities may be subject to regulation and control at the local, state, national, and international levels.

Two relatively minor failures illustrate just how embedded global utilities have become in the global information infrastructure. In 1996 a controller at the Air Force GPS control center accidentally put the wrong time into just one of the 24 satellites, and this erroneous signal was broadcast for just 6 seconds before automatic systems turned the signal off. That momentary error caused more than 100 of the 800 cellular telephone networks on the US East Coast to shut down, and some took hours or even days to recover.[112] In May 1998 "40–45 million pager subscribers lost service; some ATM and credit card machines could not process transactions; news bureaus could not transmit information; and many areas lost television service – all because of the loss of *one* satellite."[113] Clearly, space systems have become an increasingly important part of the global information infrastructure, but questions remain about how they should be regulated and protected.

Near-term space and defense issue: spectrum crowding, orbital debris, and space traffic control

The final contentious area is related to the cumulative effects of greater use of space. Current and projected use of space is creating challenges particularly in the areas of crowding of the radio spectrum for space, orbital debris, and the possible need for space traffic control.

Recent growth in commercial space activity has exacerbated crowding of the radio spectrum for space applications, and there are currently significant pressures on portions of the spectrum now allocated to military uses. These commercial pressures on the spectrum have been somewhat balanced by the use of new technologies and different orbits that lessen the effects of increased use. For example, modern satellites in GSO have only two degrees of spacing between them (versus three or more degrees in the past) for most systems providing fixed satellite services. Likewise, increasing use of NGSO for communication satellite networks has decreased the pressure on overcrowding the GSO in terms of spectrum and spacing. Current trends for the space radio spectrum do not augur major changes in the current regulatory structure. Moving the ITU to auctions for its satellite coordination/registration process would undoubtedly produce greater efficiency and generate income, but these benefits would need to be weighed against the equal access concerns of the developing world and fact that there currently seems to be little support for moving in this direction.

Orbital debris may represent the best window of opportunity for cooperative approaches to space for the United States and the global spacefaring community.[114] NASA defines orbital debris as "any man-made object in orbit about the Earth which no longer serves a useful purpose."[115] Human space activity has generated a lot of debris: there are over 17,000 objects larger than 10 cm and an estimated 100,000-plus objects 1–10 cm in size.[116] The largest single source of this debris has been intentional and unintentional satellite explosions on-orbit.[117] Orbital debris generally moves at very high speeds relative to operational satellites and thereby poses a risk to these systems due to its enormous kinetic energy.[118] Only three collisions between operational systems and orbital debris are known to have occurred thus far, but concerns about this hazard are growing due to the increasing number of operational space systems and the 5 percent growth rate in LEO orbital debris each year.[119] There is even concern about the potential for orbital debris "chain reactions" due to collisions in big LEO communication satellite constellations or due to the debris clouds that could be created by use of kinetic energy ASATs in LEO.

Since the 1980s, the United States has led the world in publicizing the risks due to orbital debris, and it has made debris-mitigation programs an increasingly important part of its overall space policy.[120] There is, however, undoubtedly more the United States could do on the orbital debris front. The United States should explore several options such as unilaterally pledging not to create space debris through testing or operations of any ASAT system, creating strict unilateral regulations that mandate debris mitigation for US commercial space operators multilateral efforts to "clean up" debris using lasers and other techniques, and creating strict multilateral regulations for debris mitigation. These and other creative approaches should be explored vigorously in order to ensure that man's increasing use of space does not impose unacceptable risks on this activity.

Finally, due again to the increasing use of space, the United States must consider the need for and implications of space traffic control systems (STCS) that could be analogous to current air traffic control systems. The idea for such a system is obviously related to the orbital debris problem discussed above, but it goes well beyond just this problem to include a wide range of factors such as: how space traffic might coordinate for and be slotted into specific orbital positions, how space traffic would be located and tracked, sanctions and liability for non-compliance and collisions under an STCS, and how such a regime might be established and funded. As with many space-related issues, the technology to at least begin implementing such a system appears to be closer at hand than is the political will to begin down this path. For example, the Missile Defense Agency's Midcourse Space Experiment Space-Based Visible (MSX ISBV) launched in April 1996 was for many years the only space-based surveillance capability; it found some "150 objects in the last three years that were completely lost" and demonstrated the potential value of space-based sensors to an STCS.[121] Likewise, GPS positioning signals could be used to locate many space systems very accurately and a transponder-like system aboard space systems could automatically provide this data in response to queries from the STCS.[122] On the political side of the equation, however, the United States must consider very carefully how its objectives in space might benefit or be harmed via the creation and operation of an STCS. It is not obvious that an air traffic control model is the appropriate regime for space, or that the political and financial costs of creating and operating such a system (many of which would likely be borne by the United States) would be outweighed by its benefits.

Notes

1 Unfortunately, the programs within the vMFP have not thus far remained constant from year to year and have not always covered all major space systems, reducing the utility of this measure for tracking national security space expenditures over time.
2 Marcia S. Smith, "U.S. Space Programs: Civilian, Military, and Commercial," (Washington, DC: Congressional Research Service, August 9, 2005).
3 Congressional Budget Office, "The Long-Term Implication of Current Plans for Investment in Major Unclassified Military Space Programs," (Washington, DC: Congressional Budget Office, September 12, 2005).
4 Ibid., p. 5.
5 Several of these individuals were quite prolific; the following list represents a few of their best known works: Alfred Thayer Mahan, *The Influence of Sea Power upon History, 1660–1783* (Boston: Little Brown, 1890); Julian S. Corbett, *Some Principles of Maritime Strategy*, edited by Eric J. Grove (Annapolis: Naval Institute Press, 1988. First published 1911); Giulio Douhet, *The Command of the Air*, edited by Richard H. Kohn and Joseph P. Harahan (Washington, DC: Office of

Air Force History, 1983. First published 1921); William Mitchell, *Winged Defense: The Development and Possibilities of Modern Airpower – Economic and Military* (New York: Dover, 1988. First published 1925); and John A. Warden III, *The Air Campaign: Planning for Combat* (Washington, DC: National Defense University Press, 1988). On the importance of these works see, Jon Tetsuro Sumida, *Inventing Grand Strategy and Teaching Command: The Classic Works of Alfred Thayer Mahan Reconsidered* (Washington, DC: Woodrow Wilson Center Press, 1997); Philip S. Meilinger (ed.), *The Paths of Heaven: The Evolution of Airpower Theory* (Maxwell AFB: Air University Press, 1997); and David R. Mets, *The Air Campaign: John Warden and the Classical Airpower Theorists* (Maxwell AFB: Air University Press, April 1999).

6 Virtually all of these concepts are applied throughout the Chief of Staff-directed year-long study by Air University that is published as *SPACECAST 2020* (Maxwell AFB: Air University Press, 1994).

7 In 1997 then-CINCSPACE Gen. Howell M. Estes III attempted to remedy the lack of a comprehensive space power vision or theory by commissioning Dr. Brian R. Sullivan to write a book on space power theory. This project was taken over by James Oberg and published as *Space Power Theory* (Washington, DC: Government Printing Office, 1999). On the enduring nature of strategy and problems with developing space power theory, see also Colin S. Gray and John B. Shelton, "Spacepower and the Revolution in Military Affairs: A Glass Half-Full," in Peter L. Hays, James M. Smith, Alan R. Van Tassel, and Guy M. Walsh (eds.), *Spacepower for a New Millennium: Space and U.S. National Security* (New York: McGraw-Hill, 2000), pp. 239–258; and Colin S. Gray, *Modern Strategy* (Oxford: Oxford University Press, 1999), pp. 243–267. The recent publications by Barry R. Watts, *The Military Uses of Space: A Diagnostic Assessment* (Washington, DC: Center for Strategic and Budgetary Assessments, 2001); Steven Lambakis, *On the Edge of Earth: The Future of American Space Power* (Lexington: University Press of Kentucky, 2001); Bob Preston, Dana J. Johnson, Sean J.A. Edwards, Michael Miller, and Calvin Shipbaugh, *Space Weapons: Earth Wars* (Santa Monica: RAND, 2002); M.V. Smith, *Ten Propositions Regarding Spacepower* (Maxwell AFB: Air University Press, 2002); Benjamin S. Lambeth, *Mastering the Ultimate High Ground: Next Steps in the Military Uses of Space* (Santa Monica: RAND, 2003); and especially Everett C. Dolman, *Astropolitik: Classical Geopolitics in the Space Age* (London: Frank Cass, 2002), will undoubtedly go a long way toward filling the yawning space power theory gap in the literature.

8 The Air Force developed the aerospace concept during the 1950s. It was forcefully advocated by Chief of Staff Gen. Thomas D. White following *Sputnik* and formed a foundational component of Air Force thinking about space. In 1992 Chief of Staff Gen. Merrill A. McPeak moved away from this concept by adding the words "air and space" to the Air Force mission statement. During the 1997–2001 tenure of Gen. Michael E. Ryan as Chief of Staff, however, the Air Force returned to the aerospace concept. For this period the Air Force emphasized that there are physical differences between the atmosphere and space but defined *aerospace* as a seamless operational medium comprised of both physical domains. See, for example, Air Force Doctrine Document 2-2 *Space Operations* (Maxwell AFB: Air Force Doctrine Center, August 23, 1998), p. 1; or Gen. Michael E. Ryan and the Honorable F. Whitten Peters, *Global Vigilance, Reach & Power: America's Air Force Vision 2020* (Washington, DC: Department of the Air Force, 2000). As reflected in recent speeches and the November 27, 2001 edition of AFDD 2-2, Gen. John P. Jumper, the current USAF Chief of Staff, has chosen to return to using "air" and "space" as separate words rather than continuing to use the term "aerospace."

Let me start off by talking a little bit about air and space versus aerospace. I carefully read the Space Commission report. I didn't see one time in that report, in its many pages, where the term "aerospace" was used. The reason is that it fails to give the proper respect to the culture and to the physical differences that abide between the physical environment of air and the physical environment of space.

We need to make sure we respect those differences. So, I will talk about air *and* space. I will respect the fact that space is its own culture, that space has its own principles that have to be respected. And when we talk about operating in different ways in air and space, we have to also pay great attention to combining the effects of air and space because in the combining of those effects, we will leverage this technology we have that creates the asymmetrical advantage for our commanders.

(Prepared remarks of Gen. John P. Jumper at Air Force Association National Symposium, Los Angeles, November 16, 2001. Available at: www.aef.org/symposia/jump1101.asp)

For a comprehensive analysis of the aerospace concept's deep roots in air power theory, see Maj. Stephen M. Rothstein, "Dead on Arrival? The Development of the Aerospace Concept, 1944–1958," Master's thesis, School of Advanced Airpower Studies (Maxwell AFB: Air University Press, 2001). For a critique of the aerospace concept, see Lt. Col. Peter Hays and Dr. Karl Mueller, "Going Boldly – Where? Aerospace Integration, the Space Commission, and the Air Force's Vision for Space," *Aerospace Power Journal* 15(1) (2001): 34–49.

9 Lt. Col. Dennis M. Drew, "Of Leaves and Trees: A New View of Doctrine," *Air University Review* 33(2) (1982): 40–48; Lt. Col. Charles D. Friedenstein, "The Uniqueness of Space Doctrine," *Air University Review* 37(1) (1985): 13–23; and Col. Kenneth A. Myers and Lt. Col. John G. Tockston, "Real Tenets of Military Space Doctrine," *Airpower Journal* 2(4) (1988): 54–68.

10 The Air Force published AFM 1–6, *Military Space Doctrine*, on October 15, 1982 and its release was designed to coincide closely with the establishment of Air Force Space Command on September 1, 1982. For a detailed critique of AFM 1–6, see Peter L. Hays "Struggling towards Space Doctrine: U.S. Military Space Plans, Programs, and Perspectives during the Cold War" PhD dissertation, Fletcher School of Law and Diplomacy, Tufts University, 1998, pp. 400–422.

11 Friedenstein, pp. 21–22.

12 Myers and Tockston, p. 59. A more up-to-date and outstanding blueprint for developing space doctrine is provided by Maj. Robert D. Newberry, *Space Doctrine for the Twenty-First Century* (Maxwell AFB: Air University Press, 1998).

13 Frederick Jackson Turner first presented his ideas in a paper, "The Significance of the Frontier in American History," at a gathering of historians in Chicago, site of the World's Columbian Exposition, an enormous fair to mark the four-hundredth anniversary of Columbus's voyage.

14 Cited in "New Perspectives on the West," available at: www.pbs.org/weta/thewest/people/s_z/turner.htm, May 20, 2001.

15 Ibid.

16 The original *Star Trek* series aired 1966–1968. Creator Gene Roddenberry sold the series concept to the NBC network as *Wagon Train in Space*. Science fiction has probably been more important in shaping our perceptions of space power than any other single factor. Gerard K. O'Neill, *The High Frontier: Human Colonies in Space* (New York: Morrow, 1977); and the report of the National Commission on Space, *Pioneering the Space Frontier: An Exciting Vision of Our Next Fifty Years in Space* (New York: Bantam Books, 1986). See, for example, Sen. Bob Smith, "The Challenge of Space Power," *Airpower Journal* 13(1) (1999): 32–39.

17 Gerard K. O'Neill founded the Space Studies Institute at Princeton University in 1977 and is probably most famous for *The High Frontier*. Representative works in this tradition include James E. Oberg and Alcestis R. Oberg, *Pioneering Space: Living on the Next Frontier* (New York: McGraw-Hill, 1986); and Robert Zubrin, *Entering Space: Creating a Spacefaring Civilization* (New York: Jeremy P. Tarcher/Putnam, 1999). Carl Sagan co-founded the Planetary Society in 1980 and was one of the most famous and articulate spokesmen for planetary science; *Pale Blue Dot: A Vision of the Human Future in Space* (New York: Random House, 1994) was one of the last major books before his death in 1996. Frank White, *The Overview Effect: Space Exploration and Human Evolution* (Boston: Houghton Mifflin, 1987).

18 McDougall, p. 4.

19 Ibid., pp. 460–461.

20 See David N. Spires, *Beyond Horizons: A Half-Century of Air Force Space Leadership* (Washington, DC: Government Printing Office, 1997); and Curtis Peebles, *High Frontier: The United States Air Force and the Military Space Program* (Washington, DC: Air Force History and Museums Program, 1997) on the military space system incremental improvements highlighted by the Gulf War.

21 Robert S. Dudney, "Washington Watch: The New Space Plan," *Air Force Magazine* 81(7) (1998), available at: www.afa.org/magazine/0798watch.html, December 13, 2000.

22 Ibid.

23 Gen. Howell M. Estes III, "The Promise of Space Potential for the Future," prepared remarks to the US Space Foundation's 1997 National Space Symposium, Colorado Springs, April 3, 1997. Available at: www.defenselink.mil/speeches/1997/s19970403-estes.html, December 11, 2000.

24 Gen. Richard B. Myers, "Integrating Space in an Uncertain Era," prepared remarks to the Air Force Association Space Symposium, Los Angeles, November 13, 1998. Available at: www.aef.org/symposia/myers.html, December 13, 2000.

25 Quoted in Peter Grier, "The Investment in Space," *Air Force Magazine* 83(2) (2000), available at: www.afa.org/magazine/0200investment.html, December 11, 2000.

26 Gen. Richard B. Myers, "Space Superiority Is Fleeting," *Aviation Week & Space Technology*, January 1, 2000, available at: www.awgnet.com/avaiation/newmillen/aw54.htm, December 13, 2000.

27 Gen. Eberhart did not put much emphasis on commercial space developments and did not even mention the term COG in his March 8, 2000 testimony to the Senate Armed Services Committee Strategic Subcommittee, available at: www.spacecom.af.mil/usspace/cinc8mar00.htm; or in his April 4, 2000 keynote speech to the US Space Symposium, available at: www.spacecom.af.mil/ usspace/cinc0404.htm. USSPACECOM picked up the CND mission in 1999 and became responsible for CNA on October 1, 2000. SPACECOM never came to grips with how to organize to perform these new missions and it is not yet clear how STRATCOM will organize for them.

28 John A. Tirpak, "The Fight for Space," *Air Force Magazine* 83(8) (2000), available at: www.afa.org/magazine/august2000/0800space.html, December 13, 2000.

29 Quoted in William B. Scott, "Cincspace: Focus More on Space Control," *Aviation Week & Space Technology*, 15 November 2000, available at: www.infowar.com/MIL_C4I/00/mil_c4I_111500b_j.shtml, April 19, 2001.

30 Futron Corporation, *Satellite Industry Guide* (Bethesda, MD: Futron Corporation, 1999).

31 US GDP figures are available from the DoC's Bureau of Economic

Analysis, available at: www.bea.doc.gov/bea/ARTICLES/2001/08august/
0801GDP.pdf. Aerospace does not qualify as a separate subset within manufacturing but manufacturing of *all* durable goods accounts for only 9.4
percent of GDP.

32 The "*Fortune* 500 List," available at: www.fortune.com. The *Fortune* 500 list for
2001 is based on data from 2000. Most major aerospace corporations do not
report space-only revenues or categorize this part of their business in consistent ways.

33 Ibid. and *Space Commission Report*, p. 11.

34 $572.9 billion for the energy sector versus $186.1 billion for the aerospace
sector. And, again, this represents *all* the revenues from aerospace corporations rather than the fraction attributable to space activities.

35 "Comparison of Aerospace Market Valuation to Top 25 U.S. Companies –
End of FY 1999," slide 31 of "Space Industry 2000 Study" presentation by
Industrial College of the Armed Forces.

36 CBO, "Investment in Major Military Space Programs," p. 17.

37 Government Accountability Office, "Defense Acquisitions: Assessment of
Selected Major Weapon Programs," Report to Congressional Committees,
GAO-06–391, March 2006, p, 54. Under the Nunn-McCurdy Act (10 USC
2433), Congress must be notified when a major defense acquisition program
experiences a cost increase of 15 percent or more. If the increase is greater
than 25 percent, the Secretary of Defense must certify to Congress that the
program is essential to national security and adequately managed, no feasible
alternatives exist, and the new cost estimates are reasonable; otherwise,
funding for the program may be suspended.

38 Ibid., p. 14.

39 Honorable Ronald M. Sega, Under Secretary of the Air Force, "Space Posture
Statement to Strategic Forces Subcommittee of House Armed Services Committee," March 16, 2006, p. 9.

40 CBO, "Investment in Major Military Space Programs," p. 5.

41 GAO, "Assessments of Major Weapon Programs," pp. 119–120.

42 Ibid., pp. 27–28.

43 Sega, "Space Posture Statement to Strategic Forces Subcommittee of House
Armed Services Committee," p. 13.

44 GAO, "Assessments of Major Weapon Programs," pp. 111–112.

45 Section 912 of the FY 2006 NDAA, Military Satellite Communications, states:

(a) Findings – Congress finds the following:

   (1) Military requirements for satellite communications exceed the capability of on-orbit assets as of mid-2005.

   (2) To meet future military requirements for satellite communications,
   the Secretary of the Air Force has initiated a highly complex and revolutionary program called the Transformational Satellite Communications System (TSAT).

   (3) If the program referred to in paragraph (2) experiences setbacks that
   prolong the development and deployment of the capability to be provided by that program, the Secretary of the Air Force must be prepared to implement contingency programs to achieve interim
   improvements in the capabilities of satellite communications to meet
   military requirements through upgrades to current systems.

(b) Development of Options – In order to prepare for the contingency
   referred to in subsection (a)(3), the Director of the National Security

Space Office of the Department of Defense shall provide for an assessment, to be conducted by an entity outside the Department of Defense, to develop and compare options for the individual acquisition of additional Advanced Extremely High Frequency space vehicles, in conjunction with modifications to future acquisitions under the Wideband Gapfiller System program, that will accomplish the following:

(1) Minimize nonrecurring costs.
(2) Improve communications-on-the-move capabilities.
(3) Increase net centricity for communications.
(4) Increase satellite throughput.
(5) Increase user connectivity.
(6) Improve airborne communications support.
(7) Minimize effects of a break in production.
(8) Minimize risk associated with gaps in functional availability of on-orbit assets.

(c) Analysis of Alternatives Report – Not later than April 15, 2006, the Director of the National Security Space Office shall submit to Congress a report providing an analysis of alternatives with respect to the options developed pursuant to subsection (b). The analysis of alternatives shall be prepared taking into consideration the findings and recommendations of the independent assessment conducted under subsection (b).

46 Office of Science and Technology Policy, "Fact Sheet: U.S. Space-Based Positioning, Navigation, and Timing Policy," (Washington, DC: The White House, December 15, 2004), p. 3.
47 Ibid., pp. 6–7.
48 Ibid., pp. 3–4. The selective availability (SA) feature of GPS introduces errors into the standard position service signal, normally reducing its accuracy from approximately 16 meters to approximately 100 meters. In September 2007 the White House announced that SA capabilities will no longer be part of future GPS modernizations such as GPS III.
49 CBO, "Investment in Major Military Space Programs," p. 9.
50 Ibid., pp. 9–11.
51 *Wall Street Journal*, February 11, 2006 asserts that the restructuring of FIA will add $8 billion to the program costs and raise the total above $20 billion.
52 Report of the Defense Science Board/Air Force Scientific Advisory Board Joint Task Force on Acquisition of National Security Space Programs (Washington, DC: Defense Science Board, May 2003), p. 6; GAO, "Assessments of Major Weapon Programs," pp. 101–102. Through the end of 2005, the SBIRS program had triggered four reports to Congress under the Nunn-McCurdy Act; the program breached its cost estimates by 25 percent in 2001, by 15 percent in 2004, and by 25 percent twice during 2005.
53 CBO, "Investment in Major Military Space Programs," p. 12.
54 Ibid., p. 14.
55 Government Accountability Office, "Polar-Orbiting Operational Environmental Satellites: Cost Increases Trigger Review and Place Program's Direction on Hold," statement of David A. Powner, Director Information Technology Management Issues, testimony Before Subcommittee on Disaster Prevention and Prediction, Committee on Commerce, Science, and Transportation, US Senate, GAO-06–573T, March 30, 2006, pp. 3–4.
56 Ibid., p. 12.
57 Joint Publication 3–14, pp. IV-5–IV-8.

58 Air Force Doctrine Document 2-2.1, *Counterspace Operations* (Maxwell AFB: Air Force Doctrine Center, August 2, 2004), pp. 25–34.

59 CBO, "Investment in Major Military Space Programs," p. 16.

60 Section 911 of the FY 2006 NDAA, Space Situational Awareness Strategy and Space Control Mission Review, states:

(a) Findings – The Congress finds that –

   (1) the Department of Defense has the responsibility, within the executive branch, for developing the strategy and the systems of the United States for ensuring freedom to operate United States space assets affecting national security; and

   (2) the foundation of any credible strategy for ensuring freedom to operate United States space assets is a comprehensive system for space situational awareness.

(b) Space Situational Awareness Strategy-

   (1) REQUIREMENT – The Secretary of Defense shall develop a strategy, to be known as the "Space Situational Awareness Strategy," for ensuring freedom to operate United States space assets affecting national security. The Secretary shall submit the Space Situational Awareness Strategy to Congress not later than April 15, 2006. The Secretary shall submit to Congress an updated, current version of the strategy not later than April 15 of every odd-numbered year thereafter.

   (2) TIME PERIODS – The Space Situational Awareness Strategy shall cover –

      (A) the 20-year period from 2006 through 2025; and

      (B) three separate successive periods, the first beginning with 2006, designed to align with the next three periods for the Future-Years Defense Plan.

   (3) MATTERS TO BE INCLUDED – The Space Situational Awareness Strategy shall include the following for each period specified in paragraph (2):

      (A) A threat assessment describing the perceived threats to United States space assets affecting national security.

      (B) A list of the desired effects and required space situational awareness capabilities required for national security.

      (C) Details for a coherent and comprehensive strategy for the United States for space situational awareness, together with a description of the systems architecture to implement that strategy in light of the threat assessment and the desired effects and required capabilities identified under subparagraphs (A) and (B).

      (D) The space situational awareness capabilities roadmap required by subsection (c).

(c) Space Situational Awareness Capabilities Roadmap – The Space Situational Awareness Strategy shall include a roadmap, to be known as the "space situational awareness capabilities roadmap," which shall include the following:

   (1) A description of each of the individual program concepts that will make up the systems architecture described pursuant to subsection (b)(3)(C).

   (2) For each such program concept, a description of the specific capabilities to be achieved and the threats to be abated.

(d) Space Situational Awareness Implementation Plan-

(1) REQUIREMENT – The Secretary of the Air Force shall develop a plan, to be known as the "space situational awareness implementation plan," for the development of the systems architecture described pursuant to subsection (b)(3)(C).

(2) MATTERS TO BE INCLUDED – The space situational awareness implementation plan shall include a description of the following:

(A) The capabilities of all systems deployed as of mid-2005 or planned for modernization or acquisition from 2006 to 2015.

(B) Recommended solutions for inadequacies in the architecture to address threats and the desired effects and required capabilities identified under subparagraphs (A) and (B) of subsection (b)(3).

(e) Space Control Mission Review and Assessment –

(1) REQUIREMENT – The Secretary of Defense shall provide for a review and assessment of the requirements of the Department of Defense for the space control mission. The review and assessment shall be conducted by an entity of the Department of Defense outside of the Department of the Air Force.

(2) MATTERS TO BE INCLUDED – The review and assessment under paragraph (1) shall consider the following:

(A) Whether current activities of the Department of Defense match current requirements of the Department for the current space control mission.

(B) Whether there exists proper allocation of appropriate resources to fulfill the current space control mission.

(C) The plans of the Department of Defense for the future space control mission.

(3) REPORT – Not later than 180 days after the date of the enactment of this Act, the Secretary of Defense shall submit to the congressional defense committees a report on the results of the review and assessment under paragraph (1). The report shall include the following:

(A) The findings and conclusions of the entity conducting the review and assessment on (A) requirements of the Department of Defense for the space control mission, and (B) the efforts of the Department to meet those requirements.

(B) Recommendations regarding the best means by which the Department may meet those requirements.

(4) SPACE CONTROL MISSION DEFINED – In this subsection, the term "space control mission" means the mission of the Department of Defense involving the following:

(A) Space situational awareness.

(B) Defensive counterspace operations.

(C) Offensive counterspace operations.

61 Theresa Hitchens, Michael Katz-Hyman, and Victoria Samson, "Space Weapons Spending in the FY 2007 Defense Budget," (Washington, DC: Center for Defense Information, March 8, 2006).

62 Ibid. The NFIRE program, in particular, has been the subject of a great deal of attention and criticism after MDA revealed that the satellite would include a sensor-projectile that would approach and likely strike a target missile in space. This portion of the experiment was eliminated but Congress, in language

accompanying the 2006 Defense Appropriations Act, encouraged MDA to reconsider. NFIRE is scheduled for launch later in 2006 and it is too late to restore the projectile but the MDA budget justification document said planning would begin next year for a follow-on mission that includes the kill vehicle. MDA is requesting $10.8 million for NFIRE in 2007; the 2006 budget for the effort is $13.5 million.

63 See the Space Posture statements presented to Congress by the two Under Secretaries of the Air Force in March 2005 and March 2006. Mr. Teets resigned in March 2005 and was replaced by Dr. Sega in July 2005.

64 CBO, "Investment in Major Military Space Programs," p. 16.

65 For a more complete listing of all the US government organizations involved in setting space policy see Klotz, p. 54.

66 Office of the Joint Chiefs of Staff, "Chronology of Changes in Key West Agreements, April 1948–January 1958," (Washington, DC: Joint Chiefs of Staff Historical Section, February 7, 1958). The best analysis of Army–Air Force rivalry over developing intermediate-range ballistic missiles during the 1950s is Michael H. Armacost, *The Politics of Weapons Innovation: The Thor-Jupiter Controversy* (New York: Columbia University Press, 1969).

67 Peter L. Hays, "Struggling Towards Space Doctrine," pp. 161–172.

68 Ibid., p. 169. DoD Directive 5160.32 is reprinted in John M. Logsdon (ed.), *Exploring the Unknown*, Vol. II, pp. 314–315.

69 The key player during this period was the Director of Defense Research and Engineering (DDR&E), Herbert F. York. York explains his decisions in *Race to Oblivion: A Participant's View of the Arms Race* (New York: Simon & Schuster, 1970).

70 Current DoD space policy is dated July 9, 1999 but a new DoD Directive on space policy is in final coordination as of this writing.

71 Under the current organizational structure, OSD space policy falls under the Under Secretary of Defense for Policy (Amb. Eric Edelman). The title DoD Space Architect was changed to National Security Space Architect in 1998. In Summer 1998, Maj. Gen. Robert S. Dickman, the first Space Architect, moved on to become Director, Office of Plans and Analysis and System of Systems Architect at the NRO. His replacement as the new National Security Space Architect was Brig. Gen. Howard J. Mitchell. For an analysis of the bureaucratic politics environment facing the Space Architect see Joan Johnson-Freese and Roger Handberg, "Searching for Policy Coherence: The DoD Space Architect as an Experiment," *Joint Forces Quarterly* 16 (1997): 91–96.

72 For strident positions on what the Navy's role in space should be see Commanders Randall G. Bowdish and Bruce Woodyard, US Navy, "A Naval Concepts Based Vision for Space," and Commander Sam J. Tangredi, US Navy, "Space is an Ocean," *U.S. Naval Institute Proceedings* 125 (January 1999), pp. 50–53. Hollywood has undoubtedly played a pervasive if subtle role in shaping popular conceptions of future space operations. For example, Navy traditions and command structures predominate in *Star Trek*.

73 A comprehensive work on von Braun's programs in both Germany and the United States is Frederick I. Ordway and Mitchell R. Sharpe, *The Rocket Team* (New York: Thomas Y. Crowell, 1979). For more in-depth analysis of von Braun's efforts in Germany, see Michael J. Neufeld, *The Rocket and the Reich: Peenemünde and the Coming of the Ballistic Missile Era* (New York: Free Press, 1995).

74 George Kistiakowsky, *A Scientist at the White House: The Private Diary of President Eisenhower's Special Assistant for Science and Technology* (Cambridge: Harvard University Press, 1976) provides a first-hand account from one of the principals involved in creating the NRO. For additional analysis on the political factors leading to the creation of the NRO see Burrows, *Deep Black*; Richelson, *America's*

Secret Eyes, and Day, *Eye in the Sky*. On the 1992 declassification of the NRO see Bill Gertz, "The Secret Mission of NRO," *Air Force Magazine* 76 (1993): 60–63.

75 The Office of Missile and Satellite Systems within the Office of the Secretary of the Air Force was established on August 31, 1960 as the "cover" organization for the NRO. The NRO itself was formally established on September 6, 1961. See Richelson, *America's Secret Eyes*, pp. 46–47; see also one of the first official explanations of the NRO's roots by NRO historian Gerald Haines, "The National Reconnaissance Office: Its Origins, Creation, and Early Years," in Day, *Eye in the Sky*, pp. 143–156.

76 On the subtle but pervasive links between NTM and the political process of verification see Robert Joseph DeSutter, "Arms Control Verification: 'Bridge' Theories and the Politics of Expediency," PhD dissertation, University of Southern California, 1983.

77 Not surprisingly, the NRO's starring role in shaping these space policy developments is not well documented. On the political factors influencing STS design and the rationale behind its 60 by 15 feet cargo bay see John M. Logsdon, "The Decision to Develop the Space Shuttle," *Space Policy* 2 (1986): 103–119. Former Secretary of the Air Force (and NRO Director) Edward ("Pete") Aldridge vented his frustrations with NASA in struggling to develop the CELV and take some spysats off the STS in "Assured Access: 'The Bureaucratic Space War,'" Dr. Robert H. Goddard historical essay, Office of the Secretary of the Air Force, n.d. The CELV evolved into the Titan IV.

78 Construction of the NRO's consolidated headquarters generated a considerable amount of public scrutiny during the period of 1994–1996 because it was contemporaneous with revelations that the NRO had accumulated a "slush fund" of some $4 billion in unspent funding. In the ensuing political fallout, both NRO Director Jeffrey K. Harris and Deputy Director Jimmie D. Hill ended up losing their jobs. The need to reorient and perhaps reorganize the intelligence community (IC) in order to deal more effectively with changed objectives and missions in the post-Cold War world is a central theme of many recent analyses. See, for example, Bruce D. Berkowitz and Allan Goodman, *Best Truth: Intelligence and Security in the Information Age* (New Haven: Yale University Press, 1999). The most recent official US government study on these issues, however, argued against making any sweeping changes in current IC organization or missions; see *Preparing for the 21st Century: An Appraisal of U.S. Intelligence* (Washington, DC: Commission on the Roles and Capabilities of the United States Intelligence Community, March 1, 1996).

79 See Lt. Col. (retired) Frank W. Jennings, "Doctrinal Conflict Over the Word Aerospace," *Airpower Journal* 4 (1990): 46–58 for the roots of the aerospace concept. It is important to note that DoD and the other Services have never accepted the Air Force's aerospace concept.

80 According to the Air Force's August 23, 1998 formulation of the aerospace concept in AFDD 2-2 *Space Operations*:

> The aerospace medium can be most fully exploited when considered as a whole. Although there are physical differences between the atmosphere and space, there is no absolute boundary between them. The same basic military activities can be performed in each, albeit with different platforms and methods. Therefore, space operations are an integral part of aerospace power. Space power is the capability to employ space forces to achieve national security objectives. Used effectively, space power enhances America's chances to succeed across the broad range of military operations. Space power is derived from the exploitation of the space environment by a variety of space systems. A key element of space power is

the people who operate, maintain, or support these systems. Space affords a commanding view of operations and provides an important military advantage. At the level of basic aerospace doctrine, the principles that govern aerospace operations are the same for air and space.

For an interesting perspective on the aerospace concept that argues that its roots are explicitly linked to air power theory and are easily traced to Gen. Hap Arnold's 1944 vision of the Air Force's future see, Maj. Stephen M. Roth-stein, "Dead on Arrival? The Development of the Aerospace Concept, 1944–1958," Master's thesis, School of Advanced Airpower Studies, June 1999.

81 The cancellation of MOL is particularly telling since this indicated the Air Force could not get approval for military personnel in space even to conduct highest priority intelligence-gathering activities.

82 Brig. Gen. (retired) Earl S. Van Inwegen, "The Air Force Develops an Operational Organization for Space," in Hall and Neufeld, *The Air Force in Space* (Washington, DC: USAF History and Museums Program, 1998) pp. 134–143.

83 There are, of course, a virtually unlimited number of ways in which space viewpoints can be delineated and grouped together. These four camps are presented from a US national security perspective; they could also be used for analysis at the global security level. There are also many strands of thought within any of these camps, and some of them might even be contradictory. The four camps presented here are similar to Lupton's four space doctrines discussed above and are derived from the schools of thought about space weaponization discussed in Hays and Mueller, "Going Boldly – Where?" For a six-part typology of perspectives on space weaponization see Karl P. Mueller, "Totem and Taboo: Depolarizing the Space Weaponization Debate," *Astropolitics* 1(1) (2003): 4–28. The growing importance of commercial space activity adds a new dimension that few of the traditional approaches seem well prepared to incorporate or even address. For a groundbreaking analysis that advocates using economic criteria to separate traditional military space functions from more regulatory functions that would be performed by a new US Space Guard (modeled after the Coast Guard), see Lt. Col. Cynthia A.S. McKinley, "The Guardians of Space: Organizing America's Space Assets for the Twenty-First Century," *Aerospace Power Journal* 14(1) (2000): 37–45.

84 Senator Bob Smith, "The Challenge of Space Power," *Airpower Journal* 13(1) (1999): 33. Prominent space hawk groups include High Frontier, the Heritage Foundation, and the Center for Security Policy.

85 *Space Commission Report*, p. x. Most US space policy, military space doctrine, and military officers probably fall into this camp.

86 Lt. Col. Bruce M. DeBlois, "Space Sanctuary: A Viable National Strategy," *Airpower Journal* 12(4) (1998): 41–57. This article is one of the most comprehensive and persuasive expositions of the space dove camp.

87 The single-best and most up-to-date source on this threat are the "High Altitude Nuclear Detonations (HAND) against Low Earth Orbit Satellites ('HALEOS')" slides (hereinafter called HALEOS slides) Advanced Systems and Concepts Office, Defense Threat Reduction Agency, available at: www.dtra.mil/about/organization/haleos.ppt, April 2001. One of the key findings of this study is: "One low-yield (10–20 kt) high-altitude (125–300 km) nuclear explosion could disable – in weeks to months – *all* LEO satellites not specifically hardened to withstand radiation generated by that explosion" slide 4. This study estimated replacement costs in excess of $50 billion for all the systems potentially disabled by a HAND (Globalstar, OrbComm, Iridium, Teledesic, Skybridge, weather satellites, commercial imaging and mapping satellites, and research systems such as the International Space Station and

the Hubble Space Telescope), slides 17–22. Other detailed open-source discussions of these phenomena are found in Samuel Glasstone and Philip J. Dolan, *The Effects of Nuclear Weapons*, 3rd edition, (Washington, DC: Department of Defense and Department of Energy, 1977), pp. 350–353, 383–385, 474–478, and 514–540; Lupton, *On Space Warfare*, pp. 71–75; Bruce G. Blair, *Strategic Command and Control: Redefining the Nuclear Threat* (Washington, DC: Brookings Institution, 1985), appendix C, "Electromagnetic Pulse," and appendix D, "Satellite Vulnerability to System-Generated EMP," 321–331; Ashton B. Carter, "Communication Technologies and Vulnerabilities," in Ashton B. Carter, John D. Steinbruner, and Charles A. Zraket (eds.), *Managing Nuclear Operations* (Washington, DC: Brookings Institution, 1987), pp. 217–281; Watts, *Military Use of Space*, pp. 19 and 98–102; and Tom Wilson, "Threats to United States Space Capabilities," staff paper for the Commission to Assess United States National Security Space Management and Organization, available at: www.space.gov/commission/support-docs/article05/ article05.html.

88 Kenneth F. McKenzie Jr., "The Revenge of the Melians: Asymmetric Threats and the Next QDR," *McNair Paper 62* (Washington, DC: Institute for National Strategic Studies, National Defense University, 2000), p. 38. Factors that might devalue HAND in the minds of terrorists is the absence of mass casualties and lack of powerful visual imagery such as produced by the September 11, 2001 terrorist attacks on the World Trade Center and the Pentagon.

89 Damage caused to satellites in line of sight by prompt X-rays decreases with the inverse square of the distance between the satellite and the explosion $(1/R^2)$. HALEOS, slide 9.

90 The cumulative effects are caused when unstable nuclear fission fragments decay, emitting electrons that are trapped in Earth's magnetic field and increase the peak radiation flux in parts of LEO by three to four orders of magnitude. The predicted lifetime of LEO satellites drops precipitously following a HAND; for example, an Iridium satellite would go from having a 72-month normal lifetime to about one month. The lower Van Allen radiation belt would remain "excited" for six months to two years. See Watts, *Military Use of Space*, p. 99; and HALEOS, slides 9–12.

91 Although not directly analogous, consider the range of predictions concerning the Y2K computer software problem. Different initial assumptions can drive even the best models to widely divergent results; this is especially true when attempting to model systemic effects for complex, interdependent systems. Wilson lists "Reliable Threat Analyses" as his first strategy for enhancing satellite survivability, p. 41.

92 Glasstone and Dolan, *The Effects of Nuclear Weapons*, p. 45; and Paul B. Stares, *Militarization of Space: US Policy 1945–1984* (Ithaca: Cornell University Press, 1985), pp. 107–108. The major tests and series included: The Hardtack Series above Johnson Island in the Pacific consisting of Teak (August 1, 1958, 77 km altitude), and Orange (August 12, 1958, 43.5 km); the Argus Operation in the South Atlantic in September 1958 consisting of three 1–2 kiloton bursts from 200–483 km altitude; and the Fishbowl Series above Johnson Island consisting of Starfish Prime (July 9, 1962, 399 km, 1.4 megatons) and three subsequent submegaton devices in October and November of 1962. Significant ground communication disruptions were recorded in Hawaii (1,125 km away) following the Starfish Prime detonation, and this explosion also eventually caused the failure of seven satellites in LEO at the time of the test.

93 Stares, *Militarization of Space*, p. 107.

94 Watts, *The Military Uses of Space*, p. 101.

95 This escalatory ladder is embedded in Office of the Secretary of Defense, *Proliferation: Threat and Response* (Washington, DC: Department of Defense,

November 1997); and is fleshed out in Peter L. Hays, Vincent J. Jodoin, and Van R. Tassel (eds.), *Countering the Proliferation and Use of Weapons of Mass Destruction* (New York: McGraw-Hill, 1998). The latest version of *Proliferation: Threat and Response* was published in January 2001.

96 Anti-western groups or states might view the destruction of the global information infrastructure as a desirable outcome regardless of (or perhaps because of) their level of connectivity with this infrastructure. Positive security assurances are commitments to come to the aid of states that have forsworn WMD but are being threatened with their use. Negative security assurances are commitments not to use WMD against states that do not have them. See Ronald F. Lehman, "Reassurance and Dissuasion: Countering the Motivation to Acquire WMD," in Peter L. Hays, Vincent J. Jodoin, and Van R. Tassel (eds.), *Countering the Proliferation and Use of Weapons of Mass Destruction* (New York: McGraw-Hill, 1998), pp. 89–120.

97 At a minimum, it would seem prudent for the United States to specify both the types of "purposeful interference" that would trigger an automatic response and to spell out what the response options might include in order to strengthen the rather milk-toast statement in its most recent space policy: "The United States will view purposeful interference with its space systems as an infringement on its rights." Office of Science and Technology Policy, "Fact Sheet: National Space Policy" (Washington, DC: The White House, October 6, 2006).

98 Wilson, "Threats to United States Space Capabilities," p. 43.

99 Though not as precise as military reconnaissance satellites, the five operating US private sector satellites, Ikonas 2 (GeoEye), QuickBird (Digital Globe), Orbview 3 (GeoEye), WorldView 1 (Digital Globe), and GeoEye-1 (GeoEye) can produce imagery with resolution down to 0.5 meter. They face significant foreign competition from systems such as the French SPOT system (satellite pour l'observation de la Terre), the Indian Remote Sensing satellites (marketed by Space Imaging), and EROS (an Israeli–US joint venture). Ikonos 1 failed to reach orbit following its launch from Vandenberg AFB on April 27, 1999. The first commercial one-meter resolution remote sensing system became operational following the successful launch of the Ikonos 2 satellite aboard a Lockheed Martin Athena II booster from Vandenberg AFB on September 24, 1999. See Vernon Loeb, "Spy Satellite Will Take Photos for Public Sale," *Washington Post*, September 25, 1999, p. 3. The Ikonos System can provide resolutions of 0.82-meters ground sample distance per pixel for digitized panchromatic images and 4 meters for multispectral images. Earth-Watch's QuickBird 1 was launched aboard a Cosmos 3M expendable launch vehicle from Plesetsk on November 21, 2000 but failed to reach the proper orbit and was presumed destroyed. Jason Bates, "QuickBird Loss Hits Firm, Remote Sensing Industry," *Space News*, December 4, 2000, p. 3. QuickBird 2 successfully reached orbit on October 18, 2001 and offers 0.61-meter resolution. ImageSat (formerly West Indian Space) – a joint venture between two Israeli companies (Israel Aircraft Industries and Electro-Optics Industries) and Core Software of Pasadena, California – successfully launched its EROS A1 0.82-meter resolution system (that is derived from Israel's Ofeq 3 spysat) aboard a Start 1 ELV from Svobodni, Siberia, on December 5, 2000. ImageSat "does not believe it is subject to U.S. jurisdiction regarding export licenses." See "West Indian Space Changes Name to ImageSat, Announces Product Offerings," *SPACEandTECH Digest*, available at: www.spaceandtech.com/digest/sd2000-22/sd2000-22-007.shtml, January 23, 2001; and ImageSat's website, www.imagesat-intl.com. In December 2000, the National Oceanic and Atmospheric Administration awarded the first two half-meter resolution licenses to Space Imaging and EarthWatch. The half-meter licenses "contain a provision that calls for a 24-hour delay from collection of an image to distribution to a customer." See Jason

Bates, "U.S. Approves Licenses for Two Imaging Satellites with Half-Meter Resolution," *Space.com*, December 18, 2000, available at: www.space.com/businesstechnology/business/satellite_licenses_ 001212.html, December 19, 2000. In 2002 EarthWatch became DigitalGlobe. EarthWatch's corporate roots go back to WorldView Imaging (the first licensee for a high-resolution commercial system) and Ball Aerospace. Space Imaging's lineage includes the Eosat, Lockheed Martin, and Raytheon corporations. In December 2005 Space Imaging and Orbimage merged to form GeoEye. In addition, from the 1950s until losing a $4.5 billion NRO Future Imagery Architecture contract to Boeing in 1999, Lockheed had been the NRO's primary spysat contractor. See Tim Smart, "Lockheed Loses Big U.S. Contract," *Washington Post*, September 8, 1999, E1. OrbImage is a corporate affiliate of the Orbital Sciences Corporation. According to the *Commercial Space Opportunities Study* (CSOS), the Air Force spends $10 million annually on commercial imagery (this includes the innovative Eagle Vision activities); the report recommends that spending be increased to $80 million annually for each year in the FYDP. Headquarters Space and Missile System Center, Developmental Planning Directorate and Headquarters Air Force Space Command, Directorate for Plans and Programs, *Final Report: Commercial Space Opportunities Study* (Los Angeles AFB: Commercial Exploitation Planning Office, February 16, 2000), pp. 3–13. On the security implications of high-resolution commercial remote sensing see Steven Livingston and W. Lucas Robinson, "Mapping Fears: The Use of High-Resolution Satellite Imagery in International Affairs," *Astropolitics* 1(2) (2003): 3–25; Yahya A. Dehqanzada and Ann M. Florini, *Secrets for Sale: How Commercial Satellite Imagery Will Change the World* (Washington, DC: Carnegie Endowment for International Peace, 2000); Gerald M. Steinberg, "Dual Use Aspects of Commercial High-Resolution Imaging Satellites," *Security and Policy Studies* 37 (1998); and Vipin Gupta, "New Satellite Images for Sale," *International Security* 20 (1995): 94–125.

100 "Fact Sheet: U.S. Commercial Remote Sensing Policy," (Washington, DC: The White House, April 25, 2003), pp. 3–4.

101 Ibid., p. 2.

102 Report of the National Commission for the Review of the National Reconnaissance Office, *The NRO at the Crossroads* (Washington, DC: National Commission for the Review of the National Reconnaissance Office, November 1, 2000), quotations from pp. 67, 74, and 71.

103 Report of the Independent Commission on the National Imagery and Mapping Agency, *The Information Edge: Imagery Intelligence and Geospatial Information in an Evolving National Security Environment* (Washington, DC: Independent Commission on the National Imagery and Mapping Agency, December 2000), quotations from pp. viii, 60, and 33.

104 Ibid., p. 56.

105 Ibid., p. 89.

106 Ibid., p. 90.

107 Steven Livingston, "Transparency or Opacity? Information Technology and Deception Operations," paper presented at the International Studies Association Annual Convention, Chicago, February 21–24, 2001. Livingston's quotation is from Don E. Tomlinson, "Computer Manipulation and Creation of Images and Sounds: Assessing the Impact," Washington, DC: The Annenberg Washington Program, 1993). See also Ivan Amato, "Lying with Pixels," *Technology Review* (July/August 2000).

108 Kimberly Amaral, "The Digital Imaging Revolution: Legal Implications and Possible Solutions," available at: www.umassd.edu/Public/People/KAmaral/Thesis/digitalimaging.html.

109 Dehqanzada and Florini, *Secrets for Sale*, p. viii.

110 Livingston and Robinson, "Mapping Fears," argues that the NIMA-SpaceImaging agreement was more important for removing "a potential source of dissonant information" than for providing operational security (p. 20). See also Kerry Gildea, "NIMA Extends Deal with Space Imaging for Exclusive Imagery Over Afghanistan," *Defense Daily*, November 7, 2001, p. 2; "Eye Spy," *The Economist*, November 10–16, 2001; and Pamela Hess, "DoD Won't Release Pix Until 5 Jan," *Washington Times*, November 7, 2001. In addition, the French Ministry of Defense barred SPOT Image from selling or distributing images of Afghanistan and the surrounding regions to anyone except that ministry. "Shutter Control for SPOT Over Afghanistan," *Space Newsfeed*, October 28, 2001, available at: www.spacenewsfeed.co.uk/2001/28October2001.html. These decisions have left Cypress-based ImageSat International as the only company able to provide one-meter commercial imagery of Afghanistan and the surrounding region. Barbara Opall-Rome, "U.S. Data Purchase Opens Doors for ImageSat," *Space News*, October 22, 2001, p. 6.

111 Lt. Gen. Bruce Carlson, USAF, "Protecting Global Utilities: Safeguarding the Next Millennium's Space-Based Public Services," *Aerospace Power Journal* 14(2) (2000): 37. For a more detailed discussion of why GPS does not fit exactly into existing categories of "natural monopoly," "public good," "utility," or "dual-use technology" see Scott Pace, Gerald P. Frost, Irving Lachow, David R. Frelinger, Donna Fossum, Don Wassem, Monica M. Pinto, *The Global Positioning System: Assessing National Policies* (Washington, DC: RAND Critical Technologies Institute, 1995), pp. 184–189.

112 Carlson, "Protecting Global Utilities," 38. All modern "digital compression" telecommunication protocols such as time division multiple access or code division multiple access require highly accurate timing signals to operate.

113 Ibid., 37. Emphasis in original. The PanAmSat Corporation's *Galaxy 4* satellite failed on May 19, 1998.

114 Planetary defense or the effort to track and eventually defend against potentially life threatening near-Earth objects (NEO) that might impact Earth is another high-profile window for cooperation on a space-related issue, but it does not appear to be a traditional control or regulation effort and is not discussed in this chapter. For more information about planetary defense, see, for example, "Preparing for Planetary Defense: Detection and Interception of Asteroids on Collision Course with Earth," *SPACECAST 2020*, appendix R, available at: www.au.af.mil/Spacecast/app-r/app-r.doc; *Air Force 2025*, "Planetary Defense: Catastrophic Health Insurance for Planet Earth,", research paper, available at: www.au.af.mil/au/2025/volume3/chap16/v3c16–1.htm; and Brig. Gen. S. Pete Worden, "NEOs, Planetary Defense and Government: A View from the Pentagon," available at: www.spaceviews.com/2000/04/ article2a.html.

115 "Frequently Asked Questions about Orbital Debris," NASA-Johnson Space Center, Space Science Branch, available at: http://orbitaldebris.jsc.nasa.gov/faq/faq.html.

116 China caused the single most serious debris generating event on January 11, 2007 when it tested its kinetic energy ASAT system by destroying its FengYun 1C meterorological satellite at an altitude of 869 kilometers. This test generated an estimated 150,000 pieces of debris larger than one centimeter, increased the amount of debris being tracked in LEO by 10 percent, and it is estimated that 79 percent of this tracked debris will remain in orbit until the year 2108.

117 The European Space Agency estimates that 44 percent of the catalogued orbit population (larger than 10 cm) originated from the 129 on-orbit fragmentations recorded since 1961. See European Space Agency, "Introduction to Space Debris," available at: www.esoc.esa.de/external/mso/debris.html; and the Aerospace Corporation's "What is Orbital Debris?" website available at: www.aero.org/cords/orbdebris.html. Until fairly recently, several spacefaring

states (Russia in particular) routinely blew up their satellites at the end of their useful life. Inadvertent mixing of propellant and oxidizer and over-pressurization of residual fuel or batteries are the most common causes of unintentional explosions.

118 In LEO (less than 2,000 km altitude) the average relative velocity at impact is 10 km per second. At this speed:

> An aluminum sphere 1.3 mm in diameter has damage potential similar to that of a .22-caliber long rifle bullet. An aluminum sphere 1 cm in diameter is comparable to a 400 lb safe traveling at 60 mph. A fragment 10 cm long is roughly comparable to 25 sticks of dynamite.

In GSO, average relative velocity at impact is much lower (about 200 m per second) because most objects in the geostationary ring move along similar orbits. See "What are the Risks of Orbital Debris?" available at: www.aero.org/cords/debrisks.html.

119 Aerospace Corporation, "What is the Future Trend?" available at: www.aero.org/cords/future.html. The space shuttle must infrequently (every year or two) maneuver away from known orbital debris. Critical components on the International Space Station have been designed to withstand the impact of debris up to 1 cm in diameter.

120 Historic Space Policy documents are available from the Air War College's Space Operations & Resources Gateway. Available at: www.au.af.mil/au/awc/awcgate/histpol.htm. The first emphasis on orbital debris in National Space Policy came in President Reagan's February 11, 1988 National Space Policy and by the Clinton Administration's September 19, 1996 National Space Policy, mitigation of orbital debris was a major intersector guideline:

(7) Space Debris

(a) The United States will seek to minimize the creation of space debris. NASA, the Intelligence Community, and the DoD, in cooperation with the private sector, will develop design guidelines for future government procurements of spacecraft, launch vehicles, and services. The design and operation of space tests, experiments and systems, will minimize or reduce accumulation of space debris consistent with mission requirements and cost effectiveness.

(b) It is in the interest of the USG to ensure that space debris minimization practices are applied by other spacefaring nations and international organizations. The USG will take a leadership role in international fora to adopt policies and practices aimed at debris minimization and will cooperate internationally in the exchange of information on debris research and the identification of debris mitigation options.

121 Leonard David, "Eye in the Sky to Track Space Junk," *Space.com*, November 7 2000, available at: www.space.com/businesstechnology/technology/space_trafficcontrol_001102.html. MSX/SBV operations ceased in June 2008; in the future this mission will be performed by a planned constellation of four Space-Based Surveillance System (SBSS) satellites with first lauch scheduled for 2009.

122 For a detailed discussion of STCS (especially the technical requirements for such a system), see "Space Traffic Control: The Culmination of Improved Space Traffic Operations," *SPACECAST 2020*, appendix D, available at: www.au.af.mil/Spacecast/app-d/app-d.html.

8 Space and intelligence

David Christopher Arnold

I remember the direction came down: "The word 'space' is forbidden to be used. You will not even talk about it. Nobody is interested in space. It's a nonuseful type of endeavor for the military to get into. It's a waste of money, just don't talk about it." We didn't quit thinking about it, but we quit talking about it.[1]

Lt. Gen. Thomas W. Morgan, USAF, retired

The money flowing in from the NRO and the CIA made satellite command and control work.[2]

Brig. Gen. William G. King, Jr., USAF, retired

Space-based intelligence is a field that focuses, during peacetime and hostilities, on the accumulation, analysis, and dissemination of information about the enemy, terrain, or weather in an area of interest. Intelligence analysts conduct information-gathering activities. Strategic intelligence deals with broad issues such as economics or military capabilities. Analysts look at changes in combination with known facts about the subject in question, such as geography, demographics, industrial capacities, and previous intelligence analysis.

Reconnaissance is the military term for the active gathering of information by physical observation. Examples of active gathering include patrolling by troops, ships, submarines, or aircraft, or by setting up observation posts. Reconnaissance seeks to collect information about an enemy. This includes types of enemy units, locations, numbers, and intentions or activity. Thus reconnaissance is a fundamental tactic which helps to build an intelligence picture. Reconnaissance is different from surveillance, which is usually a long-term activity, while reconnaissance is usually a short-term activity. Both reconnaissance and surveillance provide inputs to intelligence analysts. A reconnaissance satellite is an Earth-observation satellite deployed for intelligence purposes. The first successful reconnaissance satellite was the GRAB electronic intelligence satellite, used to gather information about Soviet air defense radars.[3] The United States uses imaging satellites, the subject of this chapter, both for peacetime collection of intelligence, as well as the location of targets to support combat operations. However, the most information is available on programs that existed before 1972. Therefore, this chapter uses the declassified reconnaissance satellite program called CORONA as a lens to look at policy issues in intelligence and space.

The 1950s were a scary time in American history. The leaders at the time had grown up through two world wars and one unending war in a remote corner of Asia. World politics, though described as a Cold War, teetered on the brink of a third world war. Since the "iron curtain descended across Europe," in Winston Churchill's words, the American leadership tried desperately to find ways to keep from merely killing people better.[4] For the President, the key to pulling the world back from the brink and to fortifying the US policy of deterrence was knowing the capabilities of the USSR. President Truman and later President Eisenhower, who both understood the consequences of a sneak attack because they had seen Pearl Harbor, wanted as much information as they could get. But with no certainty about Soviet capabilities, and with wild, propagandistic statements by its leadership, all strategists could do was guess at Soviet capabilities. One option was to put the United States back on a wartime footing, mobilize the nation, and try to match Soviet claims missile-for-missile and man-for-man, something no one wanted, particularly not the fiscally conservative Republican President Eisenhower. At one point, Air Force General Thomas E. Power, commander of the Strategic Air Command, suggested the service needed to build 10,000 missiles to keep up with Soviet strategic weapons development.[5] The 50-plus U-2 missions had failed to provide anything concrete in the way of facts about Soviet strategic nuclear capability, and since flying U-2 missions over the USSR was technically an act of war under international law, and ultimately provided very limited information anyway, leaders needed an alternative to close whatever gap existed between US and Soviet capabilities.

Another option for the President was to develop a system of reconnaissance satellites that could be used to gather data about the USSR from which educated guesses could be made, rather than wild speculations. And since there was no clear international law about satellite-based reconnaissance, the President pursued space technology as an important source of data. After a long and slow evolution, satellite reconnaissance closed that gap by offering the President the following advantages over aerial reconnaissance: persistence, perspective, range, speed, and access to denied areas. By the end of 1961, the new President knew the USSR had four ICBMs on alert and the missile gap was closed.[6]

Invention[7]

New and radical inventions occupied the minds of some airmen at the end of World War II. Chief among the Air Force visionaries was Gen. Hap

Arnold. The Wright brothers taught Arnold to fly and he served a distinguished career, culminating as a member of the Joint Chiefs of Staff during World War II for Army Air Forces. Although he never served in an independent Air Force, Gen. Arnold's vision and foresight created the most powerful air and space force the world has ever seen. His creative thinking gave people permission to develop radical ideas for problems.

Gen. Arnold fostered close relationships with various civilian academics, in particular, Theodore von Kármán, one of the founders of NASA's Jet Propulsion Laboratory (JPL) at the California Institute of Technology in Pasadena. As the end of World War II approached, Arnold asked von Kármán and the AAF Scientific Advisory Group – a group of leading American scientists – to develop a long-range vision for the post-war Air Force. Arnold speculated about "manless remote-controlled radar or television assisted precision military rockets or multiple purpose seekers."[8] Von Kármán's science and technology forecast, *Toward New Horizons*, laid out his vision for the post-World War II Air Force. Summarizing his report in a December 1945 letter to Arnold, von Kármán acknowledged that

> scientific discoveries in aerodynamics, propulsion, electronics, and nuclear physics, open new horizons for the use of air power.... The next ten years should be a period of systematic, vigorous development, devoted to the realization of the potentialities of scientific progress.[9]

The US Navy also had a small core of people willing to think in bold, new ways. In a simple three-page report – including the cover page – the Navy got the jump on the other Services in the new medium of space. The Navy's Bureau of Aeronautics (BuAer) issued Report R-48 in November 1945, "Investigation on the Possibility of Establishing a Space Ship in an Orbit Above the Surface of the Earth," prepared by Lt. Comdr. Otis E. Lancaster and J.R. Moore. The mission for the ship would determine its orbit, but for reconnaissance of enemy positions, "all the necessary information could be obtained in a few trips over the target." Lancaster and Moore speculated that "television or automatic photography could supply the desired information, without personnel [on board]." They found especially interesting

> a circular orbit, 22,300 miles above the surface of the earth, where the [space]ship would make one revolution per day. In this orbit, the ship may be kept over a designated point on the surface of the earth. Naturally, the higher the orbit above the surface of the earth, the more difficult it is to establish the orbit.[10]

Fortunately for the Air Force, Lancaster and Moore underestimated the technical requirements for their proposal. According to Dr. Robert Salter, a

BuAer contractor at the time who reviewed the proposal, the Navy wanted a single-stage-to-orbit vehicle, presumably to make worldwide launch from ships possible. To achieve orbit using a single stage, however, required a mass-fraction of about 0.95, "about the same mass ratio as an egg."[11] Even at the start of the twenty-first century, scientists and engineers have not been able to achieve such a lofty goal.

Some leaders in the US Air Force saw an obvious threat: if the Navy developed a successful reconnaissance satellite, it would threaten the Air Force monopoly on strategic reconnaissance.[12] Strategic aerial reconnaissance became an officially exclusive Air Force mission in the 1948 Key West inter-service agreement, though the Air Force acted both during and after World War II as though strategic aerial reconnaissance was their private sphere. The simple BuAer report only directed an examination into the possibility of a satellite, but Maj. Gen. Curtis E. LeMay, Deputy Chief of Staff for Research and Development for the Air Force, found out about the research and responded on a far-reaching scale. LeMay directed Douglas Aircraft's Research and Development (RAND) Group to investigate the possible uses of satellites. Released in May 1946, RAND's response, "Preliminary Design of an Experimental World-Circling Spaceship," offered a comprehensive look at what satellites could do for the military and suggested three missions: meteorology, communications, and reconnaissance. The report outlined four significant technologies for research and development: long-life electronics, video recording, attitude stabilization, and spacecraft design.[13]

The Air Force, enamored with aircraft, largely ignored RAND's report. In the 1940s and 1950s the Air Force reconnaissance community was wedded to long-range bombers flying reconnaissance missions: "When you wanted to take pictures of something, you just got in your airplane, went out and turned on your cameras and came back and processed the film."[14] The Air Force was so committed to an existing technology – aerial reconnaissance – that it would not nurture a radical new space-based reconnaissance technology even though reconnaissance aircraft like the RB-29, PB4Y-2, and later even the U-2, could not reach every place in the USSR to search for missiles and bombers.[15] Therefore, although the President needed an alternative to provide evidence of where the Soviets were in their missile development, the mainstream Air Force rejected space-based reconnaissance as technically crude and economically risky, continuing to champion piloted reconnaissance aircraft instead.[16]

Inside the Air Force, however, a small but vocal minority of space enthusiasts argued that the notion of spacecraft and satellites had technical practicality. Espousing the cause of continuing satellite studies at considerable risk to their careers because space was a "nonuseful type of endeavor for the military to get into," leaders such as Maj. Gen. Donald L. Putt, Air Force Deputy Chief of Staff for Research and Development, Gen. Hoyt S. Vandenberg, later Air Force Chief of Staff, and Gen. Bernard A.

Schriever, kept official interest in satellites and space programs alive.[17] As Gen. Thomas Morgan recalled later, word came down that officers should not talk about such missions. Yet, in forbidding officials even to speak the word "space," the Air Force merely acknowledged the character of the new and radical invention.[18]

In the late 1940s, RAND scientists and engineers continued to develop the basis for making satellite reconnaissance a reality. Building on the 1946 report, RAND submitted 12 detailed supplemental studies, aimed at convincing the Air Force of the usefulness of the orbiting spacecraft. Finally, in January 1948, the Vice Chief of Staff of the Air Force, Gen. Hoyt S. Vandenberg, authorized the Air Force's Engineering Division to fund further RAND studies of satellite operations. Vandenberg also issued a policy statement that staked the Air Force's claim for space operations: "The USAF, as the Service dealing primarily with air weapons – especially strategic – has logical responsibility for the satellite."[19] Although the Air Force did not have a formally approved space program at the time, some Service leaders had an interest in the possibilities of space and never retracted Gen. Vandenberg's statement. Unfortunately, the Air Force had not yet started building a satellite system even though RAND had been talking about it since 1946. Building a satellite for any reason, let alone strategic reconnaissance, endured an extended infancy in the post-war years.

The first detailed technical report in the evolution of satellite reconnaissance was RAND's 1951 report, "The Utility of a Satellite Vehicle for Reconnaissance." RAND took a major step by stating the technical and engineering possibilities for a reconnaissance satellite employing television techniques for data readout to ground stations. RAND found its earlier satellite advice largely ignored, but its scientists and engineers did not give up. Their engineers advocated television because they recognized that the satellite must transmit large amounts of data and television seemed the easiest way to do it. In addition, in the early 1950s the reentry and recovery capabilities needed to return an object from outer space through atmospheric heating did not exist. Very heavy copper heat sinks were the only heat-reducing system available. Their weight put a tremendous strain on the launch and recovery capabilities at the time, so engineers tried to keep satellites as small as possible.[20]

RAND's Robert Salter used personal connections at RCA to conduct some research at NBC studios in Hollywood, and figured out some early parameters of satellite reconnaissance. On "the basis of a couple of martinis," Salter did some simulated satellite photography. A year later, when Salter moved over to Lockheed, he took a few highly accurate pictures of the Earth and put them on easels. Salter degraded the photos to the level that he assumed matched the quality from space. Using image orthicons, vacuum tubes used in some television cameras, he scanned the pictures in a simulated satellite path, and then transmitted them up to Mount Wilson, near Los Angeles. Rebroadcasting the pictures back through the atmosphere, Salter recorded them using a kinescope because magnetic tape recording had not reached the television studio. A group of photo interpreters looked at the pictures to see what they could see.[21] Salter thus proved the viability of space-based reconnaissance of the Earth.

Development

In 1951 Air Research and Development Command (ARDC) authorized the RAND Corporation to make specific recommendations for the start of developmental work on a reconnaissance satellite system. RAND recommended in Project Feed Back that as a matter of "vital strategic importance" the Air Force should begin studying the "use of an efficient satellite reconnaissance vehicle."[22] Then a Major, William G. King, Jr., read the Project Feed Back reports that addressed such matters as orbital mechanics and satellite photography of the Earth. The RAND studies, and King's own earlier observation of ballistic missile launches at White Sands Army Proving Ground in New Mexico and the Joint Proving Grounds in Florida, convinced him that the United States could perform some military missions in space. King subsequently helped establish a system program office, known as the Advanced Reconnaissance System Program Office. After briefing the secretary of the Air Force, King received $2 million for systems concept studies of military satellites.[23]

RAND sold ARDC on the idea of satellites by proposing them as a system designed to test the feasibility of building a satellite, complete with estimates of costs, development time, and critical design criteria. In early 1953 ARDC's program office for the Advanced Reconnaissance System recommended acceptance of the RAND proposal for letting a system design contract within one year. Unfortunately, from December 1953 to January 1954, ARDC did little beyond documenting the reconnaissance satellite proposal as Project Number 409-40, "Satellite Component Study," engineering project MX-2226, ARDC M-80-4, Project 1115; tentatively assigning it the designation Weapon System 117L; and giving it the unclassified name "Advanced Reconnaissance System."

Although convinced of the utility of a satellite for reconnaissance, King still had to sell it to the Air Force as an operational system, not just a fancy research project. Strategic Air Command (SAC), the Air Force's nuclear command responsible for President Eisenhower's doctrine of massive retaliation, expressed interest in the uses of the Advanced Reconnaissance System. In mid-1955, King traveled from ARDC headquarters in Dayton, Ohio to SAC headquarters in Omaha, Nebraska to speak about the Advanced Reconnaissance System.

At the start of the meeting, someone yelled "attention," and everybody stood up. In walked SAC Commander-in-Chief Gen. LeMay with "half-a-dozen of his horse holders." King often used his wry sense of humor to introduce his presentation by saying, "I'm here to tell you about the Advanced Reconnaissance System, the ARS. The arse. Now, some of you people don't know your 'ARS' from a hole in the ground, and I'm going to straighten you out." He usually got a laugh, but this time, with the hard-nosed, cigar-chewing Gen. LeMay sitting right in front of him, King decided he had better not use his "arse" joke. The general stood up after the presentation and King knew for sure LeMay had been paying attention when LeMay walked by saying, "How did you fellows justify your TDY [travel money] to come in and tell me such crap?"[24] LeMay, a career pilot and responsible for strategic aerial reconnaissance as Commander-in-Chief of Strategic Air Command, soundly rejected the idea of satellite-based reconnaissance, which he had actually initiated in the Air Force when he served as chief of R&D for the Service. Later, LeMay did become a supporter of satellite-based reconnaissance, but at the outset satellite systems were a tough sell to the legendary Air Force leader.

Unfortunately for the space enthusiasts, the rest of the Air Force did not commit itself to fight for the embryonic satellite program either. No operational Air Force command bought into the idea of a reconnaissance satellite, so it remained up to ARDC to define and develop the entire Advanced Reconnaissance System. Even though the Air Force came to a decision on November 27, 1954 to support a reconnaissance satellite program, and even though ARDC published System Requirement No. 5, covering satellite reconnaissance, the Air Force still allocated no funds to cover development. Therefore the Advanced Reconnaissance System remained a research project until March 16, 1955, when Air Force Headquarters published General Operations Requirement No. 80, which called for the development of a satellite reconnaissance vehicle for use by a combat command.

The mainstream Air Force lacked enthusiasm for space, even though in the 1950s politics and space were nearly inseparable. The first American policy statement on space came from the National Security Council on May 20, 1955 in the classified document NSC 5520, "Draft Statement of Policy on U.S. Scientific Satellite Program." President Eisenhower took a keen interest in the discussions of the National Security Council, often attending meetings and taking an active role in the discussions, which considered three satellite programs in early 1955, including the reconnaissance satellite programs.[25] The policy statement stressed the political benefits of having the first satellite launched under international auspices as part of the International Geophysical Year (IGY). According to the National Security Council's report, DoD studies indicated that the United States could launch a small scientific satellite weighing 5–10 lbs into orbit using adaptations of existing rockets.[26] This small project fit well within President Eisenhower's vision for space. The United States had an official American IGY entry, Vanguard, but the country still had no national policy on satellite reconnaissance.[27]

Based on the RAND studies, in early 1955 the Advanced Reconnaissance System program office in Ohio put out a Request for Proposal targeted at specific companies from which the Air Force would issue development contracts. When the Request for Proposal for WS-117L went out the Air Force gave the Advanced Reconnaissance System a new program name, PIED PIPER. Lockheed, Glenn L. Martin Co., and RCA all decided to submit proposals. The Air Force offered Bell Labs and IBM, two other electronics firms, money to study the idea, although both declined.[28] The Air Force provided $1 million to Lockheed, Martin, and RCA for research, giving the corporations plenty of latitude.

Observers outside the Air Force assumed that RCA, a big electronics firm heavy in data processing, had the upper hand over the two old-school airplane companies, Martin and Lockheed. However, the airplane companies had an advantage that RCA did not: they had been involved in selling systems to the Air Force for years, so the airplane companies understood the Air Force procurement system much better than RCA did. Martin and RCA recommended an electronic video system, not too far removed from the original RAND proposal for a readout satellite, and the narrowness of their proposal hurt them both.[29] The Air Force accepted bids from neither RCA or Martin.[30]

Lockheed took a broader approach to the problem of satellite reconnaissance because of its unique combination of experienced scientists and engineers, including video and film reconnaissance systems in its proposal. Louis Ridenour, formerly of MIT's Radiation Laboratory and the first chief scientist of the Air Force, gave Lockheed's presentation to ARDC, very much impressing the review board. Lockheed included in its proposal a host of other functions for the satellite system besides visual (photo) reconnaissance, including infrared surveillance, communications surveillance, and others. Lockheed also wrote the proposal in the ARDC format for development plans, spelling out the system requirements, how they planned to develop the system, how they planned to train personnel, maintain and operate the system, and even where the tracking stations ought to be.[31] According to Robert Salter, Lockheed's proposal originated the requirement for launch facilities at Camp Cooke (later Vandenberg Air Force Base), "which Schriever liked because he wanted something up there. He received that with open arms."[32] Therefore, unlike the competition, Lockheed proposed a complete satellite system to the Air Force. Gen. Schriever recalled later "it turned out that the companies which had been engaged in aircraft weapon development were among the most competent companies in the country for undertaking this new work."[33] Either way, Lockheed clearly stood out in the competition to produce the Advanced Reconnaissance System. Still, nothing happened because Air Force Secretary James H. Douglas, Jr., felt that the engineers still "overestimated" the roles there might be for satellites in orbit.[34]

In June 1956, Gen. Schriever officially announced that the review board had selected Lockheed over RCA and Martin as the prime contractor for the PIED PIPER program, but again the reconnaissance satellite system suffered a setback. On April 2, the Western Development Division had published a development plan for WS-117L, calling for full initial operational capability by 1963 and research and development costs of $153 million,[35] which would be just over $1 billion 2005 dollars. The Air Force approved the development plan for PIED PIPER on July 24, 1956, but allocated a mere $4.7 million, 3 percent of Lockheed's estimate. In November new Air Force Secretary Donald A. Quarles told the Western Development Division to cease development on an Air Force "scientific" satellite. "Don't bend any metal," Quarles told them, knowing that the United States already had a very public scientific satellite program for the IGY, Vanguard.[36]

On October 4, 1957 everything changed. RAND, who had defined the feasibility of using a satellite for reconnaissance within current state-of-the-art technology in its April 1951 secret report, stepped back to the forefront. The "Utility" report had languished in the "back of the files" until *Sputnik*'s dramatic launch, when, said RAND space pioneer Scott J. King, "you never saw a file opened so dammed fast in your life!" King remembered people saying, " 'What the hell did RAND say back there? Let's get cracking on this thing.' "[37] Recalled General Schriever, "When Sputnik went up ... everybody was saying, 'Why the god dammed hell can't you go faster? Who's in charge here?' "[38] The Air Force finally accelerated the reconnaissance satellite program, but did not move it into the fast lane. Instead, the Service spent funds on its major strategic nuclear build-up by pushing the development of ICBMs, leaving satellites as a lower priority.

Fortunately for the development of space-based intelligence, *Sputnik*'s launch established the principle of freedom of overflight in space, eliminating a hurdle for the Eisenhower Administration's space policy. Some historians have argued President Eisenhower allowed the USSR to launch the first satellite in order to establish this critical precedent in international law. The Vanguard satellite program for the 1957–1958 IGY was an equatorial-orbiting satellite that did not travel above the USSR. If the USSR did not protest, so the Eisenhower Administration planned, then the precedent of freedom of overflight would be established. Vanguard would not track so far north that it would invade the "space" above the USSR, but could still help for freedom of navigation in orbit above the Earth, a major American concern before the first satellite launch. Discussions over the important precedent-setting event of satellite overflight even carried into the National Security Council, at which President Eisenhower was a frequent participant. According to ADM Arthur Radford, Joint Chiefs chairman:

> It is important to preserve the "Freedom of Space" concept in order not to impair our freedom of action to launch.... Since the mere

solicitation of prior consent from any nation over which the satellite might pass in its orbit might jeopardize the concept of "Freedom of Space," the proposed policy would preclude any action which would imply a requirement for such prior consent.[39]

Yet, even after *Sputnik*, domestic politics prevented major funding for the reconnaissance satellite program. The United States had yet to decide which organization should have responsibility for space. Senior Air Force leaders wanted to make sure that the government and the public thought of the Air Force as the nation's space Service. In a November 1957 speech to the National Press Club, Air Force Chief of Staff Gen. Thomas D. White, echoing the theme Gen. Hoyt S. Vandenberg first articulated in 1948, once again asserted the Air Force claim to space.

> In speaking of the control of air and the control of space, I want to stress that there is no division, per se, between air and space. Air and space are an indivisible field of operations.... Ninety-nine percent of the earth's atmosphere lies within 20 miles of the surface of the earth. It is quite obvious that we cannot control the air up to 20 miles and relinquish control of space above that altitude ... and still survive.[40]

Thereafter, White frequently used the term "aerospace" to make his point that the Air Force should be the service for all military missions above the surface of the Earth. Thus, by 1959 the Air Force had convinced itself that it should take the lead in space operations, but it had not convinced anyone else, especially not the supreme political and military authority in the United States, President Eisenhower.

When the CIA got involved, the entire nature of the reconnaissance satellite program changed. The CIA had had the U-2 for overhead reconnaissance for some time, but the aircraft had limited capability and President Eisenhower only reluctantly employed it. Eisenhower also did not want to admit publicly that the United States had an active reconnaissance satellite research and development program, particularly after Soviet Premier Nikita Khrushchev rejected the "Open Skies" proposal. So Col. Osmond J. Ritland sat down with Richard Bissell at the CIA and created the DISCOVERER cover story about an environmental satellite test program. The two had successfully created the U-2 reconnaissance aircraft and teamed up again on the reconnaissance satellite. The new reconnaissance system received a quick go-ahead for development. The Air Force now had public authorization to develop a prototype demonstration satellite capability using a Thor intermediate-range ballistic missile with an Agena upper stage, eventually aimed at providing a demonstration of launch, orbit, and recovery. So, although the Air Force started it, the CIA predominantly funded the covert reconnaissance satellite program called CORONA.[41]

Reconnaissance satellites were not just technological artifacts of the Cold War, but political artifacts as well. Privately, the CIA acted as the overall program manager, using Air Force resources and Air Force contractors to do the job, a return to the arrangements of the U-2 development program. The CIA provided the cameras and the film through their contractors at Itek and Kodak, respectively. The Air Force provided for the launch and operation of the satellites through their contractors, primarily Lockheed and Philco, and recovery of the film in buckets deorbited and caught as they returned to Earth by C-119 Flying Boxcars from the 6594th Recovery Control Group. The arrangement worked although sometimes it was awkward and other times it was downright difficult.

Innovation

American leaders from Eisenhower downwards wanted photography of the Eurasian landmass as soon as possible. Air Force leaders like Gen. Schriever believed that the advanced reconnaissance system could provide a significant boost to intelligence capabilities. In meetings between Air Research and Development Command, the Air Force Ballistic Missile Division (recently renamed from Western Development Division), and Strategic Air Command, officers agreed that the minimum essential capability to handle operational pictures from any research and development flights could be sped up by using the California launch facilities – the Satellite Test Center – for operational satellite control, reading out the data at the New Boston and Vandenberg stations, and using the Recovery Group's force in the Pacific to recover film capsules ejected from space.[42]

Innovations in space and missile design helped move the reconnaissance satellite program along even faster. Because of the Army's work in the Jupiter missile program, and scientist Carl Gazley's work at RAND, ablative reentry techniques emerged on the satellite stage, proving practical and, more important, lightweight. As physical recovery of data became an option, RAND Corporation scientists eagerly introduced a new innovation to the Air Force – recoverable satellites. In a research memorandum entitled "A Family of Recoverable Reconnaissance Satellites," Merton Davies, Amrom Katz, and others, described reconnaissance satellites that provided both an early and a continuing photographic reconnaissance capability. Davies and Katz had seen their ideas, first published in the late 1940s, go nowhere, especially not into space. Now finally, with the innovative ideas of Edward A. Miller at General Electric Co., the system included a water-recoverable vehicle, which would become the basis for the CORONA program until the 1970s.[43] The Air Force now had important information necessary to bypass the problems in development of the data readout satellite.

By the end of 1958, the interim satellite tracking network had been built, installed, and checked out, and crews anxiously awaited the first planned launch of *Discoverer 1*, set for late 1958.[44] In California, Charles Murphy was the lone CIA officer in Palo Alto in charge of setting up the reconnaissance program. On active-duty with the Air Force but assigned to the CIA, he had been a B-36 Peacemaker navigator and a key player in the development of the U-2 reconnaissance aircraft. Murphy figured out how to use the Advanced Reconnaissance System by determining the best time of day to launch to ensure adequate daylight in the USSR target area and what altitude to launch into to get the best pictures. Coordinating with analysts at the National Photo Interpretation Center, he learned about what scale meant to photo interpreters, how they liked to have targets illuminated, how much sun they needed, and the effect of shadows.

People had thought about and been meeting these needs for a long time for aerial reconnaissance, but mission planning for satellites differed because the camera system, instead of flying a few thousand feet in the air, orbited 100 miles above the Earth, making planning factors extremely different. For example, when do you launch from California to gather useful information on the other side of the world? Murphy eventually figured out the way that the intelligence community could use the CORONA camera: as a search system, not a spotting system. Using the search method, the photo interpreters discovered that many of the targets of American nuclear weapons did not lie where the planners thought they did. In fact, in Murphy's opinion, the maps were bad enough that an American nuclear strike on the USSR might not have been successful.[45] By himself, then, Murphy figured out launch windows and began turning the satellite program into a useful tool for gathering intelligence of the USSR.

Growth

Interested groups extended beyond the CIA, Air Force, and contractors. Instead of bestowing the space budget on the Air Force, President Eisenhower created the Advanced Research Projects Agency (ARPA) to prevent any one military Service from monopolizing the space budget. He ordered ARPA to manage and direct selected basic and applied research and development projects for the DoD. Eisenhower directed ARPA to focus on high-risk technology that offered the potential for a dramatic advance for traditional military roles and missions. The reconnaissance satellite certainly fit into both those categories. Secretary of Defense Neil H. McElroy formed ARPA on February 8, 1958 as a new agency for space technology research and development with complete authority for directing the expanding American military space program. Advanced Research Projects Agency Director Roy W. Johnson construed his mandate mission very broadly, considering his organization to be a "fourth military service," in effect a military Space Agency.[46] In the area of space systems other than launch vehicles, the DoD's Research and Development Board had assigned military satellites to the Air Force in 1950. Now, with the creation of the new agency in 1958, program direction came from ARPA, not the Air Force or the ARDC in Ohio.

More significant for this study of satellite reconnaissance, President Eisenhower authorized the creation of an agency whose very existence remained secret until 1992. On August 31, 1960, the DoD created the Office of Missile and Space Systems under Assistant Secretary of the Air Force Joseph P. Charyk, who reported directly to the Secretary of Defense. In January 1961, the DoD and CIA jointly rechartered the organization and renamed it the National Reconnaissance Office, classifying its existence. The NRO was a separate agency responsible for consolidation of all DoD "satellite and air vehicle overflight projects for intelligence," and "for the complete management and conduct" of these programs, thus fitting the American space program into its top-down Cold War paradigm.[47] The NRO developed techniques of procurement and program management, which the CIA dominated and which frequently involved over-spending. The NRO also took a share of the military space program, taking a huge bite out of total space activity and shrinking the amount of space work under "normal" military development procedures and control.[48] The NRO also assumed operational authority for overhead reconnaissance while being designated the prime customer for all overhead intelligence, taking away the Air Force's authority for operating the nation's reconnaissance satellites. The Air Force refocused space activities only on launching and tracking satellites, not on their final product. For that, whether the nation knew about it or not, the NRO had primary responsibility.

The American space program now had three branches, one concerned primarily with space science (NASA), one concerned primarily with military support missions (the DoD), and one concerned primarily with reconnaissance operations (the National Reconnaissance Office). Later presidents and secretaries of defense formally endorsed these divisions, which remain largely in effect today.

As the reconnaissance satellite system began to come together, the Air Force scheduled the first launch attempt for December 1958. The complete ground communications links between the tracking stations and the Palo Alto control center underwent a series of tests to ensure their operational readiness. The Air Force and Lockheed had a simple test objective for the first flight, primarily to prove that the booster and the ground support equipment worked, as well as to check out the telemetry, tracking, and control equipment. Other objectives included testing the interstation communications network and crew proficiency with the new equipment and procedures. This first launch attempt would be the Air Force's first operational experience with satellite command and control.[49] Even NASA's brief experience with the Explorer program did not compare to the ambition of the Air Force's first CORONA launch; however, the first attempt to launch a CORONA heralded bad tidings as failure followed failure for over a year.

The first attempt to launch a reconnaissance satellite vehicle on board a Thor/Agena booster came on January 21, 1959. While the Thor booster quietly sat on the launchpad, the Agena upper stage malfunctioned when small rockets that forced propellant into the rocket engine's fuel inlet prematurely fired. After inspection, the Agena turned out to be a total loss. People called this launch attempt *Discoverer Ø*.

On February 28, 1959, the Air Force finally launched *Discoverer 1*, carrying only a light engineering payload. The Air Force Satellite Control Facility presumed it to have crashed near the South Pole, even though at least one tracking station claimed to have heard it on orbit. *Discoverer 2*, the first mission to reach orbit, failed when controllers lost it on the seventeenth orbit because a bad command sequence ejected the capsule at the wrong time, forcing reentry near Spitzbergen, Norway. Efforts to recover the capsule ended when aerial reconnaissance revealed the Soviets probably got the capsule, which carried a pair of mice, not a camera. The first camera-carrying mission, *Discoverer 4*, failed to reach orbit when the Agena booster burned out prematurely. A string of failures followed when, for example, *Discoverer 5*'s camera batteries failed on orbit, the range safety officer blew up *Discoverer 10*, and *Discoverer 11*'s spin-stabilization rockets exploded during reentry.

True to the engineering culture of the day, James Plummer, Lockheed's CORONA program manager and later director of the National Reconnaissance Office, looked on every launch as a successful test, whether or not it provided reconnaissance of the USSR:

> We didn't look at those as twelve failures, we looked at those as twelve successes. We were learning new things every time in every part of the missile and the recovery and all of the things that we had to do.... If it hadn't been for President Eisenhower's support for the [program], of course it would have been cancelled.

Discoverer 14, the first completely successful photo reconnaissance satellite mission start-to-finish, included an air catch of the capsule and processed reconnaissance information. In one single flight, CORONA had accomplished more than dozens of U-2 flights.[50] CORONA continued to serve national reconnaissance needs until the 1970s.

As satellite technology matured and electronics became more reliable, the NRO began to see more success. From the initial successes of the CORONA photo reconnaissance program and the GRAB electronic reconnaissance program grew programs devoted to infrared surveillance and even more reliable mapping systems like the LANYARD system. Initially the NRO was organized into four programs, each of which was responsible for system design and development, coordination with contractors, and operations. Program A included all Air Force satellite intelligence programs, which were managed by the Air Force's Special Projects Office at the Air Force Space

Table 8.1 CORONA program milestones

Name	International designation	Date	Launch site	Vehicle	Orbit	Mass (kg)	Notes
Discoverer 1	1959-[Beta]1	February 28, 1959	WSMC	Thor Agena A	LEO	618	First polar orbiting satellite; prototype; did not carry camera or film capsule
Discoverer 2	1959-[Gamma]1	April 13, 1959	WSMC	Thor Agena A	LEO	743	Prototype; tested capsule recovery techniques; did not carry camera; capsule recovery failed
Discoverer 3	None	June 3, 1959	WSMC	Thor Agena A	N/A	753	Prototype; did not carry camera; film capsule recovery failed
Discoverer 4	None	June 25, 1959	WSMC	Thor Agena A	N/A	743	First generation low-resolution photo surveillance
Discoverer 5	1959-[Epsilon]1	August 13, 1959	WSMC	Thor Agena A	LEO	781	First generation low-resolution photo surveillance; film capsule boosted into higher orbit, decayed February 11, 1961
Discoverer 6	1959-[Zeta]1	August 19, 1959	WSMC	Thor Agena	LEO	783	First generation low-resolution photo surveillance; film capsule recovery failed
Discoverer 7	1959-[Kappa]1	November 7, 1959	WSMC	Thor Agena A	LEO	794	First generation low-resolution photo surveillance; satellite tumbled; film capsule not recovered
Discoverer 8	1959-[Lambda]1	November 20, 1959	WSMC	Thor Agena A	LEO	795	First generation low-resolution photo surveillance; film capsule recovery failed
Discoverer 9	None	February 4, 1960	WSMC	Thor Agena A	N/A	765	First generation low-resolution photo surveillance; premature first stage cut-off
Discoverer 10	None	February 19, 1960	WSMC	Thor Agena A	N/A	765	Destroyed by range safety
Discoverer 11	1960-[Delta]1	April 15, 1960	WSMC	Thor Agena A	LEO	790	Film capsule recovery failed
Discoverer 12	None	June 29, 1960	WSMC	Thor Agena A	N/A	790	Prototype; designed to test capsule recovery system; did not carry camera
Discoverer 13	1960-[Theta]1	August 10, 1960	WSMC	Thor Agena A	LEO	850	Prototype; designed to test capsule recovery system; did not carry camera; capsule successfully recovered from ocean
Discoverer 14	1960-[Kappa]1	August 18, 1960	WSMC	Thor Agena A	LEO	850	Film capsule recovered 1.2 days later; First successful photosurveillance mission

Source: NASA/JPL, "Corona Program," available at: http://msl.jpl.nasa.gov/Programs/corona.html, accessed March 22, 2006.

and Missiles System Center at Los Angeles AFB. CORONA was the nation's first photo reconnaissance satellite system, operating 1960–1972. During the series of 145 launches, CORONA satellites photographed the entire Earth's surface. That photography allowed the United States and its allies to keep track of military targets and operations in the Soviet Bloc and to understand Sino-Soviet strategic capabilities.

Program B encompassed CIA satellite programs, which were the responsibility of the CIA's Deputy Director for Science and Technology. Program C comprised the Navy's component of NRO, ocean surveillance satellites. Program D covered aerial surveillance programs, such as the U-2, prior to 1969. The Air Force's Strategic Air Command subsequently operated these aircraft until the establishment of Air Combat Command when control of the remaining U-2s reverted briefly to the NRO. Responsibility for these activities was subsequently removed from the NRO and ultimately lodged in the Defense Airborne Reconnaissance Office in late 1993.

The concept of non-interference with national technical means of verification first appeared in the Strategic Arms Limitation Talks (SALT) I Treaty of 1972. The intent of this non-interference measure is to preserve from attack or interference technical means of verifying treaty compliance, including space-orbiting means.

Style

Even before the first successful CORONA mission, the debate already had begun about who should operate satellites – contractors or airmen. The discussion over contractor or military staff would not be resolved quickly. In the 1960s, the Air Force preferred to use airmen instead of contractors to operate and maintain the satellite command and control system. The service planned for officers with engineering degrees to launch and operate the satellites and enlisted people with technical training to operate and maintain the ground equipment. In the April 1960 edition of *Air Force/Space Digest*, Gen. Terhune described the satellite operations in Sunnyvale. He described how the Air Force planned to do telemetry, tracking, and control at the remote tracking stations with staffs of NCOs, whom the Air Force considered similar to the airmen servicing fighters, bombers, and cargo planes. Civilian technicians would continue to do research and development, "but with operational systems, these jobs will belong to the men ... and women ... who wear Air Force blue."[51] To continue as the military Space Service, the Air Force planned to move ahead in the space operations business with a core of space operations and maintenance personnel, all of whom wore blue uniforms.

The contractors had a different idea about the staff for the tracking stations. The facilities requirements report for the New Hampshire station that Lockheed managers R. Smelt and R.A. Proctor signed in 1959 stated clearly that Lockheed wanted the station "operated, maintained, and supported by contractor personnel." Although the Air Force had to provide some facilities, like a building for the non-operations staff and a dining hall, permanently assigned contractor personnel reportedly could obtain support from the surrounding communities [Manchester and Nashua, New Hampshire], reducing this cost to the government.

The Air Force had no uniformed expertise in operating and maintaining satellites, so the first Air Force officers joined contractors already on console in 1959. Forrest S. McCartney, now a retired Air Force lieutenant general, joined the Air Force space program in 1960, with two other Air Force captains, Mel Lewin and Al Crews, shortly after *Discoverer 2*'s loss in Norway. The Air Force assigned the three space operations pioneers to work for Lockheed alongside Lockheed employees. Recalled Lewin, McCartney's carpool partner at the time and now a retired Air Force colonel, "I was so excited about what I was doing and enjoyed it a lot. It's not often you get in at the start of something."[52] These officers paved the way for all Air Force satellite operators, using only their engineering training to get them going because no formal training program existed for any of the new satellite flyers, contractors or uniformed personnel. Three weeks after they arrived in Sunnyvale, that original group of captains had become experts. Before actual flight-testing began, the captains worked only in the back room on the displays, supporting the Lockheed engineers on the consoles. The officer-engineers drew with grease pencils on acetate, putting viewgraphs up on screens or on closed-circuit television for the satellite controllers to see. Nobody knew anything more than the Lockheed employees did. The airmen's operational experience grew right along with them. The captains did not have the technical expertise on the vehicle itself, but in terms of operational satellite control they essentially possessed as much expertise as the Lockheed people.

Eventually the General Accounting Office stepped in and said that the government could contract for an end item, but that the contractors working alongside airmen in the satellite command and control business were actually performing illegal "personal services." In response, the Air Force decided to staff the Vandenberg station with only military and to staff the other stations with only contractors. This compromise tested the feasibility of an all-blue-suit operation, provided a comparison for the contractor-operated stations, and preserved the illusion in the Air Force of a uniformed ability to command and control satellites. The Air Force sent selected military personnel from the other stations to staff Vandenberg. In reality, Vandenberg never achieved a truly all-blue-suit status. It always had a cadre of technical advisors assigned to each work center to assist with training and to provide technical assistance as needed. Then, once Vandenberg became the only station where the military performed satellite control and maintenance functions, the personnel problem became acute because no other stations could rotate personnel to California.

Moreover, the Air Force personnel system, already unable to satisfy the personnel needs of the unique space missions, viewed the organization as an R&D outfit rather than an operational or warfighting unit. Many contractor technical advisors augmented the Air Force operations and maintenance teams because of turnover and chronically low experience levels. Once personnel issues became a nagging problem, the Air Force threw in the towel, transitioning Vandenberg to an all-contractor staff.

Personnel problems remained a headache for the Air Force leadership until well into the late 1960s. The Air Force constantly dealt with the conflicting desires of wanting to carry out operations with military technicians but not being able to because of a lack of trained personnel. With the increasing demands of the Vietnam War and the civil space program that was racing to the Moon drawing away technical personnel, the military began to run out of enlisted technicians, making the goal of an Air Force that could provide satellite command and control using uniformed staff a harder one to meet. Filling in the gaps with either contractor personnel or more technology to eliminate personnel seemed to be the only solutions. As the personnel available to ARDC shrank, the military and the contractors had no option but to work together to win the technological war.

Momentum

Technological challenges are certainly important, but large technological systems also face organizational challenges. Gen. Schriever assumed that "increased scientific and engineering competence" would not speed up the rate of technical progress unless the Air Force learned to administer its resources "more wisely and efficiently."[53] To get the reconnaissance system – satellites, boosters, and ground stations – set up as quickly as they did and to get them operational and supporting CORONA and the programs that quickly followed remains a credit to the leadership in Los Angeles. As Gen. King added later, the success of Air Force satellite command and control had as much to do with the leadership as it did with the money flowing in from the CIA.[54]

From the beginning, the Air Force had a special relationship with the CIA, one beginning with the CORONA program but evolving after the creation of the National Reconnaissance Office. President Eisenhower's approval for the reconnaissance satellite program forced the CIA and ARPA to provide all funding for reconnaissance satellite research, development, and procurement, while Gen. Schriever's Air Force Ballistic Missile Division (AFBMD) in Los Angeles acted as the executive agent for the program, running it day-to-day in an Air Force weapon system program office. President Eisenhower wanted the CIA to have complete and exclusive control of all the intelligence phases of the reconnaissance satellite operation, but left the Air Force to manage the technical aspects of the program's development. ARPA exercised general supervision over the development of the vehicle, but AFBMD handled the details, especially the provision of necessary ground facilities. The CIA participated by supervising the development of the actual reconnaissance equipment, and had responsibility for its covert procurement.[55] In fact, the relationship perpetuated was similar to that of the CIA and Air Force during the development of the U-2 reconnaissance airplane, including some of the same people. As in the U-2 program, Lockheed served as prime contractor. Only a handful of people knew anything about the true nature of the satellite program as a reconnaissance vehicle.[56]

Despite the President's preferences for security, the reconnaissance satellite program did not proceed in absolute secrecy. To increase secrecy, on August 31, 1960 Eisenhower ordered the CORONA program placed under a civilian-directed office in the Department of the Air Force. Joseph V. Charyk had the unique opportunity to bypass the chain of command when he took over leadership of the secret Satellite Reconnaissance Program Office, located behind door 4C1000 on the Pentagon's fourth floor.[57] A small group of military officers and government civilians reported directly to Charyk and he to the Secretary of Defense, bypassing even the Secretary of the Air Force under whom he nominally served. After the change of national leadership in early 1961, Charyk stayed on as Director of the NRO, leaving two years later to run the new Comsat Corporation.[58] In September 1961 the CIA and DoD formally established the National Reconnaissance Program, consisting of all satellite and aerial reconnaissance programs. An accompanying agreement between the DoD and CIA gave joint leadership responsibility to the Director of the NRO (Charyk, who also served as the Under Secretary of the Air Force), and the CIA's Deputy Director for Plans (Richard M. Bissell, Jr).[59]

A May 1962 agreement between the Director of the CIA, John A. McCone, and the Defense Secretary, which Deputy Defense Secretary Roswell L. Gilpatric signed, outlined basic policy for the NRO. The CIA and DoD agreed that the NRO would directly respond "to, and only to, the photographic and [redacted] collection requirements," as determined by the United States Intelligence Board. The CIA and DoD also agreed that the reconnaissance mission schedule would be the NRO's sole responsibility, and that the NRO would assign operational control for individual projects under the National Reconnaissance Program to the DoD.[60] In March 1964, CIA Director William Raborn and Deputy Defense Secretary Cyrus Vance signed the current NRO charter, DoD Directive TS-5105.23, the only one partially declassified. The two agencies agreed to make the NRO a separate agency within the DoD, giving it responsibility for the management of the National Reconnaissance Program. The directive gave the NRO director responsibility

for consolidation of all DoD satellite and air vehicle overflight projects for intelligence into a single program defined as the National

Reconnaissance Program, and for the complete management and conduct of this Program in accordance with policy guidance and decision of the Secretary of Defense.[61]

The NRO interpreted "conduct" to mean its charter also bestowed on it *operational* responsibility for reconnaissance satellites. In the following years, the NRO and its contractor teammates designed, built, launched, *and* operated all the nation's reconnaissance satellites.[62]

Difficulties between the CIA, the NRO, and the Air Force grew out of a period when "less than vigorous top-level leadership on the CIA side of CORONA" gave the Air Force, led by Air Force Under Secretary and therefore NRO chief Brockway McMillan, more of a lead role in the reconnaissance satellite business. The CIA accepted a freer Air Force hand as long as the CORONA program suffered no major setbacks. During 1962 the Air Force successfully orbited 17 payloads, retrieving 14 of them, including 69 percent of the film launched. When operations went downhill during 1963 with a string of launch failures and film recovery dipping to 50 percent, Dr. McMillan proposed that the Air Force completely take over the CORONA program. At that point the CIA stepped back up to the plate. Senior members in both the CIA and the White House did not intend to make the National Reconnaissance Program a strictly military responsibility, so the CIA held onto its role in satellite reconnaissance.[63] It remained too important to leave to the Air Force.

Air Force legend Gen. Schriever felt that the NRO's attempt to expand its sphere of influence not only extended beyond its original mandate, but also did more harm than good by keeping the operators out of the loop: "You can't run an operation given a black [covert] status and not bring the operators in because then they don't know what the hell's going on."[64] In effect, this separation kept them from developing an intimate knowledge about the technology about which they should have known the most. Schriever presented the flag flown onboard *Discoverer 13* to President Eisenhower, but no one felt Schriever had a need to know that *Discoverer 14* carried a camera. In fact, Schriever believed, satellite command and control suffered during the NRO's expansion. Furthermore, Schriever expressed in a 2001 interview with the author that the NRO should be a policy-making organization, not a hardware organization. Making the NRO a bureaucracy with operational responsibilities

> was the wrong move to take, because the operator was really taken out of the loop. If you look at all the 117L babies [like SAMOS, MIDAS, and others], they were practically all gobbled up by the NRO. That, plus the fact that you find people that are always trying to protect their rear ends – 'what's been' instead of 'what we can get' – and that sure as hell happened in space.

Schriever wanted to restrict the NRO's area of responsibility. Because about 90 percent of the NRO's staff came from the Air Force in the early days, the Air Force, if given a chance, could have done as good a job operationally as the NRO, or, in Schriever's words, "maybe better."[65]

To be fair, in the early 1960s the NRO did not behave like the massive organization that typifies it today. According to NRO pioneer Charlie Murphy, "If the NRO staff had tried to meet back then, I am sure we could have met in a phone booth. We were operating with ten or fifteen people."[66] In the early 1960s, most of the bureaucratic squabbles and jockeying for power went on at the highest levels of the government. The CIA officer in charge in Palo Alto, Air Force Lt. Col. Charles Murphy, says the NRO acted largely transparently to the satellite operators. The CIA and Air Force people involved in the program in California had an excellent relationship. They did not have to call back to Washington and ask permission to make decisions; they just made the decisions, even if they planned something for the first time. In fact, the program succeeded because the working people – wrench turners, engineers, and the people making decisions in the field – ignored the bureaucratic squabbles and did what they had to do to accomplish the mission.[67]

The NRO did reasonably well at gathering strategic intelligence while delaying the Air Force from normalizing space operations. Because the NRO had a specific charter to acquire strategic intelligence, it actively kept other organizations, particularly the Air Force, whose legal mandate included acquiring strategic reconnaissance, out of the space operations arena.[68] In recent years the NRO has implemented a series of actions declassifying some of its operations. The organization was declassified in September 1992, followed by revelation of its headquarters location in Chantilly in 1994. In February 1995 the CORONA photo reconnaissance program, in operation from 1960 to 1972, was declassified and 800,000 CORONA images were transferred to the National Archives and Records Administration. In December 1996 the NRO announced for the first time, in advance, the launch of a reconnaissance satellite.

By the 1970s, many of the technical issues had been solved with respect to the ability to perform satellite-based reconnaissance. Engineers continued to push the technological envelope, inventing new and better ways to acquire reconnaissance data from space. As they did so, the programs grew bigger and more expensive but always continued to provide useful reconnaissance data for national security decision-makers.

Implications

The story of the CORONA program reflects many of the salient issues that the satellite-based reconnaissance program has faced and overcome and to some extent still faces today. The road to invention of this radical new technology was a difficult one, more often slowed by political than technological roadblocks. When the traffic jam finally broke open in October 1957, satellite-based reconnaissance started moving in the fast lane. But almost as soon as President Eisenhower had given the CORONA program the green light, crash followed crash. Although program manager James Plummer referred to the string of failures as engineering successes, Eisenhower's perseverance gave the program a chance to succeed because, after the U-2 shoot down and the show trial of pilot Francis Gary Powers, there was no alternative to penetrate the secrecy of the USSR. The President had to know what was going in the USSR and CORONA was his only option. Would a president or subordinate allow the same string of failures to befall a program today in the absence of a grave threat to the existence of the United States? Or is the Global War on Terror enough of an existential threat to allow multiple failures of a space-based program in a fiscally constrained environment? There is no doubt that satellite-based reconnaissance was a remarkable invention, maybe even a revolution in military affairs.[69] Space did close the missile gap in 1960 – there should be no doubt about that – and the technology has advanced to the point where formerly strategic systems can also be used for tactical purposes. Now the challenge for the intelligence community is to figure out how to work together to make such data available to tactical users.

During the second phase of the CORONA program, satellite reconnaissance endured a long and tortured development. Domestic politics over control of the space budget largely prevented the development of a comprehensive satellite reconnaissance program; there was no major technological obstacle to overcome at the time. A similarly Byzantine arrangement applies to the current space budget as interest groups like the CIA, NRO, and Air Force still control pieces and parts of the space budget but no single agency controls it all.

The biggest problem that arose initially during the development phase was the reluctance by the traditional military to accept the potential engineers offered with satellite reconnaissance. Even leaders who initially pushed the concept of space-based reconnaissance, like Gen. LeMay, were reluctant to accept it later on when it offered obsolescence for airplane-based reconnaissance or failed to deliver on its promises after initial difficulties. Of course, CORONA's first successful missions revealed not only the potential of the system but yielded incredible intelligence coups. By the time photo reconnaissance had become a fact, euphemistically referred to as "national technical means," the initial barriers had been broken down. The US reduced investments in non-space-based reconnaissance programs such that the pendulum swung in the opposite direction. However, with the realization that space-based programs were not sufficient in and of themselves, air-breathing reconnaissance platforms like the U-2 remained in the inventory. One of the most important tasks for leaders, therefore, is to figure out what the balance is between space and air-breathing systems.

Furthermore, as the number of major satellite contractors has dwindled to just a handful, new concerns arise. During this phase, the CORONA program underwent development by a large airplane contractor, Lockheed. The company proved to have the capability, know-how, skill, and financial wherewithal (when the Air Force would not fund satellite development, Lockheed used its own funds to sustain the program[70]) to provide a revolutionary product for national security. The company has gone on to be involved in national security space ever since. In modern terms, the CORONA program cost just over $1 billion to develop and field, with the result being a national program with satellites, boosters, and a ground system. The President judged these to be acceptable costs given the perceived risk. The decision reflects his political choices for a technological system. Are similarly large programs with huge satellites, boosters, and a ground system still needed today? Could a system of distributed, smaller satellites provide the same information that today's large multibillion-dollar systems do? And if it could, should the decision-makers choose smaller distributed systems over larger systems of fewer satellites?

At the time of SALT I's acceptance of national technical means as a legitimate use of space and space systems, the United States and the USSR were the only two nations that possessed space-based reconnaissance systems. Space was as yet uncluttered and secrets remained to those two nations. However, modern decision-makers have to deal with multiple nations that invest in space. Nations as diverse as China, France, Israel, India, and Japan, all have "remote sensing" satellites. In addition, several companies offer pictures from space of varying degrees of quality.[71] The implications for US space-based reconnaissance of multiple players in the overhead arena, some of whom are not even nation-states but multinational corporations, are vast.

Examination of the CORONA program's technological style raises some questions that are relevant to modern satellite intelligence programs. To what extent have contractors become indispensable to space operations in the modern intelligence program? By the end of the 1960s, the Air Force had given up on uniformed operations and maintenance, and fallen back on the original personnel plan, having contractors perform the vital national service of satellite command and control. What are the implications of this personnel choice, for example, for the Law of Armed Conflict? The engineers inside and outside the Air Force, whether contractors, civilians, or military, built the AFSCF because they had unbounded faith in each other that they could get the job done and that they had the technology to make satellite reconnaissance work. And in the middle of the Cold War, they also had unbounded faith in their country, knowing that the system they built supported the national space program, an international source of prestige for the United States. Those with special access worked with extra diligence, knowing that their work was considered essential to the country's security in the dangerous days of the Cold War.

Nevertheless, this aspect of technological style has exerted a kind of soft determinism on space professional development in the military, leading the 2001 Rumsfeld Space Commission to remark that the lack of space leadership was a critical limiting factor to a continued American advantage in space capabilities. For example, the commission found that about one-third of the officers commanding space wings, groups, and squadrons have "extensive" space experience, while the remaining two-thirds have less than 4.5 years in space-related positions.[72] One former member of Gen. Schriever's team felt that the heavy reliance on contractors was a terrible development in military space. Said the father of the solid-fueled Minuteman ICBM, Col. Edward N. Hall, reliance on contractors for systems engineering and technical direction was "the biggest mistake ever made."[73] By continually deferring to "relatively naïve" groups of scientists and encouraging them to believe they possessed a unique competence in the development of complex weapon systems, Hall believed the Air Force destroyed its own ability to carry out its assigned task of defending the nation.[74] However, the heavy reliance on contractors has not necessarily been a negative in the development of satellite reconnaissance. As retired Gen. McCartney put it in a November 2000 interview,

> I think the smartest, most aggressive, best motivated people are blue suiters. I believed it when I was in the Air Force and believe it now, and wish I [were] still in the Air Force. So, it's not a question of are they capable of doing the job. The question is whether that is the best use of the resource.[75]

As the former CIA official who headed development of the U-2 and A-12 spy planes, and the first generation of spy satellites, Richard Bissell, put it in his posthumously published book: "It is no exaggeration to say that what was accomplished in this period of less than ten years [U-2 development to Corona's first successful flight in 1960] was a revolution in intelligence collection."[76] Although the United States continued to fly reconnaissance missions both in the air and in space for the remainder of the Cold War, the two provided unique and distinct capabilities. Satellites provided repetitive and broad views of surveyed territory, while piloted reconnaissance aircraft remained available for deployment to accessible "areas of interest" at any time. But there should be no doubt that satellites came to be the more important of the two systems, if only because from the start the Soviets chose to ignore satellites' violation of the overflight of their territory.[77]

Notes

1 Lt. Gen. Thomas W. Morgan, USAF (retired), interview by Dr. James C. Hasdorff, March 20–21, 1984, Jemez Springs, NM, transcript, p. 32 (Air Force Historical Research Agency, Maxwell AFB, AL) (hereafter AFHRA),

K239.0512–1576. Morgan was the recently retired commander of the Air Force's Space and Missile Systems Organization in Los Angeles.

2 Brig. Gen. William G. King, Jr., USAF (retired), telephone interview by author, tape recording, October 25, 2000.

3 For more on the development of the GRAB electronic intelligence satellites, see Reid D. Mayo "Conceiving the World's First Signals Intelligence Satellite," in Robert A. McDonald (ed.), *Beyond Expectations – Building An American Reconnaissance Capability: Recollections of the Pioneers and Founders of National Reconnaissance* (Bethesda, MD: American Society for Photogrammetry and Remote Sensing, 2002), pp. 129–138.

4 Bernard A. Schriever, interview by author, Washington, DC, tape recording, June 27, 2001.

5 Heather A. Purcell and James K. Galbraith, "Did the US Military Plan a Nuclear First Strike for 1963?", *American Prospect* 19 (1994): 88–96, available at: http://utip.gov.utexas.edu/jg/archive/1994/STRIKEF2.pdf, accessed January 18, 2006.

6 Jeffery T. Richelson, "From CORONA to LACROSSE: A Short History of Satellites," *Washington Post*, February 25, 1990, B1.

7 Models are generally accepted research methods in the social sciences but not within the humanities. However, one model for the study of large technological systems has been developed and articulated over the last half century by historian Thomas Parke Hughes. In his extensive body of work, Hughes successfully articulated a model of systems development that has been applied to projects as varied as Boston's Central Artery/Tunnel (also known as "The Big Dig") (see Thomas P. Hughes, *Rescuing Prometheus* (New York: Pantheon Books, 1998)) and the US Air Force's system of satellite command and control (see David Christopher Arnold, *Spying from Space: Constructing America's Satellite Command and Control Networks* (College Station: Texas A&M University Press, 2005)). For more on the specifics of Hughes' model, see especially Thomas Parke Hughes, *Networks of Power: Electrification in Western Society, 1880–1930* (Baltimore: Johns Hopkins University Press, 1983) and Thomas Parke Hughes, "The Evolution of Large Technological Systems" in Wiebe Bijker, Thomas P. Hughes, and Trevor Pinch, *The Social Construction of Technological Systems: New Directions in the Sociology and History of Technology* (Cambridge, MA: MIT Press, 1989). Thus, this chapter is not a history of the National Reconnaissance Program, nor is it a history of the NRO, but a look at the social construction of a system for satellite-based reconnaissance. This chapter will not be a strict chronological approach to the subject because of the often-overlapping phases of systems development.

8 Arnold to von Kármán, November 7, 1944, in Dik A. Daso, *Architects of American Air Supremacy: Gen. Hap Arnold and Dr. Theodore von Kármán* (Maxwell AFB: Air University Press, 1997), p. 319.

9 Von Kármán to Arnold, December 15, 1945, in Dik A. Daso, *Architects of American Air Supremacy: Gen. Hap Arnold and Dr. Theodore von Kármán* (Maxwell AFB: Air University Press, 1997), pp. 321–322.

10 Lt. Comdr. Otis E. Lancaster and J.R. Moore, ADR Report R-48, "Investigation on the Possibility of Establishing a Space Ship in an Orbit Above the Surface of the Earth," November 1945, Jet Propulsion Laboratory Archives, 5-492. Provided to the author by Dr. Rick Sturdevant, Air Force Space Command History Office, Peterson AFB (hereafter AFSPC/HO). Lancaster and Moore may have heard about the 1945 Arthur C. Clarke article suggesting a geostationary orbit for communications satellites and then expanded on it.

11 Robert Salter, interview by Martin Collins and Joseph Tatarewicz, July 29, 1986 and July 7, 1987, transcript, p. 14 (RAND Oral History Collection, National Air

and Space Museum Archives, Suitland, MD) (hereafter RAND/NASM). Mass fraction is an efficiency factor for total propulsion. The mass of the propellants of the rocket divided by the total mass of the rocket gives mass fraction (MF). The space shuttle has an MF of approximately 0.82. To achieve orbit using a single-stage vehicle requires a high mass fraction, currently only achievable with small objects, not large useful satellites.

12 Robert Frank Futrell, *Basic Thinking in the United States Air Force, 1907–1960*, vol. 1 of *Ideas, Concepts and Doctrine* (Maxwell AFB: Air University Press, 1989), pp, 196–200.

13 Salter interview, July 29, 1986, pp. 45–47.

14 Merton Davies, interview by Joseph Tatarewicz, December 12, 1985, transcript, p. 35 (RAND/NASM). For a frank discussion of "occupationalism" in the United States Air Force, see Carl H. Builder, *The Icarus Syndrome: The Role of Air Power Theory in the Evolution and Fate of the U.S. Air Force* (New Brunswick, NJ: Transaction Publishers, 1996).

15 Frederic C.E. Oder, telephone interview by author, tape recording, October 10, 2001.

16 For more on aerial reconnaissance, see William E. Burrows, *By Any Means Necessary: America's Secret Air War in the Cold War* (New York: Farrar, Straus and Giroux, 2001) and John T. Farquhar, "A Need to Know: The Role of Air Force Reconnaissance in War Planning, 1945–1953" PhD dissertation, Ohio State University, 1991.

17 Morgan interview, March 20–21, 1984, p. 32.

18 Hughes, "Evolution of Large Systems," pp. 58–59; Morgan interview, March 20–21, 1984; Bruno Augenstein, interview by Martin Collins and Joseph Tatarewicz, July 28, 1986, transcript, p. 28 (RAND/NASM).

19 Maj. Gen. L.C. Craigie, Air Force Director of Research and Development, SUBJ: Satellite Vehicles, January 16, 1948, with incl. by Gen. H.S. Vandenberg, Vice Chief of Staff of the Air Force, SUBJ: Statement of Policy for a Satellite Vehicle, January 15, 1948, to Brig. Gen. A.R. Crawford, Chief, Engineering Division, Air Materiel Command, in Joseph W. Angell, Jr., *USAF Space Programs, 1945–1962* (Washington, DC: USAF Historical Liaison Office), Tab A, AFSPC/HO.

20 Davies interview, December 12, 1985, p. 36. This problem disappeared in the late 1950s with the introduction of ablative reentry vehicles, not only quite feasible but also very lightweight.

21 Salter interview, July 29, 1986, p. 48.

22 J.E. Lipp and Robert M. Salter, "Project Feed Back Summary Report," RAND Corporation, Report R-262, March 1, 1954, Defense Technical Information Center, Fort Belvoir, VA (hereafter DTIC), AD354297.

23 Air Force Space Command History Office, "Brigadier General William G. King, Jr.," *Air Force Space and Missile Pioneers* (Colorado Springs: Air Force Space Command, 2000), p. 11. Topeka, Kansas native William G. King, Jr., received an ROTC commission at Kansas State University, entering active duty during World War II and serving as an anti-aircraft artillery officer in the Pacific Theater. After the war, this early architect of the Air Force's Advanced Reconnaissance System – if such a term should even apply to individuals – returned to Kansas State, completing his BS in 1946. He later received a Master's degree from the University of Chicago in business administration, with an emphasis on the management of research and development. Gen. King retired from the Air Force in April 1971, beginning an eighteen-year career with Aerojet General Corporation, where he served as both vice president and general manger of the Space Surveillance Division of the Electronics Systems Division, and as Director of the Defense Support Program, the nation's ballistic missile early warning satellites.

24 King interview, October 25, 2000. When King returned to Ohio, his boss, Lt. Gen. Howell M. Estes, Jr., a LeMay protégé, gave King "a stern talking to." King's "sales trips," however, continued.

25 See Fred I. Greenstein, *The Hidden Hand Presidency: Eisenhower as Leader* (New York: Basic Books, 1982), especially pp. 129–131.

26 National Security Council, NSC 5520, "Draft Statement of Policy on U.S. Scientific Satellite Program," May 20, 1955, in John M. Logsdon (ed.) *Exploring the Unknown: Selected Documents in the History of the U.S. Space Program*, Volume I, Organizing for Exploration, NASA SP-4407 (Washington: Government Printing Office, 1995), pp. 308–309.

27 For more on the debate over the principle of overflight, see especially Walter A. McDougall's Pulitzer Prize-winning history, ... *the Heavens and the Earth: A Political History of the Space Age* (New York: Basic Books, 1985).

28 Salter interview, July 29, 1986, pp. 48–49.

29 Ibid, p. 49; Davies interview, December 12, 1985, p. 31.

30 Augenstein interview, July 28, 1986, p. 33; Air Force Space Command History Office, "James W. Plummer," *Space and Missile Pioneers* (Colorado Springs: Air Force Space Command, 2000), p. 16. Lockheed's president, L. Eugene Root, on the other hand, enlisted some of the best minds in the burgeoning space industry to help his company's proposal. A former RAND employee, Root recruited Robert Salter, Bruno Augenstein, William Fry, and Sidney Brown from the original team of RAND "World-Circling Spaceship" engineers and scientists. Augenstein eventually became chief scientist of Lockheed's satellite program, and then moved on to the DoD, working under Harold Brown, Defense Director of Research and Engineering during the Kennedy and Johnson administrations. James W. Plummer, a Salter protégé, later served as Under Secretary of the Air Force and Director of the NRO in the Nixon and Ford administrations. Other engineers and scientists joined Lockheed and together they made a difference for the airplane manufacturer trying to make its way into the realm of space contractors.

31 Salter interview, July 29, 1986, p. 49.

32 Ibid.; James S. Coolbaugh, "Genesis of the USAF's First Satellite Programme," *Journal of the British Interplanetary* Society 51 (1998): 293; Oder, electronic mail to author, SUBJ: Early 117L, January 28, 2000.

33 Maj. Gen. Schriever, Commander, Air Force Ballistic Missile Division, April 24, 1958, in *Astronautics and Space Exploration*, Hearings before the Select Committee on Astronautics and Space Exploration, 85th Congress, 2nd session (Washington: Government Printing Office, 1958), p. 677.

34 James H. Douglas, Jr., interview by Hugh H. Ahmann, 13–14 June 1979, Chicago, Ill., transcript, 126, AFHRA, K239.0512–1126; Osmond J. Ritland, interview by Lyn R. Officer, March 19–21, 1974, Solano Beach, Calif., transcript, 231–235, AFHRA, K239.0512–722.

35 Western Development Division, *WS-117L Advanced Reconnaissance System Development Plan*, April 2, 1956, AFHRA, K243.8636-39.

36 Oder interview, October 10, 2001. For more on the Vanguard program, see Constance McLaughlin Green and Milton Lomask, *Vanguard: A History* (Washington: NASA, 1970) SP-4202.

37 Scott J. King, interview by Martin Collins and Joseph Tatarewicz, July 22, 1987, transcript, p. 14 (RAND/NASM).

38 Schriever interview, June 27, 2001.

39 Arthur Radford, Chairman, Joint Chiefs of Staff, Memorandum for the Secretary of Defense, "U.S. Scientific Satellite Program," May 24, 1955, in *National Security Council, Minutes of Meetings*, A:III:0403. See also Walter A. McDougall, ... the *Heavens and the Earth*, especially chapter 7.

40 General Thomas D. White, "At the Dawn of the Space Age," *Air Power Historian* 5 (1958): 17. This article is a reprint of a speech he gave to the National Press Club on November 29, 1957.
41 Ritland interview, March 19–21, 1974, p. 238a.
42 MEMO, Maj. Gen. James H. Walsh, Assistant Chief of Staff, Intelligence, SUBJ: SAMOS, to Secretary of the Air Force Dudley C. Sharp, Jr., June 24, 1960, AFHRA, K243.012-33.
43 M.E. Davies and A.H. Katz, R.W. Buchheim, T.F. Burke, R.T. Gabler, T.B. Garber, C. Gazley, E.C. Heffern, J.R. Huntzicker, H.A. Lieske, D.J. Masson, "A Family of Recoverable Reconnaissance Satellites," RAND Research Memorandum RM-2012, November 12, 1957; Davies interview, December 12, 1985, p. 36; Augenstein interview, July 28, 1986, pp. 32–33; Amron Katz, interview by Martin Collins and Joseph Tatarewicz , July 24, 1986, transcript, p. 41, Rand Oral History Collection, National Air and Space Museum, Archives Division, MRC 322, Washington, DC.
44 For more on the development of the satellite command and control system, see David Christopher Arnold, *Spying from Space.*
45 Colonel Charles Murphy, USAF (retired), telephone interview by author, September 26, 2001, tape recording.
46 Roy W. Johnson, Director, Advanced Research Projects Agency, in *Astronautics and Space Exploration,* Hearings before the Select Committee on Astronautics and Space Exploration, 85th Congress, 2nd session (Washington: Government Printing Office, 1958), p. 1171.
47 Cyrus Vance, Deputy Secretary of Defense, DoD Directive Number TS 5105.23, "National Reconnaissance Office," March 27, 1964, in John M. Logsdon (ed.) *Exploring the Unknown,* pp. 373–374. With the exception of this directive, the DoD prohibited using the terms National Reconnaissance Office, National Reconnaissance Program, or NRO in any document. Any reference had to use the phrase "Matters under the purview of DoD TS-5105.23."
48 John L. McLucas, "The U.S. Space Program since 1961: A Personal Assessment," in R. Cargill Hall and Jacob Neufeld (eds.), *The U.S. Air Force in Space: 1945 to the 21st Century* (Washington, DC: USAF History and Museums Program, 1998), pp. 85–87.
49 *SENTRY Program Status Report,* December 10, 1958, AFHRA, K243.012-36.
50 Thomas C. Reed, *At The Abyss: An Insider's History of the Cold War* (New York: Ballantine Books, 2004), p. 183.
51 Charles H. Terhune, Jr., "In the 'Soaring Sixties' Man is on His Way – Up," *Air Force/Space Digest* (1960): 71. Ellipses in the original.
52 Colonel Melvin Lewin, USAF, retired, telephone interview by author, December 21, 2001, tape recording.
53 US Air Force, "Air Force Systems Command," in *Air Force and Space Digest* (1962), pp. 165.
54 King interview, October 25, 2000.
55 MEMO, Richard Bissell, CIA, to Maj. Gen. Jacob E. Smart, Assistant Vice Chief of Staff, USAF, SUBJ: "Distribution of Responsibilities for CORONA," November 25, 1958, NRO, 1/A/0008.
56 MEMO for the Record, Gen. A.J. Goodpaster, April 21, 1958 NRO, 2/A/0043; MEMO of Conference with the President [on] February 7, 1958; Gen. A.J. Goodpaster, February 10, 1958, NRO, 2/A/0040. Present at the conference besides Eisenhower and Goodpaster were Dr. James Killian, the President's science advisor, and Edwin "Din" Land, president of Polaroid and responsible for the development of the CORONA camera.
57 NRO's main Pentagon office, in rooms behind door 4C1000, earned them the somewhat pejorative and medieval nickname "Forcey One Thousand."

58 For a brief overview, see David J. Whalen, "Billion Dollar Technology: A Short Historical Overview of Origins of Communications Satellite Technology, 1945–1965," in Andrew J. Butrica (ed.), *Beyond the Ionosphere: Fifty Years of Satellite Communication* (Washington: NASA, 1997), SP-4217, especially pp. 110–114.

59 R. Cargill Hall, *The NRO at Forty: Ensuring Global Information Supremacy* (Chantilly, VA: NRO, 2000), pp. 2–3; Cyrus Vance, Deputy Secretary of Defense, DoD Directive Number TS 5105.23, "National Reconnaissance Office," March 27, 1964, in John M. Logsdon (ed.) *Exploring the Unknown*, pp. 373–374.

60 John A. McCone, Director of the CIA, and Roswell L. Gilpatric, Deputy Secretary of Defense, "Agreement between Secretary of Defense and the Director of Central Intelligence on Responsibilities of the National Reconnaissance Office," May 2, 1962, NRO, 2/A/0036.

61 DoD Directive TS-5105.23, in John M. Logsdon (ed.) *Exploring the Unknown*, vol. 1, pp. 373–375.

62 Hall, *The NRO at Forty*, p. 4.

63 CIA Office of Special Projects, *History*, June 1, 1973, NRO, 2/A/0075.

64 Schriever interview, June 27, 2001.

65 Ibid.

66 Charles L. Murphy, "Commanding and Controlling Corona," in Robert A. McDonald (ed.), *Beyond Expectations – Building an American National Reconnaissance Capability: Recollections of the Pioneers and Founders of National Reconnaissance* (Bethesda, MD: American Society for Photogrammetry and Remote Sensing, 2002), p. 279.

67 Murphy interview, September 26, 2001.

68 Schriever interview, June 27, 2001.

69 For more on satellite reconnaissance and the concept of revolution in military affairs, see David Christopher Arnold, "The Revolutionary Bureaucrats: Independent Inventors Roaming Free In The 1950s Revolutionize Reconnaissance," unpublished paper, 2005.

70 James W. Plummer, telephone interview by author, September 8, 2001, tape recording.

71 For example, for an overview of the history of the Indian remote sensing program, see Shubhada S. Savant and Santhosh K. Seelan, "India's Remote Sensing Program – A Historical Perspective," *Quest: The History of Spaceflight Quarterly* 12(4) (2005): 26–33. For an overview of commercial remote sensing satellites, see Incigul Polat-Erdogan, "A Brief History of Commercial High-Resolution Satellite Remote Sensing," *Quest: The History of Spaceflight Quarterly* 12(4) (2005): 34–43.

72 "Report of the Commission to Assess United States National Security Space Management and Organization," January 11, 2001, p. 43.

73 Col. Edward N. Hall, interview by Jack Neufeld, Washington, DC, July 11, 1989, transcript, AFHRA, K239.0512-1820.

74 Edward N. Hall, "Epitaph," August 29, 1958, Ballistic Missiles Division, ARDC, AFHRA, K243.0122-7; Hall interview, July 11, 1989, p. 5.

75 McCartney interview, November 9, 2000.

76 Richard M. Bissell, *Reflections of a Cold Warrior: From Yalta to the Bay of Pigs* (New Haven, CT: Yale University Press, 1996), p. 93.

77 Ibid., p. 139.

9 The acquisition process

Acquiring technology for space and defense

Steve G. Green, Kurt A. Heppard, and Robert L. Tremaine

Introduction

With the end of the Cold War, the United States faced an unclear future regarding spacecraft and their delivery systems. The nation bore the responsibility of preserving its access to space while meeting a growing demand in both military and commercial sectors. This demand resulted in the Department of Defense (DoD) spending more than $18 billion annually to develop, acquire, and operate satellites and other space-related systems.[1]

Given the invaluable strategic potential of space systems, it comes as no surprise that investment in them continues to be significant. But making sure that major space systems and capabilities meet critical operational requirements without exceeding budget limitations is a major challenge. This challenge has forced the space community to confront the adequacy of its current space launch and acquisition strategies and ultimately resulted in several comprehensive reviews that fundamentally changed the way space systems are bought.

In addition, over the last few years, the Global War on Terror (GWOT) has validated and even amplified the need for space systems. It has forced the space community to confront the efficacy of its current space launch and acquisition strategies. At the same time, growing congressional interest and the presence of certain technological obstacles served to spotlight key areas that deserved additional study. This attention was amplified by the fact that the DoD's space system acquisitions have experienced problems over the past several decades that have driven up costs by hundreds of millions, even billions, of dollars; stretched schedules by years; and increased performance risks. In some cases, capabilities have not been delivered to the actual "warfighter" even after decades of development.[2]

This chapter characterizes the substantive changes to space acquisition. It provides background information by illustrating the unique nature of space systems acquisition while giving insight into the major players involved. To better comprehend the current space acquisition approach, an understanding of its policy evolution provides an excellent foundation for appreciating the need for change. In particular, two high-level reports served as a catalyst – the Report of the Commission to Assess United States National Security Space Management and Organization, January 2001; and the Report of Defense Science Board/Air Force Scientific Advisory Board Task Force on Acquisition of National Security Space, May 2003. Exposure to the acquisition process is provided because understanding how space systems are acquired, relative to how other major DoD systems are procured, is vital to appreciating space policy implementation. This will be followed by an explanation of pivotal policy changes that have been implemented to address key systemic acquisition issues. Finally, acquisition strategy implementation will be described as examples of how the space community is trying to better meet operational space requirements while continually improving the space acquisition process.

The changing space acquisition landscape

The DoD's current space network includes an impressive array of satellite constellations, as well as their related ground-based systems, launch vehicle systems, and associated terminals and receivers. These systems are used to perform strategic intelligence, surveillance, and reconnaissance functions; perform missile warning; provide communication services to the DoD and other government users; provide weather and environmental data; and provide positioning and precise timing data to US forces as well as national security, civil, and commercial users.[3]

In essence, DoD space acquisition programs can be grouped into four system types. The acquisition characteristics of space system types 1, 2, and 3 are similar to each other, but differ from space system type 4. These types can be seen in Table 9.1.

This difference results in two primary acquisition models that fit most National Security Space systems – the Small Quantity System model and the Large Quantity Production-Focused model. This chapter is focused on the Small Quantity Systems model associated with types 1, 2, and 3, which

Table 9.1 Space system type examples

1 Space-based systems	Satellites
2 Ground-based systems	Satellite command and control (C2), launch C2, ground station payload data processing stations, space surveillance stations, command and control systems
3 Satellite launch vehicle systems	Boosters, upper-stages, payload processing facilities, space launch facilities, ground support equipment
4 User equipment	Hand-held user terminals, data reception terminals, user terminals

Source: NSS Acquisition Policy 03-01, 2004, p. 7.

are predominately satellite programs, along with their ground stations and boosters, and are usually bought in quantities of ten or fewer.

These types of programs generally do not have the luxury of building on-orbit prototypes before a winner is selected for a production contract. Also, the incredibly expensive nature of satellites and associated launch costs prohibits traditional "fly-offs" where the firm that performs best wins the contract. Instead, the final selection between space system competitors normally occurs in the form of a "design-off," where the contractor with the best proof of concept wins.

The growing importance of space systems to military and civil operations requires the DoD to seek out and develop cutting-edge technologies and achieve timely delivery of capabilities while being very sensitive to budget constraints. Also, many argue that assured access to space and non-reliance on foreign governments or corporations is also important to national security. For example, the space launch market share began shifting from the United States to Europe and Asia, fueled by new foreign launch providers and international partnering agreements. The growth of commercial entities seeking launch capabilities to service commercial users led to a demand for internationally supported launch vehicles that could provide reliable service at a lower cost than American-supported vehicles.[4]

The government had significant expectations that a commercial space market would develop, particularly in commercial space-based communications and space imaging. Most parties in the government space community assumed that this commercial market would pay for portions of space system research and development and that "economies of scale," or lower costs based on quantity, would result. Consequently, government funding was reduced.[5]

But a widespread commercial space program demand never materialized. In response, domestic industry dealt with excess capacity with a series of mergers and acquisitions. In some cases critical suppliers with unique expertise and capabilities were lost or put at risk of going out of business. Successfully competing on major space programs became "life or death" for some, resulting in extreme optimism in cost estimates and program plans.[6]

Between these political and fiscal realities, acquiring the many different types of space systems for the various users was, and continues to be challenging. The impressive array of space systems and the different organizations involved in developing, producing, and operating them is referred to as the National Security Space (NSS) environment.[7] During the 1990s, several phenomena also occurred that had varying degrees of impact on the space acquisition process which included:[8]

- declining acquisition budgets
- acquisition reform with significant unintended consequences
- increased acceptance of risk
- unrealized growth of a commercial space market

- increased dependence on space by an expanding user base
- consolidation of the space industrial base.

These changes in the NSS environment took place in the face of evolving national security needs. Simultaneously, the DoD transitioned from the structured Cold War environment to the more recent global and unpredictable threat. As requirements changed and technology evolved, costs skyrocketed and completion schedules slipped significantly in many cases. Even the ability of the NSS community to effectively develop, produce, and operate space systems was openly questioned.

The 2001 Space Commission

In order to more tightly couple the national and DoD space efforts, the Report of the Commission to Assess United States National Security Space Management and Organization, often referred to as the "Space Commission," was published in 2001, identifying five matters of key importance that the members believed needed immediate attention in order to improve space systems acquisition.[9] The Commissioners strongly recommended that DoD space operators become intimately more familiar with system acquisition to bridge a known gap between two distinct cultures: acquisition and operations. The National Reconnaissance Office (NRO) had the apparent solution. The NRO's approach to acquisition and operations, referred to as "cradle-to-grave," harbored a much more symbiotic relationship between the acquirers and operators. In essence, operators worked directly with technical design teams, and consequently had much greater influence on the systems they would one day be operating.[10] The approach was also studied by the Government Accountability Office (GAO) which arrived at similar conclusions regarding process, staffing issues, and gaps in expectations.[11]

The 2003 Defense Science Board joint taskforce

In 2003, leaders of the space security community, including the Under Secretary of Defense for Acquisition, Technology and Logistics (ATL), Secretary of the Air Force (SECAF), and Under Secretary of the Air Force/Director of the NRO chartered a study on NSS programs to be jointly accomplished by the Defense Science Board (DSB) and the Air Force Scientific Advisory Board (AFSAB).[12] This formidable taskforce, comprised of government and industry leaders of the space community, investigated the systemic issues related to space systems acquisition. Their focus included all aspects ranging from requirements definition and budgetary planning, through staffing and program execution. The joint taskforce was also given the charter to recommend improvements to the acquisition of space programs from program initiation to sustaining operations.[13]

Over the course of this study, the team members identified four chronic problems associated with space acquisition. Their conclusions subsequently helped drive several major process changes. The following list summarizes, roughly verbatim, the joint taskforce's key findings:[14]

- *US national security is critically dependent upon space capabilities and that dependence will continue to grow.* Pressing requirements exist to monitor activities and events throughout the world, transfer massive quantities of data, and project force on a global scale. The taskforce felt that in all likelihood this requirement would increase in the future.

- *The space industrial base is adequate to support current programs, although there are long-term concerns.* Nearly every mission area in national security space was in transition, with concurrent development of entirely new satellite systems or major block upgrades. These factors led to concerns that the industrial base would not be adequate to support the required range of activities. They also felt that the aerospace industry is characterized by an aging workforce, with a significant portion of this force eligible for retirement. Developing, acquiring, and retaining top-level engineers and managers will be a continuing challenge, particularly since a significant fraction of engineering graduates are foreign students.

- *The taskforce found five basic reasons for the significant cost growth and schedule delays in national security space programs.* Any of these could have a significant negative effect on the success of any program. When taken in combination, these factors could have a negative compound effect on achieving program success.

  1 *Cost had replaced mission success as the primary driver in managing space development programs.* When dealing with space-related activities, a single engineering flaw or workmanship error can result in catastrophe. Mission success in the space program has historically been based on an unrelenting emphasis for high quality. The taskforce felt that the change of emphasis from mission success to cost has resulted in excessive technical and schedule risk as well as a failure to make responsible investments to enhance quality and ensure mission success.

  2 *Unrealistic estimates lead to unrealistic budgets and unexecutable programs.* The task-force felt that during program formulation, advocacy tends to dominate and a strong motivation exists to minimize program cost estimates. Independent cost estimates and government program assessments had proven ineffective in countering this tendency. The taskforce found that most programs at the time of contract initiation had a predictable cost growth of 50–100 percent.

  3 *Undisciplined definition and uncontrolled growth in system requirements increased cost and schedule delays.* As space-based support had become

more critical to our national security, the number of users had grown significantly. As a result of conflicting desires and needs, the taskforce felt requirements had proliferated.

4   *Government capabilities to lead and manage the space acquisition process have seriously eroded.* The taskforce felt that this erosion ironically could be traced back, in part, to acquisition reform environment of the 1990s. The authority of program managers and acquisition officials appeared to have eroded to the point where it reduced their ability to succeed on development programs. In addition, the taskforce felt that widespread shortfalls existed in the experience level of government acquisition managers, with too many inexperienced personnel and too few seasoned professionals working directly in space acquisition.

5   *Industry had failed to implement proven practices on some programs.* Successful development of space programs requires strong leadership and rigorous management processes both in industry and in government. The taskforce felt that it was of paramount importance that industry leadership assures that programs are implemented utilizing proven management and engineering practices.

Space systems acquisition and the NSS Policy Directive 03-01

Over the years, the manner in which the DoD acquires major weapon systems had evolved and adapted to changes in threat, technology, and political will, but the change in space systems acquisition was relatively swift and dramatic. To best grasp the gravity of this change, it is useful to understand how the DoD buys its major weapon systems.

The Defense Acquisition System exists to manage the nation's investments in technologies, programs, and product support necessary to achieve the National Security Strategy and support the US Armed Forces. In that context, the DoD's continued objective is to rapidly acquire quality products that satisfy user needs with measurable improvements to mission capability at a fair and reasonable price. The fundamental principles and procedures that the DoD follows in achieving those objectives are described in DoD Directive 5000.1 and DoD Instruction 5000.2.[15]

For space acquisition programs, current DoD acquisition directives (the 5000 series) have been augmented with the NSS Acquisition Policy Directive 03-01, entitled "Guidance for DoD Space System Acquisition Process," to create an acquisition policy environment that fosters efficiency, flexibility, creativity, and innovation.[16] This was a significant change in policy. In fact, using the words "flexibility, creativity, and innovation" in the same context as major systems acquisition had previously only been associated with highly classified programs where the pursuit of breakthrough technologies in an environment of security limitations and strict need-to-know rules increased responsibility and decreased oversight.[17] Along with the findings and recommendations of the aforementioned DSB/AFSAB joint taskforce report, high-profile news items and other congressional commissions and inquiries all combined to prompt the development of NSS Acquisition Policy Directive 03-01.

The purpose of NSS 03-01 was to provide acquisition process guidance for the DoD organizations that were part of the NSS team. NSS is defined as the combined space activities of the DoD and National Intelligence Community (IC). This policy described the streamlined decision-making framework for all DoD space systems. The policy also defined a DoD space acquisition program as a program that can be found as "a table" within the President's Budget submission that had total expenditures for research, development, test, and evaluation (RDT&E) of more than $365 million in a fiscal year or, for procurement, of more than $2.19 billion. It should be noted that highly sensitive classified programs are not included in this policy.[18]

NSS 03-01 also recognized that the acquisition of space systems is different than other major system procurement. For example, aircraft have the ability to test, land, fix, and re-fly. In a space environment, satellite launches and orbit placement must be failure-free and precise the first time and every time. There is no room for error. Repairing satellites in orbit is virtually impossible. Failed satellites become space junk along with the other 10,000 pieces of debris 4 inches or larger orbiting the Earth today. Because of prohibitive costs as indicated earlier, space systems usually do not have on-orbit prototypes to serve as a precursor decision point before a winner is selected for a production contract. As a result, the government reviews and assesses a series of comprehensive "design-offs" rich in modeling and simulation for component, subsystem, and system testing and integration. Many key components are built and exposed to harsh testing environments similar in nature to the operational conditions they will face, including wide ranges and swings in temperature, shock, vibration, rattle and roll. Ultimately, from a life-cycle perspective, more resources are dedicated earlier in space acquisition programs since failure in space is not an option.

Significant cost growth and schedule delays in many critical space system programs have caused senior DoD and IC leadership to question our nation's ability to acquire and sustain NSS systems. The recent series of problems came at a time when our nation had become increasingly reliant on space systems to perform military and intelligence operations.[19]

The space acquisition process

To address the unique nature and complications of space systems acquisition, NSS 03-01 provides an ideal roadmap designed to facilitate the development, production, and deployment of space systems. NSS 03-01 contains an acquisition continuum similar to DoD Directive 5000.1. Its sequence of events follows the same systems engineering context that begins with understanding the user's requirements via an initial capabilities document (ICD) and ends with a cost-effective solution that meets the users' needs. It employs most of the same tools, including analysis of alternatives (AoAs), concept of operations (CONOPS), and architectural views to help narrow system choices and ensure interoperability with other systems. It contains a series of "timed" technical reviews (e.g. system requirements reviews, system design reviews, preliminary design review, critical design reviews) to verify progress and identify issues. It also requires high-level process plans including test and evaluation management plans (TEMP), systems engineering management plans (SEP), etc. At the same time, NSS 03-01 is different, beginning with discrete events within the acquisition timeline. Modeled after the NRO's Directive 7, NSS 03-01 has a definitive acquisition continuum characterized by:[20]

- Pre KDP A: Concept studies
- Phase A: Concept development
- Phase B: Preliminary design
- Phase C: Complete design
- Phase D: Build and operations.

Second, NSS 03-01 uses structured reviews like Key Decision Points (KDPs) to assess whether a program is ready to proceed into the next phase through the use of very well-defined and distinctive criteria, customized for each phase. KDPs also ensure senior-level involvement early in the acquisition process, and serve as timely and focused independent assessments before proceeding into the next acquisition phase. The Milestone Decision Authority (MDA), whose selection is based on the dollar threshold of the space acquisition program, expects proof that a program is ready to proceed forward into the next phase. The MDA is usually very cautious in authorizing a program to move ahead. Third, NSS 03-01 does not normally have a low rate initial production (LRIP) decision since, as mentioned earlier, space system quantities are generally relatively small (ten or fewer). Fourth, there is limited logistics infrastructure since satellite repair is problematic. Ground systems represent the extent of any required spare parts or repair facilities. The acquisition continuum that captures the overall process is illustrated in Figure 9.1. Progression through the five distinct stages of the DoD space acquisition process defined in NSS 03-01 provides a means of progressively refining broadly stated mission needs into well-defined, system-specific requirements and ultimately into operationally effective, suitable, and survivable systems.

Acquisition strategy

While a generic acquisition process exists, an acquisition strategy can vary greatly based on the specific needs of a program. For all DoD system

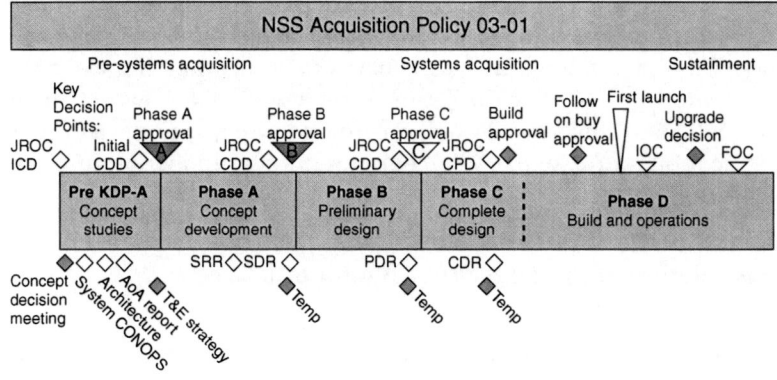

Figure 9.1 NSS Acquisition Policy Directive 03-01, Appendix 1, page 7.

acquisitions, "evolutionary acquisition" is now the preferred strategy for rapid acquisition of mature technology. An evolutionary approach delivers capability in increments, recognizing at the beginning the need for future capability improvements. The objective is to balance needs and available capabilities with resources and to place capabilities into the hands of the user, as quickly as possible. Evolutionary acquisition also requires constant collaboration between the user, tester, and developer. Not surprisingly, all weapon and support systems become much more viable when the user works side-by-side with the designers and the developers from the outset. The evolutionary acquisition approaches include[21]

• Spiral development: a development process where a desired capability is identified, but the end-state requirements are not known at program initiation. Requirements are refined through demonstration, risk management, and continuous user feedback. Each increment provides the best possible capability, but the requirements for future increments depend on user feedback and technology maturation.

• Incremental development: in the incremental development process, a desired capability is identified, an end-state requirement is known, and that requirement is met over time by developing several increments, each dependent on available mature technology.

Like any acquisition program, specific criteria must be met to proceed through the phases of the process. Acquisition Decision Memorandums (ADMs) are the documents that describe the entry and exit criteria for a particular acquisition phase. Entrance criteria describe in general terms what accomplishments are necessary to obtain a successful decision at each major phase, and they apply to all programs. Exit criteria are program-specific accomplishments that must be satisfactorily demonstrated before a program can progress further. Exit criteria are normally selected to track progress in important technical, schedule, or management risk areas. They serve as gates that, when successfully passed or exited, demonstrate that the program is on track to achieve its final program goals and should be allowed to continue with additional activities within an acquisition phase or to be considered for continuation into the next acquisition phase.

Recommendation of the DSB/AFSAB joint taskforce and current state of affairs

These examples of improvements to the acquisition strategy did not occur by chance. They were the products of select groups of highly qualified professionals that have studied the issues and made specific recommendations on how to not only improve, but to transform the way our space acquisition process works. As noted earlier, the key issue is that despite our nation's continued dependence on its space assets, space acquisitions have experienced problems that have driven up costs by hundreds of millions, even billions, of dollars and stretched schedules by years.

The previously described position of the DSB/AFSAB Joint Taskforce was that there were significant, systemic problems in the acquisition of NSS systems and they felt that there was urgent change required to prevent recurring problems from happening to future space programs. The panel recommended several immediate actions that related to process improvements including better, more realistic cost estimates as well as structural issues that address chain of command and other authority concerns.[22]

From the interchange with the DoD and the GAO, and through testimonies to various congressional committees and subcommittees that directly addressed issues involving space acquisitions, many recommendations emerged. For example, a 2005 GAO study on program management best practices, surveyed the DoD's major weapon program managers, including some managing space programs. It cited the following as "top" obstacles to achieving successful outcomes:[23]

- funding instability (approximately 36 percent)
- requirements instability (13 percent)
- staffing problems (8 percent)
- excessive oversight (7 percent)
- inexperienced leadership (7 percent).

In addition, a June 2005 GAO study of space programs was consistent with these views and found that for major space systems, "technologies [being] immature at program start" was the key issue. Further, in delving deeper into the root causes behind these problems, they found that competition for funding has given programs the incentive to produce optimistic cost and schedule estimates, over-promise on capability, suppress bad news, and forsake the opportunity to identify potentially better alternatives. They also found that the DoD starts its space programs "too early"; specifically before it has assurance that the capabilities it is pursuing can be achieved within available resources and time constraints, and it allows new requirements to be added well into the acquisition phase, a course of action that can further increase technology challenges. The report also mentioned that many officials working within the space community agreed that these were key underlying causes of acquisition problems.[24]

Constituents' widely held perception of "problems," or more specifically, judicious spending, has always been very important to law-makers, and a constant challenge to manage. But equally as difficult to accomplish, for those involved with major systems acquisition, is repairing the reputation of the acquisition corps that was severely damaged by apparent mismanagement and high-profile scandals. Repairing the perception that the management and oversight of space acquisitions has sufficient checks and balances to ensure acquisition integrity was no small task. However, many of the systemic issues involving individuals have been repaired by initiatives predominantly aimed at process improvement. In short, most experts feel that the overwhelming majority of the people involved in space acquisitions are ethical and hard-working even though these people, individually and collectively, operate in a complex and, at times, conflicting world of law, regulations, and practices that is demanding and time-sensitive.[25]

These experts also concluded that acquisition professionals should not be encumbered with more rules and regulations since that would not prevent a determined insider from illegal behavior. Their efforts should be enabled by a work environment that encourages and rewards integrity and mutual respect. Processes and oversight that reinforce key principles such as open communications, distribution of authorities, and ethical and respectful behaviors will not only assist in fairness to all, but will encourage the sharing of best business practices, improve decision-making, and create an environment within which transgressions will be more difficult to hide.[26]

Summary

The constant high-level attention that space acquisitions attract underscores the importance of its role in national defense and policy formulation. Also, the significance of learning from its triumphs and failures is a lesson that any serious student of defense policy should embrace. As the process of space system design, production, and operation experienced dramatic change, the nation found itself positioned for more efficient and effective system acquisition in general. The DoD's new space acquisition oversight processes will help increase insights into gaps between requirements and resources as well as ensure that technology maturity is adequately assessed before major acquisitions proceed.[27] Improvement to space acquisitions will not only guarantee America's access to space, but will boost taxpayers' confidence that their money is being efficiently and effectively spent.

A robust NSS program is critical to the security of the United States and will remain a challenging endeavor, requiring the nation's most competent acquisition personnel, both in government and industry.[28] Learning about space acquisition policy evolution provides an excellent foundation for appreciating why, and when, change is needed. Most importantly, appreciating current space acquisition policy will best prepare the nation to successfully acquire future space systems.

Notes

1 Government Accountability Office, *Defense Acquisitions: Improvements Needed in Space Systems Acquisition Management Policy*, GAO-03–1073 (Washington DC: Government Printing Office, September 15, 2003), p. 3.

2 Government Accountability Office, *Defense Acquisitions: Space Systems Acquisition Risks and Keys to Addressing Them*. GAO-06–776R (Washington, DC: Government Printing Office, June 1, 2006), p. 1.

3 Government Accountability Office, *Defense Acquisitions: Improvements Needed in Space Systems Acquisition Management Policy*, p. 2.

4 Robert K. Saxer, J.M. Knauf, L.R.Drake, and P.L. Portanova, "Evolving Expendable Launch Vehicle System: The Next Step in Affordable Space Transportation," *Program Manager* Vol. XXXI, No. 2 (March/April 2002): 3.

5 Report of Defense Science Board/Air Force Scientific Advisory Board Taskforce on Acquisition of National Security Space (Washington, DC: Government Printing Office, May 2003), p. 10.

6 Ibid., p. 11.

7 Department of Defense, National Security Space Acquisition Policy, "Guidance for DoD Space System Acquisition Process," Number 03-01 (Washington, DC: US Department of Defense, December 27, 2004), p. 1.

8 Report of Defense Science Board and Air Force Scientific Advisory Board Taskforce on Acquisition of National Security Space, p. 1.

9 Report of the Commission to Assess United States National Security Space Management and Organization (Washington, DC: Government Printing Office, January 11, 2001), p. 9.

10 Ibid., p. 26.

11 Government Accountability Office, *Defense Acquisitions: Improvements Needed in Space Systems Acquisition Management Policy*, p. 10.

12 Report of Defense Science Board/Air Force Scientific Advisory Board Taskforce on Acquisition of National Security Space, p. i.

13 Ibid.

14 Ibid., pp. 1–4.

15 Defense Acquisition Guidebook, "Forward" (Defense Acquisition University), available at: http://akss.dau.mil/dag/DoD5000.asp?view=document (accessed February 13, 2007).

16 Department of Defense, National Security Space Acquisition Policy, "Guidance for DoD Space System Acquisition Process," p. 1.

17 Joe Mazur, Jr., "Acquisition Secrets from the National Reconnaissance Office," *Air Force Journal of Logistics* Vol. XXVII, No. 4 (2003): 2.

18 Department of Defense, National Security Space Acquisition Policy, "Guidance for DoD Space System Acquisition Process," p. 1.
19 Report of Defense Science Board/Air Force Scientific Advisory Board Taskforce on Acquisition of National Security Space, p. 8.
20 Department of Defense, National Security Space Acquisition Policy, "Guidance for DoD Space System Acquisition Process," p. 5.
21 Ibid., p. 9.
22 Report of Defense Science Board/Air Force Scientific Advisory Board Taskforce on Acquisition of National Security Space, pp. 4–5.
23 Government Accountability Office. *Best Practices: Better Support of Weapon System Program Managers Needed to Improve Outcomes*, GAO-06–110 (Washington, DC: Government Printing Office, November 30, 2005), p. 44.
24 Government Accountability Office. *Defense Acquisitions: Incentives and Pressures That Drive Problems Affecting Satellite and Related Acquisitions*, GAO-05–570R (Washington, DC: Government Printing Office, June 23, 2005), p. 4.
25 Report of Defense Science Board Taskforce on Management Oversight in Acquisition Organizations (Washington, DC: Government Printing Office, March 2005), p. 3.
26 Ibid.
27 Government Accountability Office, *Defense Acquisitions: Improvements Needed in Space Systems Acquisition Management Policy*, p. 10.
28 Report of Defense Science Board/Air Force Scientific Advisory Board Task Force on Acquisition of National Security Space, p. 33.

10 The civil sector

Michael Gleason and John M. Logsdon

Introduction

Civilian space activity carried out by the US government contributes to US national security through its relevant scientific and technological accomplishments. It also contributes to the *international prestige* and *national pride* of the United States that result from its achievements. Looking at the civilian space program through this lens provides an important perspective on the development and implementation of civilian space policy. It also provides one justification for the continuing large government investments in the US civilian space program, both in absolute terms and in comparison to the investments of other leading countries. For example, in 2008 the United States' share of global government spending on civilian space activities was approximately 60 percent of the world's total.

This chapter provides an overview of the evolution of civilian space policy and activities. It discusses the processes through which space policy is developed, and the government structures through which the civilian aspects of that policy are implemented. It concludes with an overview of those current and emerging issues in the civilian space sector of particular relevance to national security interests.

The evolution of civilian space policy and programs

After almost five decades of operations, the US civilian space program has seen triumph – 12 Americans walked on the Moon between 1969 and 1972 – and tragedy – three Apollo astronauts died during a launchpad test in January 1967 and 14 astronauts died in the January 1986 and February 2003 space shuttle accidents. From its peak budgets during the build-up needed to send people to the Moon, the resources allocated to NASA rapidly decreased after 1970, to approximately 20 percent of their highest level, and have stayed relatively flat as a share of the overall Federal budget for more than 35 years. NASA believed that it would be able to continue to pioneer both robotic space science and human exploration, and it has had difficulty adjusting its expectations to the resources it could command; there has been a continuing tendency to try to do too much with too little. It has also had difficulty adjusting to a post-Cold War mindset where it was not in competition with the Soviet Union for space leadership. Nonetheless, NASA has continued for most of its existence to symbolize American power to the rest of the world and to serve as a source of pride for the American public.

Americans to the Moon

In 1959 NASA set human missions to the Moon and eventually other planets as its long-term goal and began very preliminary planning on what the technical and financial requirements of such an undertaking might be. The White House caught wind of this planning, and in December 1960, just a month before leaving office, President Eisenhower was briefed on what NASA was thinking. His reaction was strongly negative. Eisenhower's space policy since the creation of NASA in 1958 as a reaction to the launch of *Sputnik 1* had been marked by modest ambitions and a cautious entry into all realms of civilian space activity; the President wanted to contain the ambitions of enthusiastic space advocates, and setting out on a journey to the Moon was well beyond what he thought justified, given the high costs involved.

Space policy was not a high priority issue at the start of John F. Kennedy's Administration in 1961. That was to change less than three months later, on April 12, 1961, when the Soviet Union became the first country to place a human, Yuri Gagarin, into orbit and return him successfully to Earth. Once again the United States was second in a highly visible area of technological accomplishment.[1] Again, the media and Congress demanded a US response. President Kennedy decided that the United States could not continue to cede space leadership to the Soviet Union. On April 20, he asked his top space adviser, Vice President Lyndon Johnson, to get answers to the following questions:

> Do we have a chance of beating the Soviets by putting a laboratory in space, or by a trip around the moon, or by a rocket to land on the moon, or by a rocket to go to the moon and back with a man? Is there any other space program that promises dramatic results in which we could win?[2]

The technical answer to these questions came from Wernher von Braun, now working for NASA after being employed by the US Army in the period 1945–1960, who told the Vice President that with its existing rocket, the Soviet Union had a good chance of being first to accomplish everything on the President's list except sending a man to the Moon and back. That would require both the Soviet Union and the United States to develop a new, more powerful rocket; not surprisingly, von Braun thought that the United States could do that before the USSR.

The political rationale for a decision to compete with the Soviet Union came in a May 8 memorandum signed by Secretary of Defense Robert McNamara and the man Kennedy had chosen to head NASA, James Webb. They recommended an across-the-board acceleration of the US space effort, with its focus being a lunar-landing mission. "It is man, not machines, that captures the imagination of the world," they told the President. They added, "Dramatic achievements in space symbolize the technological power and organizing capability of a nation." For this reason, "*this nation needs to make a positive decision to pursue space projects aimed at enhancing national prestige*" (emphasis in original) because space achievements and the prestige they conferred were "part of the battle along the fluid front of the cold war."[3]

Kennedy accepted these recommendations, and on May 25, 1961 told a joint session of Congress that it was "time for this nation to take a clearly leading role in space achievement which in many ways may hold the key to our future on Earth." As the centerpiece of his speech, he said "I believe that this nation should commit itself to achieving the goal, before this decade is out, of landing a man on the Moon and returning him safely to Earth."[4]

This was the beginning of Project Apollo, which led to the spectacular *Apollo 8* circumlunar flight at Christmas-time 1968, the *Apollo 11* first lunar landing on July 20, 1969, and five subsequent sorties to the lunar surface. The Apollo program achieved its basic objective – beating the Soviet Union to the Moon. (The Soviet Union did attempt both flights around the Moon and a lunar landing, but a combination of a late start, insufficient funding, and poor political and management decisions led it to abandon its effort in the early 1970s.)

Apollo also defined the lasting character of the US civilian space program, one centered on developing large and complex systems for human space flight. In the aftermath of the Kennedy speech, NASA created a new "field center" in Houston, Texas (now the Johnson Space Center) to manage its human flight activity, and a new launch center on Merritt Island, Florida (now the Kennedy Space Center) to accommodate the massive rockets needed to launch people to the moon. Apollo required a peaceful, but war-like mobilization of financial resources; in the months following the President's speech, NASA's budget was increased by 89 percent, and by another 101 percent in 1962. At its peak, the NASA budget was over 4 percent of the overall Federal budget. (It has been less than 1 percent in recent decades, as will be discussed below.) When NASA began operations in October 1958, it had approximately 8,000 employees; at the peak of Apollo, NASA employed over 34,000 people, working with over 375,000 industrial and academic contractors. This space industrial and academic base in future years contributed to both civilian and national security space and defense efforts; Apollo also resulted in increased enrollments in university science and engineering programs.

In addition to Apollo, during the 1960s NASA undertook an increasingly ambitious space science effort and did pioneering work in developing the technologies that enabled the development of a commercial satellite communications industry and of operational meteorological satellites. With a robust overall NASA budget, these efforts were not in conflict for scarce resources, unlike the situation in the 35 years since Apollo. By the time Neil Armstrong stepped on the lunar surface, it was clear that the United States had achieved the "clearly leading role in space achievement" that Kennedy had called for eight years earlier.

What do you do after you have been to the Moon?

The people of the United States and their government have been willing, in the 35 years since the end of the Apollo program, to continue a substantial civilian space program, but only at a level of funding that has forced NASA to constantly operate on the edge of viability in terms of achieving its ambitious objectives. Even so, the program is still seen by national leaders as an important element of US global leadership and national power. In addition, government civilian space programs provide essential operational services such as meteorological observation and remote sensing of the Earth's surface. Government organizations to promote and regulate private sector space activities have emerged in recent years.

Two things are remarkable about the pattern of resources allocated to NASA over its history. The one most usually remarked upon is the rapid build-up of funding in the early 1960s in support of Project Apollo. Equally remarkable, however, is the rapid decrease in financial resources allocated to NASA between 1965 and 1974, and even more so the stability of that allocation over the past 35 or more years. It is impossible to escape the conclusion that, whatever the specific content of the NASA program at a particular time, the American public and their leaders, through the political process, have consistently decided to allocate less than 1 percent of the annual Federal budget to the civilian space program. This decision has been made, and reinforced, as the Federal budget for each successive fiscal year has been assembled in the White House and approved or modified by Congress. Within that allocation, national leaders have expected NASA to carry on a successful program of human spaceflight as well as its other activities. The result, as the Columbia Accident Investigation Board observed with respect to the 2003 *Columbia* space shuttle accident, has been an agency striving to "do too much with too little."[5]

The first step in the process of formulating a policy to guide the civilian space program after the end of the Apollo program was the creation in February 1969 of the Space Task Group, chaired by Vice President Spiro T. Agnew. This group was charged by President Richard M. Nixon with preparing "definitive recommendations on the direction which the US space program should take in the post-Apollo period." In its September

15, 1969 report, the Space Task Group set out several ambitious options for the future and proposed, "as a focus for the development of new capability," that "the United States accept the long-term option or goal of manned planetary exploration with a manned Mars mission before the end of the century as the first target."[6] Intermediate objectives recommended by the report included a series of increasingly larger space stations in low-Earth orbit, continuation of lunar exploration, and a reusable logistics vehicle – called a space shuttle – to reduce the costs of regular operations in space.

Accepting the Space Task Group's recommendations would have meant accepting a long-term national commitment to a robust program of human spaceflight, with repeated trips to the Moon and, eventually, forays to Mars. This was not at all what Richard Nixon and most of his advisers had in mind for the post-Apollo space effort. On March 7, 1970, the White House released a presidential statement on the future of the US space program. The statement was cast both as a response to the Space Task Group report and as an evaluation of where the civilian space program fit into the country's future. Its message was clear:

> Space expenditures must take their proper place within a rigorous system of national priorities. What we do in space from here on in must become a normal and regular part of our national life and must therefore be planned in conjunction with all of the other undertakings which are important to us.[7]

To Richard Nixon and his advisers, the rationale that had been accepted at the start of the Apollo program – that national prestige itself was an adequate justification for an expensive human spaceflight program – was not an acceptable justification for an expansive post-Apollo space program, especially given their desire to reduce Federal spending overall and the resource demands of winding down US engagement in Southeast Asia. They did not want to put an end to human spaceflight and its associated contributions to national self-image and prestige, but they were unwilling to set an ambitious, Apollo-like goal to guide that effort.

The first program that NASA had hoped to gain post-Apollo approval to develop was a 12-person space station launched by the large rocket developed for Apollo, the Saturn V. In addition, NASA proposed developing a fully reusable Earth-to-orbit launch vehicle called the space shuttle to carry crew and supplies to the space station and to carry out other space operations. When the Nixon Administration, as part of its decision to reduce the NASA budget, refused to approve the space station, NASA, in the fall of 1970 deferred – not canceled – its space station plans and directed its shuttle contractors to design a vehicle capable of carrying pieces of a space station into orbit. Thus the current strong link between the space shuttle and International Space Station programs actually has its roots in decisions taken more than 35 years ago. In the interim, NASA attempted to get White House approval to develop the space shuttle.

In 1971, there was intense debate within the Executive Branch and its advisers of whether to approve such a development. While Nixon's scientific and budgetary advisers favored further reductions in the NASA budget and even the cancellation of the *Apollo 16* and *Apollo 17* missions, the President's long-time confidant Caspar (Cap) Weinberger disagreed. He told the President in an August 1971 memorandum that further reductions of the NASA budget and cancellation of the remaining Apollo missions

> would be confirming in some respects, a belief that I fear is gaining credence at home and abroad: That our best years are behind us, that we are turning inward, reducing our defense commitments, and voluntarily starting to give up our super-power status, and our desire to maintain world superiority.

Nixon wrote on the memo "I agree with Cap."[8]

It was arguments such as these, linking a continued strong civilian space program to national image, which led in January 1972 to approval of space shuttle development. Even so, the shuttle that emerged from the 1970–1971 policy process was a product, according to the Board that investigated the 2003 *Columbia* accident, of "a series of political compromises that produced unreasonable expectations – even myths – about its performance," with a "technically ambitious design [that] resulted in an inherently vulnerable vehicle."[9]

In order to make the case that the investment in developing the space shuttle was cost effective, NASA had to gain the agreement of the civilian leadership of the military and intelligence communities that when it became operational, the space shuttle would be the only launch vehicle for almost all national security payloads, including the most critical classified payloads. In order to gain this agreement, NASA had to design a shuttle with specific performance characteristics, including the size of its payload bay and its ability to maneuver upon reentry. Meeting these requirements increased the technological risks of the vehicle and increased its operating costs.

NASA spent the rest of the 1970s developing the space shuttle and readying it for its first flight, which took place (more than two years behind schedule) on April 12, 1981.

Soon after that first flight, the new NASA leadership set its two top priorities as: bringing the shuttle to operational status as soon as possible; and getting presidential and congressional approval to develop a shuttle-launched space station. NASA had deferred its space station ambitions in 1970, but they re-emerged as soon as the shuttle began flying. In 1970, NASA had decreed that any shuttle that might be developed needed to be capable of launching a modular space station and no alternatives to using the space shuttle in this role were considered at the inception of the space station program.

Also in 1981, after only two shuttle flights, President Ronald Reagan approved a formal policy statement saying that the space shuttle would be the primary space launch system for both US military and civil government missions. This policy was reinforced in a 1982 statement of National Space Policy, which said that "completion of transition to the Shuttle should occur as expeditiously as possible" and that "government spacecraft should be designed to take advantage of the unique capabilities of the STS [Space Transportation System, another designation for the space shuttle]."[10]

The US Air Force, as the launch agent for both military and intelligence spacecraft, early on recognized the dangers of this "all eggs in one basket" policy. Soon after the shuttle was declared operational on July 4, 1982, after only four flights, the Air Force began to argue that the risks and costs of the system could be a detriment to its ability to perform its launch responsibilities for critical national security payloads. Most of those payloads had been designed since the late 1970s so that they could only be launched on the shuttle.

Beginning in 1983, the Air Force campaigned for approval of a backup to the shuttle in order to provide assured access to space for such payloads. NASA fought this move. The dispute between the Air Force and NASA reached the White House in early 1985, where it was decided in favor of the Air Force. This decision led to the development of the Titan IV expendable launch vehicle, which was capable of launching the largest military and intelligence spacecraft. After the January 1986 *Challenger* accident, the Titan IV became the primary launcher for large national security missions, and those spacecraft that had been intended for shuttle launch had to be redesigned at high cost. In addition, the Department of Defense (DoD) had agreed to bear the multi-billion dollar costs of creating launch facilities for the space shuttle at Vandenberg Air Force Base on the California coast, so the shuttle could be launched into polar orbit. After the *Challenger* accident, the White House decided that this facility would not be used. These actions created tensions between the military and intelligence space programs and the civilian space effort that persist even today. In addition, the White House in 1986 decided that the space shuttle would no longer be marketed as a launch vehicle for commercial payloads such as communication satellites, and existing shuttle launch contracts were voided. This left NASA as the only user of the shuttle, and assembly and servicing of a space station as its primary mission.

President Reagan had approved space station development in 1984, over the opposition of the DoD, which saw no military value in the project; at that point the facility was to be completed within a decade. The primary public justification for the space station was the scientific and technological potential of research carried out in a well-equipped laboratory in the microgravity space environment of Earth orbit. However, the US decision was also influenced by the fact that the Soviet Union, after abandoning its efforts to send people to the Moon, had launched since the early 1970s a series of rudimentary stations. For example, one aerospace company ran a television advertisement portraying a Soviet space station and asking "Shouldn't we be there too?"

From its inception, the space station program ran into cost overruns and schedule delays, and its first elements were not launched until 1998, with a crew taking up permanent occupancy only in November 2000. After its 1988 return to flight following the *Challenger* accident, the space shuttle flew a variety of missions, launching various spacecraft ranging from probes to Jupiter to the Hubble Space Telescope (to which it also flew a 1993 servicing mission to correct the effects of an incorrectly ground mirror). The shuttle also served as a surrogate space station, as its crew carried out various microgravity experiments during missions lasting as long as 16 days. But from 2000 onwards, the shuttle's primary mission has been space station assembly and servicing – the link between the shuttle and the space station created almost 30 years earlier remained unbreakable.

The space station, and to some degree, the space shuttle programs reflected an important policy shift in the US approach to linking space activities to broader foreign policy interests. While Apollo had been a unilateral demonstration of US power, in the post-Apollo period the White House decided that the space program should become an instrument of US leadership in international partnerships. Rather than leading through singular achievements, the goal became a demonstration of US capability as the managing partner, together with other countries, in complex technological undertakings. While the robotic space science program had from its inception been characterized by international cooperation, only in 1969 did NASA begin to court international partnerships in its human spaceflight program. European countries and Canada contributed elements to the space shuttle program, and Japan joined them in 1984 as a partner in the space station. After the end of the Cold War and the collapse of the Soviet Union, Russia also joined the station partnership in 1993, and the program became known as the International Space Station (ISS). As ISS assembly began in 1998, 16 countries were members of this partnership.

By the start of 2003, NASA's future looked relatively stable. In the preceding two years, cost overruns in the US portion of the ISS had appeared, threatening the program's schedule and causing tensions between the United States and its ISS partners. But it seemed as if NASA had gotten the ISS under managerial control, and the space shuttle was flying regularly and expected to remain in service until at least 2020. There had been several attempts to develop a replacement for the shuttle, but at the turn of the century none was in sight. The robotic space science program was moving forward with a series of impressive missions in the 1990s, and even more ambitious missions were being planned.

However, on February 1, 2003 the space shuttle orbiter *Columbia* broke up upon reentry into the atmosphere after a 16-day mission; all seven crew members died. In the aftermath of the *Columbia* accident, there was a searching re-examination of the goals and content of the NASA program. The result of this re-examination was a fundamental shift in US civilian space policy; that shift will be discussed below.

Actors, policies, and organizations

The civilian space sector is composed of a variety of organizations and actors concerned with creating and implementing US civilian space policy. This section provides an overview of the actors involved in creating that policy, the process by which policy decisions are reached, and the organizations that implement space policy and operate the nation's civilian space program. Starting at the top, the White House formulates policy, Congress provides oversight and funding, and on occasion writes White House policy into law, and multiple executive branch organizations implement laws, policies, and funded programs.

NASA is the biggest and most well known organization charged with implementing US civilian space policy. However, the National Oceanographic and Atmospheric Administration (NOAA), which is part of the Department of Commerce (DoC), also plays a significant role, although its activities are less well known. There is also an Office of Space Commercialization in the DoC; it is based within NOAA. In addition, the Federal Communications Commission, the Office of Commercial Space Transportation, part of the Federal Aviation Administration within the Department of Transportation (DoT), and the US Geological Survey (USGS) of the Department of the Interior (DoI) have roles in the civilian space sector. The State Department (DoS) is involved in international space interactions.

As discussed previously, the civilian space sector is responsible for the US human spaceflight program as well as robotic space and Earth science missions. Human spaceflight garners the largest budget, the most human resources, and the most prestige. Although the science program's research accomplishments are substantial (two researchers have won the Nobel Prize for their space-based work), the human space program generates the most public attention and enthusiasm. Although they co-exist within the same overall organization, there have been tensions between NASA's human spaceflight and robotic science efforts since NASA began operations in 1958. While the overall NASA budget was at a high level during Apollo, these tensions were muted. Since 1969, they have increased as NASA tries to balance its limited budget between the two sets of activities.

The civilian space program supports US national security directly via the prestige it provides the United States and indirectly as it supports the US economic, military and social instruments of power by helping stimulate US superiority in science, technology, and engineering. Furthermore, the civilian space sector is intertwined with the military space sector in significant areas. The terms "dual-use" or "multi-use" describe technology and programs used for both civilian and military applications, such as Global Position Satellites or polar-orbiting meteorological satellites.

International cooperation is an important feature of the US civilian space program. It is called for in the National Aeronautics and Space Act of 1958, and is epitomized by the 16-country ISS program. The civilian space program has enjoyed a significant amount of success with partner nations in its scientific, exploratory, and human spaceflight missions, but questions about the pay-offs for such partnerships are growing and barriers to cooperation, such as US export control regulations, are getting more difficult to overcome.

This section first discusses the actors that develop formal statements of space policy and determine space budgets in the White House. The contributions of Congress to the policy process are then discussed. A detailed overview of how civilian space policy is implemented through various executive branch organizations follows.

Actors that generate policy

The president determines national space policy within the legal boundaries and funding provided by Congress. A number of bodies within the White House provide policy advice for the president, including the National Security Council (NSC), the Office of Science and Technology Policy (OSTP), the Office of Management and Budget (OMB), and the National Economic Council (NEC). The Administrator of NASA reports directly to the president and thus can provide his own views on space policy issues to the chief executive. Cabinet-level officials and their representatives from the DoD, DoS, DoC, DoT, DoI, Homeland Security Departments, the Director of National Intelligence, and the Central Intelligence Agency also provide input to the president, as necessary, to guide the formulation of US national space policy, usually in the context of their participation in one of the councils noted above.

The US Congress also shapes national space policy through legislation, such as the National Aeronautics and Space Act of 1958 or various authorization or appropriation bills. Significant space policy-related legislation is relatively infrequent, however, so Congress exercises its policy role mostly through oversight of the executive branch and by controlling the purse strings.

The White House

The 1958 NASA Act created a National Aeronautics and Space Council at the White House level to coordinate interagency space policy. The Council was to be chaired by the president and to have a small staff. President

Eisenhower made little use of the Council. John F. Kennedy proposed making the Vice President the Council chair and the Congress agreed. Under Vice President Lyndon B. Johnson, the Council played a seminal role in developing space policy during the Kennedy administration. Under President Johnson and President Nixon, the Council had limited influence, and Nixon abolished it in 1973. During the late 1970s and early 1980s, the lead role on space policy within the Executive Office of the President was assigned to the OSTP, which was created in 1976. In 1982, President Ronald Reagan shifted that role to the NSC. Congress re-established a National Space Council in 1988, and during the term of President George H.W. Bush (1989–1993) the Council and its staff played an active role in space policy development. During this period the Council was chaired by Vice President Dan Quayle. President Bill Clinton again abolished the Council in 1993. Rather, President Clinton created a National Science and Technology Council under the auspices of the OSTP. This Council oversaw the development of all science and technology policies, including space, during the Clinton administration. In 2001 President George W. Bush reassigned primary responsibility for developing space policy back to the NSC and its staff, who work in close collaboration with OSTP space staff. As this summary suggests, there is no single accepted approach to the management of space policy development at the level of the president, and all concerned agencies participate regardless of which organization has the lead role. Within the Executive Office of the President, only a small number of staff members are dedicated to the space policy area. Their work is crucial to the development of options for consideration by policy-level officials in the various bodies described above.

Whatever the lead organization for developing space policy, the process leading to policy decisions is similar. Representatives of the concerned executive agencies, with whatever White House organization has been designated as lead playing a coordinating role, work together to draft a new policy or revise an existing one. This process usually begins with mid-level officials, and gradually involves more senior people until a draft policy is agreed upon and sent to agency heads for their comments and ultimate approval. Only rarely does the policy statement reach the president and his most senior aides until there is agreement on its contents, although on occasion disagreements cannot be resolved at levels lower than agency heads and the president.

The result of this process is usually a formal statement of either a general National Space Policy or a policy related to a specific area of space activity, such as Earth observation, position, navigation and timing, space exploration, or space transportation. Elements of these policy statements may be classified, but there is almost always an unclassified fact sheet issued that summarizes the policy.[11]

Space policy decisions are also frequently made in the context of the annual process through which the president prepares his budget for submission to the Congress. A small staff within the Office of Management and Budget (OMB) has year-round contact with NASA and other space-active agencies, and reviews agency budget proposals when they are submitted to the White House, usually in September. Between September and the time final decisions are made on the budget at the end of the year, there is intense interaction between the OMB and NASA on one hand and the OMB and the White House policy-makers on the other to shape a space budget that matches presidential priorities and fiscal realities.[12]

US Congress

Relatively few members of Congress give focused attention to space policy issues; rather, subcommittees within the Senate and House are designated to deal with space issues, and it is their senior members who become space specialists. Other members of Congress become involved with space issues if they are particularly important in their states or districts. Authority is also divided between the authorizing subcommittees, which provide a policy framework for space activities and oversee their implementation, and appropriations subcommittees, which decide on the funding to be made available for space. With respect to authorization, in the Senate the Subcommittee on Science, Technology, and Space of the Committee on Commerce, Science and Transportation concerns itself with issues surrounding such executive branch organizations as NASA, NOAA, the National Science Foundation (NSF), and OSTP. In the House of Representatives, the Subcommittee on Space and Aeronautics of the Committee on Science takes the lead role concerning issues related to the US civilian space program, including NASA and commercial space activities within the DoT and DoC.

Subcommittees of the Senate and House Appropriations Committees review space funding requests and appropriate funds to agency budgets. These subcommittees do not only deal with space issues; they also appropriate funds for several other, often unrelated, executive branch agencies.

A new space policy with an emphasis on space exploration

In the aftermath of the February 1, 2003 *Columbia* accident, there was a searching re-examination of the goals and content of the NASA program. The result of this re-examination was a fundamental shift in US civilian space policy, one that has given NASA an open-ended mission, to "implement a sustained and affordable human and robotic program to explore the solar system and beyond."

In the immediate aftermath of the shuttle accident, the Columbia Accident Investigation Board was created by NASA. As its investigation evolved, the Board decided that, to get to the roots of the accident, it had to broaden its focus from the physical causes of the accident to the policy, organizational, and management context within which the accident had occurred. In its August 2003 report, the Board also discussed "future directions for the United States in space." The Board observed that there had been a "lack, over the past three decades, of any national mandate providing NASA a compelling mission requiring human presence in space," and suggested that this lack, and the consequent absence of polit-ical will to develop a replacement for the aging space shuttle, constituted "a failure of national leadership."[13]

President George W. Bush and his key advisers took this criticism to heart, and in the fall of 2003 a small group of individuals from NASA and the Executive Office of the President developed a proposal for returning NASA's focus to exploration beyond Earth orbit, rather than to continue with an emphasis on utilization of the ISS and on repetitive space shuttle flights to the ISS. The President approved this proposal and announced what has come to be called the "Vision for Space Exploration" in a speech at NASA Headquarters on January 14, 2004. The key elements of the Vision include:

- completion of the ISS
- safe flight of the space shuttle until 2010
- development and flight of a Crew Exploration Vehicle (CEV) no later than 2014
- return to the moon no later than 2020, and stay on the lunar surface for increasingly longer periods of time
- extend human presence across the solar system and beyond, begin-ning with human voyages to Mars
- promotion of international and commercial participation in explo-ration.[14]

This new policy direction lays out a major implementation challenge for the nation's principal civilian space agency, NASA. If public and political support for the Vision for Space Exploration remains stable in the next decade, the United States will be well on its way back to the Moon.[15]

Implementing US civilian space policy

This section provides an overview of the organizations that implement space policy and operate the nation's civilian space program. The most visible and important organization is NASA, but NASA is a research and development agency. Operational responsibility for civilian space efforts is exercised by other agencies or by the private sector once a particular capa-bility is demonstrated. For example, NOAA in the DoC is responsible for operational meteorological satellite programs, and the USGS in the DoI is responsible for the Landsat Earth observation satellite program. Other organizations, such as the Office of Commercial Space Transportation in the DoT and the Office of Space Commercialization in the DoC, do not operate space systems but also are involved in implementing space policy, as is the Federal Communications Commission.

National Aeronautics and Space Administration[16]

NASA headquarters

NASA is the largest organization in the US civilian space program and conducts the most visible US space activities, such as human spaceflight with the space shuttle and ISS, and space science with such programs as the Hubble Space Telescope, robotic missions to Mars, and other space and Earth science missions.

NASA is an independent agency that reports directly to the White House. NASA headquarters is located in Washington, DC and there are nine other installations, known as "field centers," located around the United States, plus the Jet Propulsion Laboratory (JPL), which is managed for NASA by the California Institute of Technology, but in its other respects is a tenth NASA center. The NASA workforce fluctuates but is comprised of approximately 19,000 civilian service employees. NASA grants and contracts also support a large workforce in universities across America and in the aerospace industry.

There are four Mission Directorates within NASA headquarters: Exploration Systems, Space Operations, Science, and Aeronautics Research. The NASA headquarters is responsible for liaison with the White House, other Executive Branch Agencies, Congress, NASA's international partners, the media, and the general public. Through its mission directorates, it develops the projects and programs, and associated budgets that NASA's field centers are responsible for implementing. Figure 10.1 is a NASA organizational chart (current as of November 2008).[17]

NASA's FY 2009 budget request was $17.6 billion; this was 0.6 percent of the overall Federal budget requested by President George W. Bush in February 2008 (see Table 10.1). Within this budget, NASA was asked to fly the space shuttle for the minimum number of missions required to complete assembly of the ISS, estimated to be ten more flights, carry out a robust program of space and Earth science, support aeronautical research, and get started on developing a replacement for the space shuttle and on the new Vision for Space Exploration. Most outside observers agreed that there were inadequate resources to carry out all these missions effectively.

The Johnson Space Center

The Johnson Space Center (JSC), located in Houston, Texas is the lead center for all US human spaceflight, including space shuttle and ISS

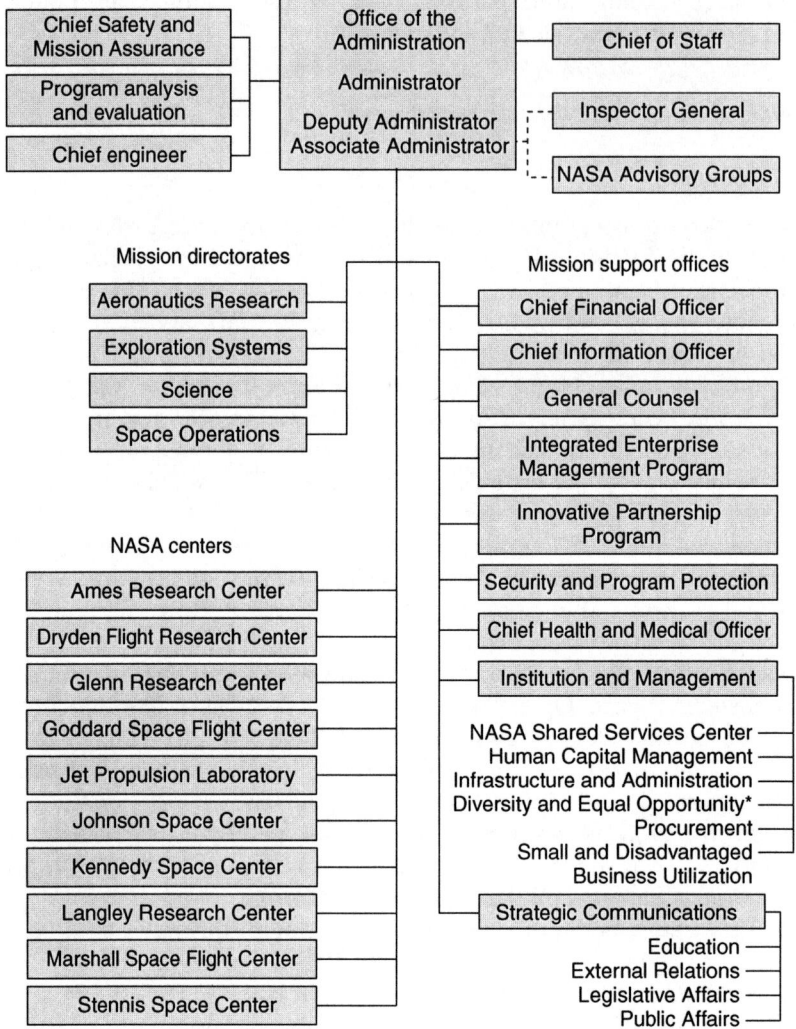

Figure 10.1 NASA organizational chart (source: NASA).

activities, and is responsible for astronaut training. The Mission Control Center (MCC) directs all space shuttle missions after their launch and manages all activity onboard the ISS. JSC will manage the development of the spacecraft that will replace the space shuttle as the means for carrying American astronauts to space, the CEV, which is to be called Orion. About 110 astronauts are currently eligible for flight assignments.

Table 10.1 NASA budget (millions of dollars)

Area of activity	FY 2009 request
Space operations (shuttle and ISS)	5,775
Science	4,441
Exploration systems	3,500
Aeronautics research	446
Cross-agency support programs	3,300
Education	116
Inspector General	36
Total, discretionary budget authority	17,614

Source: NASA.

Kennedy Space Center

The Kennedy Space Center (KSC), located near Cocoa Beach, Florida, is the only launch base for US human spaceflight, including Apollo program launches, current space shuttle launches, and most likely future CEV launches. (Launches in the Mercury and Gemini programs took place at the adjacent Cape Canaveral Air Force Station.) KSC prepares the spacecraft and launch vehicles for each mission, operates each countdown and manages end-of-mission landing recovery activities. KSC also coordinates all expendable vehicle launches carrying NASA payloads at Cape Canaveral Air Force Station in Florida, Vandenberg Air Force Base in California, or elsewhere. KSC also prepares ISS hardware for launch.

Marshall Space Flight Center

Marshall Space Flight Center (MSFC), located in Huntsville, Alabama, is responsible for key space launch and propulsion system development. In other words, MSFC develops large rockets for human exploration, from the Saturn 1B and Saturn V used during Apollo to the new Crew Launch Vehicle and Cargo Launch Vehicle, which will be known as Ares-1 and Ares-5. It will also manage robotic lunar exploration missions and the development of a new lunar lander for human missions to the Moon.

Stennis Space Center

Stennis Space Center (SSC), located in southern Mississippi, is NASA's primary center for rocket engine testing and is America's largest rocket test complex.

Ames Research Center

Ames Research Center (ARC) in Mountain View, California, is a leader in information technology research with a focus on supercomputing, networking, and intelligent systems as well as nanotechnology, fundamental space biology, biotechnology, aerospace and thermal protection systems, and human factors research. Ames also conducts research on the effects of gravity on living things and the nature and distribution of stars, planets, and life in the universe.

Goddard Space Flight Center

Goddard Space Flight Center (GSFC), located in Greenbelt, Maryland, a suburb of Washington, DC, operates numerous scientific spacecraft including the Hubble Space Telescope, making GSFC the largest organization in the US engaged in researching the Earth, the solar system, and the universe through satellite-based observations. GSFC also manages the operational Space and Ground Network that supports the Human Spaceflight Program, as well as Earth-orbiting missions, international, commercial, classified, and unclassified national missions.

Jet Propulsion Laboratory

The Jet Propulsion Laboratory (JPL) at the California Institute of Technology, located in Pasadena, California is a federally funded research and development center managed and staffed by Caltech for NASA. JPL is responsible for interplanetary, deep space scientific and exploratory missions. Recent JPL missions include the Mars Exploration Rovers Spirit and Opportunity, the Deep Impact spacecraft which blasted a crater in a comet, and the Cassini mission to Saturn and its moon Titan. JPL is also responsible for management of NASA's Deep Space Network, a global network of antenna complexes for controlling deep space spacecraft and retrieving data from them.

Dryden Flight Research Center

Dryden Flight Research Center (DFRC), located at Edwards Air Force Base, California, is NASA's primary installation for flight research. Dryden is chartered to conceive and conduct experimental flight research for integrated flight and propulsion controls; advanced optical sensors and controls; viscous drag reduction; advanced configurations; high-altitude, long-endurance aircraft; remotely piloted vehicle technology; hypersonic vehicle experiments; high-speed research for civilian transportation; atmospheric tests of advanced rocket and air-breathing propulsion concepts; instrumentation systems; and flight loads predictions. In carrying out this mission, Dryden operates some of the most advanced research aircraft in the country. It also serves as a backup landing site for the space shuttle and a facility to test and validate design concepts and systems used in development and operation of manned spacecraft.

Glenn Research Center

Glenn Research Center (GRC), located in Cleveland, Ohio, is engaged in research, technology, and systems development programs in aeronautical propulsion, space propulsion, space power, space communications, and microgravity sciences in combustion and fluid physics.

Langley Research Center

Langley Research Center (LARC), located in Hampton, Virginia, was founded in 1917 and was the nation's first civilian aeronautical research facility. Langley leads NASA initiatives in aviation safety, quiet aircraft technology, small aircraft transportation, and aerospace vehicles system technology. It supports NASA space programs with atmospheric research and technology testing and development. More than half of Langley's research is in aeronautics. Its workforce is comprised of 1,906 civilian servants and 1,400 contractors for a total workforce of approximately 3,300.

Other executive branch implementing agencies

National Oceanic and Atmospheric Administration

NOAA is a significant actor in the US civilian space sector and directly supports US national security through its operation of civilian and military meteorological satellite programs, space environment monitoring, and its international space cooperation activities.

NOAA is a element of the DoC, with its headquarters located in the DoC in Washington, DC. NOAA conducts research and gathers data about the oceans, atmosphere, space, and sun. It provides these services through five major organizations: the National Weather Service, the National Ocean Service, the National Marine Fisheries Service, NOAA Research, and the National Environmental Satellite, Data and Information Service (NESDIS). NOAA also issues the licenses required for private-sector Earth-observation missions.

NESDIS is the nation's primary source of space-based meteorological and climate data and is the world's largest environmental space organization.[18] NESDIS headquarters is in Silver Spring, Maryland, and it has facilities in: Suitland, Maryland; Wallops, Virginia; Fairbanks, Alaska; Asheville, North Carolina; and Boulder, Colorado. NESDIS develops, acquires, and operates US civilian environmental satellites which are used for weather forecasting, climate monitoring and other environmental applications such as fire detection, ozone monitoring and sea surface temperature measurements. The NESDIS budget request for FY 2009 was $1.16 billion. Its workforce is comprised of approximately 750 employees.

The NOAA–NESDIS satellite system is composed of geostationary operational environmental satellites (GOES) and polar-orbiting environmental satellites (POES). They are operated from the NOAA Satellite Operation Facility in Suitland, Maryland. GOES spacecraft produce the satellite photos the public associates with television weather forecasts and Internet satellite weather maps. There are two operational POES spacecraft and two operational GOES spacecraft. Non-fully operational POES and GOES spacecraft, with degraded sensor capabilities, but still providing useful data, are also part of the system.

NOAA's Space Environment Center (SEC) is the lead national and international warning center for disturbances in the space environment that can affect people and equipment. SEC performs critical space weather operations, jointly staffed by NOAA and the US Air Force, to provide forecasts and warnings of solar and geomagnetic activity to users in government, industry, the private sector, and the public. It also conducts research to understand the space environment. The SEC is located in Boulder, Colorado.

NOAA also plays a lead role for the United States in international space cooperation, especially with regard to space-based Earth observation. In July 2003, NOAA organized the first-ever Earth Observation Summit. The result was agreement on creating a Global Earth Observation System of Systems (GEOSS). On February 16, 2005, 55 countries endorsed this plan for comprehensive, coordinated, and sustained observation of the Earth. GEOSS will allow scientists and policy-makers in many different countries to design, implement, and operate integrated, compatible Earth-observation systems. It will link existing satellites, buoys, weather stations, and other observing instruments. The Strategic Plan for the US Integrated Earth Observation System (IEOS) will become the US component of the GEOSS. The Interagency Working Group on Earth Observations is co-chaired by the White House OSTP, NASA, and NOAA and reports to the NSTC's Committee on Environment and Natural Resources (CENR).

The above summary of NOAA activities highlights the multi-use nature of civilian space-based Earth observation systems and their vital contribution to US national security. Moreover, the military, intelligence, and commercial space sectors benefit significantly from the basic space services provided by NOAA.

US Geological Survey

USGS is an agency of the DoI. Its headquarters are located in Reston, Virginia. It is currently responsible for the operation of the Landsat-5 and Landsat-7 satellites. As of late 2008, Landsat-5 was likely to fail soon, after 24 years of operations; Landsat-7 was launched in 1999. NASA developed and launched these spacecraft, and the USGS operates the satellites and manages the data the satellites provide. Landsat spacecraft operations are performed at the GSFC by contractors under USGS supervision. The USGS uses Landsat data sales to partially fund operations but neither satellite's sensors are currently fully operational, and as the quality of data has degraded, sales have fallen. There are plans to launch a Landsat 8 satellite, but that would happen only after a data gap of several years. The United States has as a policy objective of continuity in Landsat-type data, and a Landsat Data Continuity Mission is contemplated to bridge this gap.[19]

Office of Space and Advanced Technology

The Office of Space and Advanced Technology (SAT) handles international space issues for the DoS. Among its goals are: to ensure that US space policies support US foreign policy objectives; to ensure that US international initiatives and political commitments on space are science-based, protect national security, advance economic interests, and foster environmental protection; and to enhance US space leadership and the competitiveness of the US aerospace industry. The SAT office has primary responsibility for US representation on the United Nations Committee on the Peaceful Uses of Outer Space, where a wide range of civil space issues are discussed among nations. In the 1960s and 1970s this committee developed the Outer Space Treaty and three related UN conventions which serve as the bedrock of international space law. The SAT office also represents the DoS in interagency deliberations on civil space policy issues, maintains the official US registry of objects launched into outer space, reviews export license requests for space technology, and provides support to NASA for a network of overseas emergency landing sites for the space shuttle.

Office of Space Commercialization

The Office of Space Commercialization is the principal unit for the coordination of space-related issues, programs, and initiatives within the DoC. It is housed within NOAA/NESDIS, but performs its functions for the DoC as a whole. The goal of the office is to foster an economic and policy environment that ensures the growth and international competitiveness of the US commercial space industry. The office conducts activities in three primary areas: policy development, market analysis, and outreach/education. In fulfilling these roles and functions, the Office of Space Commercialization focuses its efforts on a select group of commercial space industry sectors, including satellite navigation, satellite imaging, space transportation, and entrepreneurial space business. The office also participates in broad discussions of national space policy and other space-related issues.[20]

Regulatory agencies

The Office of the Associate Administrator for Commercial Space Transportation

The Office of the Associate Administrator for Commercial Space Transportation (AST) is part of the Federal Aviation Administration, which in turn is part of the DoT. Its mission is to ensure protection of the public, property, and the national security and foreign policy interests of the United States during a commercial launch or reentry activity and to encourage, facilitate, and promote US commercial space transportation. Established in 1984 as the Office of Commercial Space Transportation (OCST) reporting to the Secretary of Transportation, it was transferred to the FAA in November 1995.

Under its authorizing legislation, AST is given the responsibility to:

- regulate the commercial space transportation industry to ensure compliance with international obligations of the United States and to protect the public health and safety, safety of property, and national security and foreign policy interests of the United States;
- encourage, facilitate, and promote commercial space launches and reentries by the private sector;
- recommend appropriate changes in Federal statutes, treaties, regulations, policies, plans, and procedures; and
- facilitate the strengthening and expansion of the US space transportation infrastructure.

In fulfilling its responsibilities, AST issues launch licenses for commercial launches of orbital rockets and sub-orbital sounding rockets. The first US licensed launch was a sub-orbital launch of a Starfire vehicle on March 29, 1989. Since then, AST (including its predecessor, OCST) has licensed almost 200. AST also licenses the operations of non-Federal launch sites, or "spaceports." AST is also involved in setting the government regulatory framework for private human spaceflight efforts – "space tourism."[21]

Federal Communications Commission

The Federal Communications Commission (FCC) is an independent US government agency, directly responsible to Congress. The FCC was established by the Communications Act of 1934 and is charged with regulating interstate and international communications by radio, television, wire, satellite, and cable. Within the FCC, the primary organization responsible for space-related issues is the Satellite Division of the International Bureau. The primary mission of the Satellite Division is to serve US consumers by promoting a competitive and innovative domestic and global telecommunications marketplace. The division strives to achieve this goal by: (1) authorizing as many satellite systems as possible and as quickly as possible to facilitate deployment of satellite services; (2) minimizing regulation and maximizing flexibility for satellite telecommunications providers to meet customer needs; and (3) fostering efficient use of the radio frequency spectrum and orbital resources. The division also provides expertise about the commercial satellite industry in the domestic spectrum management process and advocates US satellite communication interests in international coordination and negotiations.[22]

DoD interactions with civilian space agencies

In addition to the contentious NASA–Air Force relationship with respect to the space shuttle, the DoD, usually through the US Air Force, has on other occasions worked closely with NASA and other civilian agencies on space matters. While NASA, during the 1970s and early 1980s, allocated only limited funding to advanced space transportation technology, the DoD did support a fair amount of such research and technology development related to advanced-technology crew-carrying systems. By the early 1980s, these efforts were focused on a vehicle that used air-breathing engines to accelerate to hypersonic or perhaps even orbital velocity. The Air Force program was focused on a Trans Atmospheric Vehicle (TAV), while a separate, highly classified, Advanced Research Projects Agency (ARPA) study was called Copper Canyon. In late 1985, all DoD research and development activity on hypersonic flight was consolidated into a program that became known as the National Aerospace Plane (NASP); NASA joined the DoD as a minority funder and co-manager of the NASP effort. This program was given presidential endorsement in the 1986 State of the Union Address. In his address, President Ronald Reagan spoke of an "Orient Express" that would, "by the end of the decade," be able to "take off from Dulles Airport [near Washington, DC], accelerate up to 25 times the speed of sound attaining low Earth orbit, or fly to Tokyo within two hours."[23]

DoD–NASA work on NASP continued through the late 1980s, with the DoD bearing some 80 percent of its costs. However, the NASP program struggled to achieve its technological and schedule goals. A 1988 Defense Science Board report concluded that the program's advocates had been overly optimistic in their initial promise of an early flight demonstration. A year later, the Air Force withdrew funding from the program, and the White House, in 1989, approved a stretch-out of the program with NASA taking over full management control with a flight demonstration of the X-30 test vehicle to come only after relevant technologies had been developed. In the face of competing budget priorities and slow technological progress, the NASP program was canceled in 1992, after $1.7 billion had been spent on it. At that point, the cost of a full X-30 flight-test program was estimated at $17 billion, with another $10–20 billion to develop an operational vehicle. No flight demonstration was attempted, but the program left a technological legacy for future advanced space transportation efforts.

In 1994 a Presidential Decision Directive established the National Polar-Orbiting Operational Environmental Satellite System (NPOESS). At the time, the US operated completely separate civilian and military polar-orbiting environmental satellite systems to collect meteorological, oceanographic, and space environmental data. These are POES and the Defense Meteorological

Satellite Program (DMSP), respectively. NPOESS will reduce this duplication of effort by establishing a single satellite system to replace both DMSP and POES. The NPOESS program is managed by an Integrated Program Office consisting of representatives from the DoC, DoD, and NASA.[24]

In 1998 the US Air Force transferred control of its DMSP satellites to the DoC as an interim step toward the merger of the two systems. DMSP and POES operations were consolidated at the NOAA Satellite Operation Facility in Suitland, Maryland. All DMSP spacecraft are operated by NESDIS, with coordination and oversight provided by the US Air Force Space and Missile Center. In addition, a fully capable DMSP backup satellite operations facility is operated by the US Air Force Reserve at Schriever Air Force Base, Colorado.

The NPOESS program has been troubled with large cost overruns and schedule delays. The first launch of an NPOESS satellite is not expected until 2013. Until then, the converged POES and DMSP programs will have to continue to provide the environmental data needed by both civilian and military users.

The NOAA–DoD relationship exemplified by DMSP demonstrates the link between the civilian space sector and the military space sector and again highlights the multi-use nature of space systems and technology.

Conclusion

With his January 14, 2004 speech announcing the Vision for Space Exploration, President George W. Bush set out a direction for NASA, the central US civilian space agency, that could persist for decades. The President indicated that "the fundamental goal of this vision is to advance US scientific, security, and economic interests through a robust space exploration program."

With this policy declaration, the United States hopes to rebuild the close link between its leading civilian space activity, human spaceflight, and national security interests. That link had become attenuated in the more than 30 years since the end of the Apollo program, with the failed partnership in the operation of the space shuttle and DoD opposition to the approval of a space station program.

The connection between space exploration and national security is not associated primarily with warfighting capability; it is much more linked, as it was at the start of the Apollo program, to using space achievements as an instrument of national leadership. NASA Administrator Michael D. Griffin clearly articulated this rationale in an April 2006 speech:

> The most enlightened, yet least discussed, aspect of national security involves being the kind of nation and, doing the kinds of things, that inspire others to want to cooperate as allies and partners rather than to be adversaries. And in my opinion, this is NASA's greatest contribution

to our nation's future in the world. At NASA, we beat swords into plow-shares to fulfill one of the oldest, strongest, and most persistent dreams of mankind: to know and experience what lies beyond the horizon. We have reached the point where there are no more horizons on Earth, and people everywhere know it. We see, repeatedly, that as nations and societies attain the technical capability to attempt spaceflight, first robotic and then human, they do so. And they will continue to do so. They don't go because we did, and they won't stop if we stop. They go because that is what people do, when they can.

Today, and yet not for much longer, America's ability to lead a robust program of human and robotic exploration sets us above and apart from all others. It offers the perfect venue for leadership in an alliance of great nations, and provides the perfect opportunity to bind others to us as partners in the pursuit of common dreams. And if we are a nation joined with others in pursuit of such goals, all will be less likely to pursue conflict in other arenas.

[...]

Imagine if you will a world of some future time – whether it be 2020 or 2040 or whenever – when some other nations or alliances are capable of reaching and exploring the Moon, or voyaging to Mars, and the United States cannot and does not. Is it even conceivable that in such a world America would still be regarded as a leader among nations, never mind the leader? And if not, what might be the consequences of such a shift in thought upon the global balance of economic and strategic power? Are we willing to accept those consequences? In the end, these are the considerations at stake when we decide, as Americans, upon the goals we set for, and the resources we allocate to, our civil space program.[25]

There is little doubt that the US civilian space program has served as an important instrument of US national power, both in terms of the concrete capabilities it has provided the nation and in terms of its contributions to maintaining US global leadership. It is very unlikely that a future President would commit the country to a highly mobilized effort along the lines of Project Apollo; that was perhaps a unique undertaking. But if the citizens of the United States accept as part of the country's portfolio of national activities "a sustained and affordable program of human and robotic exploration," the civilian space program can continue to be an important element of US global strength.

Notes

1 The October 4, 1957 launch of *Sputnik 1* had catalyzed action in the United States to create a new civilian space agency and give it the assignment of carrying out a variety of programs, but President Eisenhower did not believe that it was in the US interest to try to match the Soviet Union in highly visible space achievements. John F. Kennedy had a different view.

2 John F. Kennedy, "Memorandum for the Vice-President," April 20, 1961, in John M. Logsdon (ed.), *Exploring the Unknown: Selected Documents in the History of the U.S. Space Program*, Volume I, Organizing for Exploration, NASA SP-4407(Washington: Government Printing Office, 1995), p. 423.

3 James E. Webb, NASA Administrator, and Robert S. McNamara, Secretary of Defense, memorandum to the Vice President with attached "Recommendations for Our National Space Program: Changes, Policies, Goals," May 8, 1961 in Logsdon, *Exploring the Unknown*, pp. 444, 446.

4 John F. Kennedy, "Urgent National Needs," May 25, 1961, in Logsdon, *Exploring the Unknown*, p. 453.

5 Report of the Columbia Accident Investigation Board, Volume I, August 2003, p. 209.

6 Space Task Group, "The Post-Apollo Space Program: Directions for the Future," September 1969, in Logsdon, *Exploring the Unknown*, p. 524.

7 Richard M. Nixon, "Statement about the Future of the United States Space Program," March 7, 1970, in *Public Papers of the Presidents of the United States: Richard Nixon, 1970* (Washington, DC: GPO, 1971), p. 251.

8 Memorandum for the President from Caspar W. Weinberger, Deputy Director, Office of Management and Budget, via George M. Schultz, "Future of NASA," August 12, 1971 in Logsdon, *Exploring the Unknown*, p. 547.

9 Report of the Columbia Accident Investigation Board, p. 21.

10 The White House, National Security Decision Directive 42, "National Space Policy," July 4, 1982, in Logsdon, *Exploring the Unknown*, p. 591.

11 The most recent statement of National Space Policy was released on October 6, 2006, available at: www.ostp.gov/html/ US%20National%20Space%20Policy.pdf. Other policy statements on specific areas of space activity, such as remote sensing, positioning, navigation, and timing, and space transportation can also be found at www.ostp.gov.

12 The most recent NASA budget can be found by inserting the appropriate fiscal year in the following URL: www.whitehouse.gov/omb/budget/fy2007/nasa.html.

13 Report of the Columbia Accident Investigation Board, chapter 9.

14 The President's January 14, 2004 speech and associated policy statement is available at: www.whitehouse.gov/news/releases/2004/01/20040114–3.html (accessed October 27, 2006).

15 For information on the current status of NASA's exploration efforts, see www.exploration.nasa.gov.

16 More information about NASA and its activities can be found at www.nasa.gov.

17 Information about the various organizational elements of NASA can be found at www.nasa.gov/centers/hq/organization/index.html.

18 More information on NESDIS is available at www.nesdis.noaa.gov.

19 More information about the space activities of the USGS can be found at http://landsat.usgs.gov.

20 More information about the Office of Space Commercialization can be found at www.nesdis.noaa.gov/space.

21 More information about the Office of Commercial Space Transportation can be found at http://ast.faa.gov.

22 More information on the Satellite Division of the FCC can be found at www.fcc.gov/ib/sd.

23 www.c-span.org/executive/transcript.asp?cat=current_event&code=bush_admin &year=1986 (accessed October 26, 2006).

24 More information on NPOESS can be found at www.ipo.noaa.gov.

25 www.nasa.gov/pdf/146291main_NationalSpaceSymposium_new.pdf (accessed October 27, 2006).

11 Russia and China

Strategic choices in space

James Clay Moltz

Introduction

By any measure – launches, human presence in orbit, or military systems – the United States and the Soviet Union dominated what William E. Burrows has called the "first" space age.[1] Russia remains the second most capable space power, next to the United States, and still leads in a number of categories. While its space program suffered from considerable decline during the 1990s, it has made significant steps toward recovery, thanks to a combination of newly found economic resources (due to recently high world oil prices) and a political leadership (led by former President Vladimir Putin) interested in restoring the country's space prominence, both to bolster national pride and to shore up Russia's military capabilities.

While currently well behind the United States, China is emerging for the first time as a significant space power. Unique among actors in space today, China is committed – both politically and economically – to a major expansion of its program of human spaceflight, following on the success of its manned missions in 2003, 2005, and 2008. Beijing is also moving forward with commercial, scientific, and military capabilities as well. Most importantly, China's expanding space infrastructure and its growing cadre of young space engineers portend the ability to continue this progress well into the future. These trends raise a variety of possible challenges to US space interests.

This chapter analyzes Russian and Chinese space activities and their implications for the United States. Given its longer history, the chapter begins with the Russian space program, its past and current goals, and its future trajectory, including an examination of its military space policy and possible US concerns. It then considers the same issues in regard to China. A key theme is tracking the balance of interests between civilian and military programs. In both countries, military aims are currently taking a back seat to higher priority civilian goals in space. But both have the potential to develop and operate space weapons, if their priorities change in the future. How the United States chooses to respond to these two programs, and the subsequent reactions of each of these two states, will have a significant effect on the future character of international relations in space. For this reason, the chapter concludes by laying out a range of possible US policy responses and discussing their costs and benefits.

The Russian space program: a brief history

Although serious research on space questions began in Russia in the late 1800s in the works of such leading thinkers as Konstantin Tsiolkovskiy, the roots of what became the Soviet space program evolved out of a number of disparate rocket research teams formed in the 1920s by groups of engineers, inventors, and rocket enthusiasts. The Soviet government consolidated these scattered organizations in the 1930s out of a desire to tap the military potential of these technologies and a fear of trends toward war in Europe. During the Stalinist purges of the mid-to-late 1930s, many leading missile designers and engineers suffered imprisonment or death, as the government pursued so-called "enemies of the state" in political witchhunts and trumped-up show trials. Following the invasion of Nazi Germany in June 1941, however, the Soviet state began to retrieve many of those who had survived the purges from distant Gulag labor camps to assist in the development of near-term applications of missile technologies for the war effort. Although still under arrest, these scientists produced wing-borne rockets to assist aircraft in short-runway takeoffs, missile artillery units (the so-called "Katyusha" rockets), and other military applications.

Following the war, the Soviet Union brought many of these experts – some still serving prison sentences – to Germany to supervise the removal of technology and experts from Nazi factories and missile facilities, including the V-2 development complex at Peenemunde. Using German designs and their own expertise, Soviet technicians constructed their first long-range ballistic missiles and began working toward the development of extended-range systems for possible use against US forces in Western Europe and, eventually, the United States itself. The Soviet Union's great interest in missile technology stemmed not only from a realization of the emerging role of missile technology in warfare, but also from the Soviet strategic disadvantage in lacking bomber bases near US soil to counter American air bases in Western Europe and Turkey. Given the Soviets' intention of mounting their then-heavy nuclear weapons on missiles, the requirement for large, heavy-lift rockets became rooted in Soviet requirements by 1949. The first Soviet launch of the liquid-fueled R-7 (US designated SS-6) intermediate-range ballistic missile in August 1957 gave Moscow not only a possible – though rather awkward – nuclear delivery system, but also an effective space launch vehicle.

With this early advantage in launch capacity, the Soviet Union dominated the early years of spaceflight, while the United States struggled to catch up (despite its significant lead in almost all other space-related military technologies, particularly photo-reconnaissance). Following the Soviet Union's launch of the world's first orbital satellite (*Sputnik 1*) in October 1957 and again after the world's first manned flight in April 1961 (by cosmonaut Yuri Gagarin), the United States found its global leadership challenged and its international technological reputation sorely damaged. Soviet First Secretary Nikita Khrushchev's boasted that Moscow could now mass-produce intercontinental ballistic missiles "like sausages," causing Americans to fear the possible military implications of early Soviet space dominance. Responding to US nuclear tests above the atmosphere in 1958 and 1962, the Soviets reacted by conducting four nuclear tests in space in 1961 and 1962, first to study the effects of electro-magnetic pulse radiation and second to determine the ability of nuclear weapons to serve in an antiballistic missile (ABM) role. Fortunately, Moscow agreed with the United States to the Partial Test Ban Treaty (PTBT) in 1963 and halted its space nuclear tests.

From 1959 to 1971, the Soviet Union accomplished a number of space milestones before the United States: the first object to the Moon, the first two-man spaceflight, the first woman in space, the first space walk, and the first space station. The United States took up this challenge – particularly under President Kennedy and President Johnson – and eventually moved into a leadership position over the Soviet Union in various civilian, commercial, and military space technologies, capped by the high-profile US Moon landing in 1969. Nevertheless, during the 1970s, Soviet co-orbital anti-satellite weapons and long-duration manned missions continued to exceed US capabilities, and the country expanded its space communications, early warning, and military reconnaissance networks, although lagging behind US accomplishments in these latter areas. It operated three main launch sites during the Cold War: Baikonur in Soviet Central Asia (for manned, scientific, and geostationary orbital missions), Kapustin Yar in southern Russia (mostly for missile tests), and Plesetsk in northern Russia (for military payloads).

Organizationally, the Cold War Soviet space program – a conglomerate of scientific design and production facilities hidden under the innocuously named Ministry of General Machine Building – benefited from a favored status among other sectors within the Soviet state planning system. It enjoyed top-level political support and received ready access to funding, material resources, and highly qualified personnel. Even in the early 1980s, when state support began to wane, the construction of the *Mir* space station still yielded new long-duration space firsts, reaffirming the Soviet Union's supremacy in human spaceflight. As the United States struggled to return to space after the 1986 *Challenger* disaster, the Soviets continued a steady stream of manned and unmanned launches. In 1988, the Soviets successfully tested an unmanned space shuttle (*Buran*), and seemed poised to step up their space competition with the United States. However, the serious financial problems that plagued the Soviet state in the later part of President Mikhail Gorbachev's rule in the late 1980s began to chip away at these impressive programs, as funds for maintenance, operations, and new spacecraft dried up. For the next decade, Russian space enterprises would experience a serious decline in the midst of the country's spiraling financial crisis, forcing many facilities to reconstitute themselves on a commercial basis, downsize, or close. Russia also faced the ignominy of being forced to pay for the right of using the main Soviet launch facility at Baikonur, which was now located in the newly independent country of Kazakhstan.

During the 1990s, new demands for self-financing drove previously secret and protected space facilities to undertake a desperate search for foreign partners in an effort to maintain their workforces, research activities, and production lines. Aided by political reforms under Gorbachev and Russian President Boris Yeltsin, China became one new partner, although its small-scale purchases could not stem the industry's decline. Unfortunately, in the face of serious financial pressures, some enterprises made deals with states of proliferation concern (such as Iraq, Syria, and India), given Soviet-era contacts and the weakness of export controls during this transitional period. The collapse of the ruble in August of 1998 and the subsequent run on the Russian stock market and banks marked a new low. The formerly proud Russian space program now became the object of international ridicule, as its orbiting *Mir* space station sat empty for lack of funding and dangerous gaps emerged in its network of early-warning satellites.

The rise of a new Russian space program

As domestic demand evaporated, however, some of Russia's leading space and missile enterprises (Khrunichev, Energomash, and others) broke out of Cold War secrecy restrictions and began to redirect their main productive focus from hardware for the Russian military to commercial products for Western companies (Boeing, Lockheed Martin, and others). These new ventures included the US–Ukrainian–Russian–Norwegian Sea Launch consortium, the US–Russian International Launch Services (which markets the Russian Proton booster and its engines), the Russo-German Eurorockot small booster, and various cooperative projects with NASA and the European Space Agency. These new outlets – and efforts by the US government in the Department of Defense (DoD) Cooperative Threat Reduction program – helped keep Russian missile specialists from emigrating to countries of proliferation concern, sustained the bulk of the Russian space infrastructure, and allowed Russia to continue as a leading participant in international human spaceflight, space science, and space-launch services. In addition, these new contacts and the commercial processes they introduced helped change the mindset within the Russian space industry from Soviet-style planning to more market-oriented thinking. Private funding for cooperative ventures with Western companies, foreign government contracts, and space tourism now constitute a significant portion of the Russian space industry's overall budget.

As the Russian government has returned to solvency since 2002, it too has gradually increased funding for various missions – ranging from human spaceflight to military satellites and even to a select number of scientific projects (although usually conducted in cooperation with other countries). Although the new Russian Space Agency (Roskosmos) is much smaller than its Soviet predecessor, its enterprises no longer face imminent bankruptcy as they did in the 1990s. Thus, Russia will remain a key player in a variety of areas of space activity for decades to come. Russia's ability to step in and replace planned US shuttle missions after the 2003 *Columbia* disaster showed that it continues to have unique assets and experience not easily replicated by other countries.

In terms of US security interests, a sign of some positive change in the Russian space industry has come in the form of gradually expanding compliance with one of the benchmark indicators raised by critics of US–Russian space cooperation, the Missile Technology Control Regime (MTCR). Here there have been clear linkages observable between the participation of Russian enterprises in the International Space Station (ISS) and US space launch initiatives and improved nonproliferation compliance by these entities. However, those enterprises that have not been able to benefit from Western cooperation have been more likely to engage in proliferation-sensitive sales in an effort to survive. Over-supply from Soviet-era production helped fuel these exports, particularly in the 1990s, although they have still have not ceased entirely.

Moscow's final decision in 2001 to shut down and de-orbit the *Mir* station marked the symbolic end of the highly competitive and nationalistic Soviet-era space program. Given Russia's major financial and political investment in the ISS and many joint ventures with US and other Western companies in the space-launch field, it is now unlikely that Moscow will seek to establish itself as an independent rival to the United States in human spaceflight. Russia has also ceased prior efforts to re-acquire the territory of the Baikonur facility in neighboring Kazakhstan. Instead, in 2004 Moscow signed a long-term lease agreement with the Kazakhstani government to extend cooperation at the facility through 2050, with Russia paying an annual fee of $115 million. However, Moscow continues to ruminate about building a major spaceport in the Russian Far East (at Svobodniy) at some future point to replace it.

After 16 years of struggle, the Russian space industry has reconstituted itself as a leading edge of the country's twenty-first-century economy. Even though Russia's "guest cosmonaut" program met with initial US opposition, space tourism and the private marketing of space ventures may well be the wave of the future in space. The new international direction of Russia's space orientation also provides ready US access to formerly closed and often secret space enterprises. It is important to note, however, that not everyone in Moscow agrees with current Russian space policy, whose new commercial orientation entails high levels of cooperation and openness. Some Russian conservatives still complain about these "give-aways" of Russia's technological jewels and the negative strategic impact of this space "brain drain." Others point out that many Russian space facilities would have closed altogether were it not for these ties and new sources of income.

As Russia looks toward the next decade in space, it has again embraced optimistic scenarios for its future capabilities. Specific goals of Roskomos' current ten-year plan include the development of two new launch vehicles (the flexible Angara and the heavy-launch *Soyuz-2*), deployment of a six-person space ferry (the Clipper, which Russia hopes to build in cooperation with the European Space Agency), and a deep space robotic mission to the Mars moon, Phobos.[2] The Russian government has pledged to provide an estimated $1 billion each year toward the Roskomos budget,[3] with another $500 million to $1 billion coming from foreign contracts. (Although these figures seem low, Russian labor costs in 2006 were only about 25 percent of those in the United States.) Roskomos Director Anatoly Perminov explains, "It's obvious that the scale of international cooperation in space will only continue to grow."[4] Perminov notes, for example, that the Clipper – unlike the NASA-funded US space shuttle – will be paid for on a commercial basis, drawing on international partners. Despite some doubts in Moscow about high-level US enthusiasm for the ISS, Russia's plans envisage continued close cooperation with NASA and other space station partners, particularly at the operational level.

These points highlight how much the former competitive ethos of the Russian space program has changed. From what was once a coddled, heavily manipulated, and state-funded tool, like the Soviet Olympic program, the civilian space program emerged by the late 1990s as a kind of international space "Wal-Mart," which provided readily available, highly functional, and generally less expensive services than any other supplier. It was also willing to change its product line to accommodate customer tastes. Today, the Russian space program is gradually resuming its own direction and can be expected to remain active, particularly in the area of human spaceflight, which remains a high priority for Moscow.

Military and US space issues

During the Cold War, the Russian space program represented a central concern to US security interests. Its highly militarized character and well-funded research and development infrastructure meant that the United States could expect a continued quantitative and qualitative rival, as long as the Soviet economy could sustain it.

The Soviet Union developed three main types of devoted space weapons during the Cold War: ground-based, nuclear-tipped ABM systems (eventually deployed around Moscow – the so-called Galosh system), co-orbital ASATs, and ground-based lasers. After further testing of nuclear weapons in space was banned by the 1963 PTBT, Moscow shifted its attention to the conventionally armed ASAT program, which conducted some 20 tests from 1968 to 1982, demonstrating a capability to intercept non-evasive satellites in low-Earth orbit. Like the United States, the Soviet Union is believed to have "illuminated" satellites with lasers in tests during the Cold War. However, its actual capability to damage hardware remains uncertain, and the system has not been used or maintained since the Soviet break-up. Thus, Russia's readiness in most of these areas today is undoubtedly poor. Remaining space weapons capabilities – short of high-risk nuclear explosions and releases of debris – are likely residual only.

Russia's military space activities, which had to rely entirely on the state budget, have fared less well than civilian and commercial projects during the period of transition in the 1990s. Military launches dropped from 31 in 1992 to a low of four in 1999. As Brian Harvey noted, "it is remarkable that Russia has been able to maintain a military space programme at all."[5] Nevertheless, Soviet-era weapons programs could be reconstituted should Moscow again feel threatened in space, particularly if Russia's financial picture continues to improve. President Dmitry Medvedev announced vague plans in September 2008 for development of a Russian "aerospace defence" system in the wake of US–Russian disagreements over US missile defense sites planned for Poland and the Czech Republic.[6] What exactly Russia has in mind and whether it will actually proceed with such a program remains to be seen.

Key Russian military space assets developed during the Cold War include space communications, reconnaissance, tracking, and early warning satellites. In the latter area, as noted above, serious problems arose in the 1990s. The Soviet space-based, early-warning system began in the late 1970s, with a small constellation of Cosmos-series satellites in highly elliptical orbits, focused mostly on US ICBM sites. Maintaining the network required several launches per year to ensure proper coverage. This network was gradually expanded to include more sophisticated satellites in geostationary orbit. However, by the mid-1990s, gaps of sometimes as much as eight months at a time appeared, as Russia struggled to fund the new Oko and Prognoz satellites needed to keep the system operational. This deficit put Moscow's ability to deal reliably with incoming data on possible missile attacks at risk. The problem clearly heightened the chances of misinterpretation and even accidental nuclear war during this time. Fortunately, that network has now been reconstituted, and Russia has more recently begun to upgrade its electro-optical space-monitoring system in cooperation with Tajikistan.

In the early 1980s, the Soviet Union launched its competitor to the US Global Positioning System (GPS), called the Global National Satellite System (GLONASS). The system lacked the precision of GPS, but functioned adequately for its main initial consumer, Soviet ships and aircraft needing positions accurate to only about 10 meters (in an encrypted military system). The system also had a civilian counterpart with about half the accuracy. But serious gaps emerged within the network during the 1990s. These problems have reportedly been fixed, and the system is expected to achieve full operational capability soon. In addition, Russia has pledged to allow commercial use of its more precise military system in the future. As part of Russia's ten-year plan for space, a total of 24 GLONASS satellites are scheduled to be deployed by 2010. Internationally, China and India have recently signed on as users of the new system.

In organizational terms, the Russian military has begun to reorganize itself for space activity with a new Aerospace Defense Command. Russia is also undertaking a major expansion of its once-secret military launch facility at Plesetsk, upgrading its capabilities and planning for an increased number of yearly launches. Russia eventually plans to conduct all military launches from this facility, thus consolidating defensive capabilities on Russian soil. Although currency issues and hidden aspects of the military budget make exact figures difficult to calculate, the Russian military space budget likely amounted to about $1 billion in 2007.[7] It is probably somewhat higher today.

In terms of space weapons, Moscow has taken a consistent diplomatic stance since the early 1980s against the weaponization of space (as opposed to its existing militarization). While originally believed to be an effort to prevent the US Strategic Defense Initiative, the policy has been remarkably consistent since that time, and Russia has shown few signs of seeking to redeploy its own offensive systems. In the late 1990s, Russia began to join with China and many other countries at the United Nations in efforts to block the future deployment of space weapons. Russia has repeatedly backed UN resolutions calling for the Prevention of an Arms Race in Outer Space (PAROS) and supported calls for discussions at the Conference on Disarmament (CD) in Geneva on space security and arms control. More specifically, Russia has co-authored several draft space treaties that have been floated at the CD since 2001. Finally, then-President Vladimir Putin announced at the United Nations in September 2003 that Russia had adopted a "no-first-deployment" policy on offensive space weapons, in effect, challenging others (especially the United States) not to place weapons in space.[8]

Part of this new effort to promote space cooperation seems to stem from "lessons" drawn from its own experience during the Cold War. As Russian Ambassador to the CD, Leonid Skotnikov argued in June 2005, "Hopes to achieve domination in space with the use of force are illusory, and ultimately such ambitions would weaken rather than strengthen the security of all States without exception."[9] He concluded his speech by renewing Russian calls, made at a number of international venues since the late 1990s, for a treaty or other international legal arrangement that would "reliably block attempts to place weapons of any type in outer space or to use or threaten to use force against space objects."[10] Notably, in contrast to more restricted definitions used by some other states for "space weapons," this Russian concept could allow jamming of satellite signals, as long as it did not involve "force" or permanent damage to the spacecraft. Such definitions could expand or be wained altogether, however, depending on Russian reactions to possible future US deployments in space.

The Chinese space program: a brief history

China has emerged, since its first human spaceflight mission in October 2003 on its Shenzhou V, as the state with the most dynamic trajectory of any country in space. From a weak and under-funded effort that suffered tremendous hardships during the 1960s and 1970s, the Chinese space program has recently emerged as a key focus of government efforts to accelerate national development. Beijing has used the space program to make a statement domestically, regionally, and internationally about its technological prowess, creating fears of a second "space race" among some Western military analysts. To date, Beijing's main focus has been on its human spaceflight program, although its test of an ASAT weapon in January 2007 marked a sharp departure from this pattern and sets an ominous tone for the future. Currently, China's annual launch rate remains a fraction of those of the United States and Russia, rising recently to an all-time high of six launches, but it is expected to grow in the coming years in both the civilian and military sectors.

China's current space capabilities are comparable to those of the US or Soviet programs in the late 1960s or early 1970s. However, China is also benefiting from Russian help, meaning that China has the capacity as a late-coming space power to accelerate its development of space technologies more rapidly than the original countries that had to develop space systems from scratch. For these reasons, and because of China's growing demographic advantages – including an expanding pool of young, talented, and motivated space scientists and engineers – the Chinese space program merits US attention. Mismanaging US–Chinese space relations, either by over- or under-estimating Chinese capabilities, could cause problems for the United States.

The original roots of the Chinese space program date to the 1950s, when Chief of the General Staff Nie Rongzhen began coaxing leading Chinese-born scientists who had been educated in the West and were living abroad to return home to help their mother country. One of these specialists, a PhD missile engineer named Tsien Hsue Shen, returned to China after being deported by the United States in 1955 for alleged communist ties. Tsien's eventual assumption of a leading role in the space program marked a key turning point in China's organization for both missile and space activities. Helped by early technical assistance from the Soviet Union, China began expanding its educational and research facilities to train a cadre of specialists, benefiting from General Nie's ability to commandeer top university graduates into the military's Fifth Academy in charge of missiles. However, China's ongoing domestic political turmoil and the Sino-Soviet split in the early 1960s began two decades of difficulties for the country's space efforts, setting them back considerably.

After barely surviving anti-rightist campaigns in the late 1950s (which placed Western-educated scientists under suspicion), recent returnees and their students found their loyalty again questioned during the Cultural Revolution in the mid-1960s, which called for unquestioned "revolutionary" loyalty and a policy of returning power to the "masses." Under the rule of the radical Gang of Four, the country wasted its scientific talent, as young Communist Party fanatics fired leading technical specialists, who were jailed, placed under house arrest, or relocated to dismal "re-education" camps in the countryside. Despite these obstacles, Tsien managed to curry favor with the ruling "Gang of Four" and continued to lead a small group of devoted Chinese researchers, who attempted to follow US and Soviet space accomplishments by sending animals into space in the mid-1960s, beginning work on satellite technologies, and establishing plans for a manned space program. In 1970, the Chinese Air Force began selecting a team of Chinese pilots to begin training for human spaceflight. However, a combination of scientific problems, lack of funding and eventual political opposition by China's leadership caused the program's cancellation in 1972.[11] As the country struggled through yet another set of political changes, the space program experienced several more years of neglect, although progress in missile technology continued.

In early 1975 Chinese leader Mao Zedong approved work toward launching China's first communications satellite, ensuring its support through the mid-1970s. The first successful Chinese orbital satellite flight took place from the Jiuquan test site (located in the high desert of Inner Mongolia) in November 1975 aboard a Long March 2C booster, although it would take scientists several years to iron out reliable launch and recovery procedures. Following Mao's death in the fall of 1976 and the ousting of the Gang of Four, Chinese politics gradually stabilized, and ties with the outside world improved. The eventual emergence of a more pragmatic leadership under Deng Xiaoping brought a more positive attitude toward space activity, which began to be linked to the country's economic and military modernization. China's enduring political hostility with Moscow had not subsided, however, thus preventing Soviet space assistance. Following the normalization of US–Chinese political relations in 1979, some limited contacts began with the United States, with visits by small numbers of aerospace experts, including some younger scientists who are now leading Chinese space officials. Specific cooperation in the space arena included some Chinese work on small experiments for the US space shuttle.

Chinese space scientists, organized under the new Ministry of the Space Industry, continued through a trial-and-error process toward building a reliable space launch vehicle. In 1984, Chinese technology finally succeeded in launching and placing into orbit the country's first communications satellite aboard a Long March 3 rocket. Other types of satellites for remote sensing, meteorology, and military uses followed.

In 1989, the government brought together the major facilities working on space-related technologies, creating the Chinese Aerospace Corporation. To put a civilian face on its human spaceflight program, China established the China National Space Administration (CNSA) in 1993. However, the military-led Commission for Science, Technology, and National Defense (COSTIND) played an important supervisory role over all space activity until its reorganization and renaming within the newly formed Ministry of Industry and Information Technology in 2008. While the exact relationship of the players within this new structure is still unclear, all signs indicate that the military will continue to play a significant role in space activity.

Internationally, with Soviet President Mikhail Gorbachev's rapprochement with China in 1989, China gained a new and valuable partner for space activity: one with tremendous technological experience and, as importantly, the willingness to sell it. The subsequent re-establishment of Russo-Chinese industrial, scientific, and security-related ties under President Boris Yeltsin in the early 1990s opened the door for acquisition of Russian space technology. The Chinese and Russian space programs signed an official cooperative protocol in 1994, leading to subsequent Chinese visits to Moscow for extensive purchases of hardware, training, and consultations, particularly in the manned space field. These exchanges contributed greatly to China's Shenzhou spacecraft, which is based on the Soviet Soyuz module.

As China geared up for the expansion of its space program and the range of missions it would serve, diversifying its launch sites became a high priority. Beginning in 1992, the Chinese expanded the Jiuquan site in northwest China to speed up operations and allow it to handle more complex launch operations. Beginning in the late 1970s, China also began work on the Xi Chang site – in mountainous southwest China near Chengdu – to facilitate the insertion of payloads into geostationary orbit. It now handles both civilian and military payloads. Finally, China established a smaller facility on Hainan Island in the late 1980s for sounding rockets. China's space launch portfolio today represents a diverse set of capabilities. Since 1984, its Long March 1 satellite launchers have incorporated a three-stage design using liquid hydrogen in their final stage. According to Harvey,[12] Chinese human spaceflight missions are typically launched by two-stage Long March 2E boosters, which have a liquid-fueled first stage with four strap-on boosters, plus a liquid second stage. China's three-stage Long March 3B is its most powerful booster and is used for sending payloads into geostationary orbit. By 2005 China had conducted more than 80 space launches.

One of the most important recent developments has been China's establishment of a reliable space-tracking network, a prerequisite to effective manned, commercial, and military programs. While China previously depended on a fleet of ships for such tracking (much as the Soviet Union did), it now has a land station in Namibia and agreements with Sweden and

Italy to assist in tracking from other locations. Across various fields of space activity, China also has cooperative agreements with over 40 countries, including working with Brazil on high-resolution photo-reconnaissance satellites. In 2005 China established the Asia Pacific Space Cooperation Organization (APSCO) with seven countries (Bangladesh, Indonesia, Iran, Mongolia, Pakistan, Peru, and Thailand) to promote cooperation in space science and the application of space technology to economic development tasks. This organization could be used in the future to expand China's international influence in the space field.

In terms of its finances, Chinese sources put the civilian space budget at about $500 million per year in 2006. Figures for the military budget are unknown, but likely total somewhat less. These figures reflect low Chinese costs for labor, which are about 15–20 percent of those in the United States.

Prospects for US–Chinese space cooperation

Compared to the vast scale of US–Chinese commercial trade, bilateral space cooperation between the two countries remains extremely limited. Indeed, the two space programs are even less engaged than the US and Soviet space programs during the Cold War, when there was considerable cooperation in space science and even a few joint manned missions (including the highly publicized *Apollo–Soyuz* docking in 1975). Today, Chinese space activities are viewed in a highly politicized light by some members of the US Congress, which has blocked meaningful cooperation with China, such as Beijing's effort to join the ISS. The policies of the US government continue to be affected by the congressionally mandated *Cox Commission Report* of 1999, which criticized the Loral company's dealings with China in the 1990s, when it is believed that Beijing may have gained valuable military information from an exchange of commercial satellite information. US International Traffic in Arms Regulations (ITAR) now prohibit all cooperation with China in space technology due to fears of technology transfer. Supporters of the current US policy of attempting to "isolate" the Chinese space program also point to such issues as China's human rights record, its copyright infringements, its past history of aiding Pakistan's nuclear program, and its continuing pressure on Taiwan. In addition, some officials and analysts fear that China may emerge as a peer military competitor to the United States in space. Thus, the logic has been "Why give Beijing a possible step up technologically if we are going to end up *competing* with China in space?" On the other hand, supporters of cooperation ask: "are U.S. interests served by the current policy, if useful space technologies are available from other suppliers (in Russia and in Europe) and if refusing to engage the Chinese might encourage them to develop a hostile attitude about U.S. intentions in space?"

Some tentative US overtures concerned with engaging China in some forms of space cooperation are beginning. After a series of initial contacts in 2003, representatives from the CNSA and NASA held their first official discussions in 2004. In addition, the United States provided China with tracking data for space debris on the eve of China's second human space-flight mission, which sent two taikonauts aboard Shenzhou VI in October 2005.[13] In April 2006 Vice Administrator Luo Ge of the Chinese space agency met with NASA chief Michael Griffin during a visit to the United States. In October 2006 Dr. Griffin reciprocated by visiting China for a series of meetings with senior Chinese space officials. Talks are reportedly continuing, but no specific cooperative ventures have been announced.

Notably, US and European views on how to deal with China diverge. NATO allies, by and large, have supported a cooperative strategy – including allowing Chinese participation in the proposed Galileo satellite network – as the best means of ensuring a friendly China in the future.[14] In this context, some US analysts argue that current US policies are outdated (and even counter-productive), given our currently limited ability to learn about China's intentions and capabilities in space. They argue that civilian and manned space cooperation with China would allow the United States to improve its knowledge about China's space program and develop possible US leverage once China begins to see cooperation as beneficial.[15] Other points raised by pro-cooperation analysts in the United States include the possibility that isolating China will foster greater Russo-Chinese space cooperation and stimulate China's development of independent space capabilities (such as a separate space station, instead of joining the ISS).

Yet, perhaps the biggest challenge posed to the United States by China's civilian space program is one of *demographics.* China's core space scientific and engineering cadre is about two decades younger than its counterparts in the United States and Russia, which are now retiring. This dynamic younger generation of Chinese specialists could be a powerful force in developing Chinese military space programs, if it is not instead engaged in channeling its energies into cooperative civilian, scientific, and commercial space efforts. Thus, given China's unique recent success in attracting the "best and the brightest" into the space program, despite more lucrative opportunities in private industry, the United States may find itself with a competitive disadvantage if US–Chinese relations deteriorate and space relations turn into a military competition.

Military space issues and US interests

Recent US policy toward the Chinese military space program has been characterized by mistrust, stiff export controls, and fears about Chinese long-term intentions. Many US military analysts assume that China will eventually develop a major space weapons program, due to what one expert calls "strategic logic."[16] However, other analysts question China's commitment to military space applications and point out that the country's firm commitment to human spaceflight means that fewer funds will be available for military purposes. Not surprisingly, there is still limited information available about the extent of Chinese military space research and development. Chinese military writings discuss possible military operations in space in conceptual terms and conduct a great deal of analysis of US space efforts. Actual reference to Chinese space weapons is virtually non-existent.

Areas of possible concern to the United States include anti-satellite lasers, "parasite" satellites, and jammers, as well as blunter instruments such as debris-causing kinetic weapons (such as the direct-ascent ballistic missile tested in January 2007), and even nuclear blasts. However, while China is capable of using a variety of such systems against the United States, its reasons in doing so remain unclear (short of a major war or threats to its core interests). Indications of the progress of such test programs still remain limited. In the fall of 2006, reports also surfaced indicating at least one instance in which a Chinese ground-based laser had illuminated an orbiting US satellite.[17] Whether this incident was meant as a warning or indicates plans for a dedicated laser weapons program remains unclear.

In terms of military support capabilities, China's current-generation reconnaissance and remote-sensing satellites lack high resolution and cannot yet transmit digital images (relying instead on electro-optical transmissions). China supplements this data with information provided to Chinese ground stations by French, European Union, US (Landsat), and other foreign satellites. China's data relay capacity also remains limited, although there are plans to address this gap going forward. China possesses a small fleet of navigational satellites and plans eventually to orbit a complete constellation to facilitate its development of precision-guided munitions. China has acquired considerable recent experience in building and launching microsatellites and nanosatellites. While the military capabilities of such spacecraft remain unclear, their dual-use potential could provide future low-cost alternatives to large military spacecraft for specific, high-need missions. However, all of these efforts are proceeding in a gradual manner due to China's limited resources, measured goals, and preference for indigenous technology (although China continues to purchase and learn from foreign technological experience in a number of fields). As of 2005, reported gaps in China's military capabilities also included an absence of dedicated electronic intelligence satellites.[18]

As noted above, despite US reticence to engage China in space cooperation, the European Union has readily embraced Beijing. China has pledged some $236 million to participate in Galileo, which will likely allow China to provide some input as to how the system is eventually operated.[19] However, one author notes that future Chinese access to "an encrypted and potentially unjammable navigation service" would be a source of real concern to US military officials.[20]

Despite US concerns about Beijing's military programs in space, China's official policy for over a decade has strongly opposed space weaponization. China has been one of the single most determined leaders within the CD in Geneva, having led, sponsored, or co-sponsored a number of initiatives since 1994. Beijing has pressed repeatedly for international negotiation of a formal treaty to plug loopholes that exist in the current 1967 Outer Space Treaty, which calls for "peaceful uses" of space but does not explicitly ban non-WMD space weapons. Such a treaty, from the Chinese perspective, should require commitments by states "Not to place in orbit around the Earth any objects carrying any kinds of weapons" and "Not to resort to the threat or use of force against any space objects"[21] More recently, China has sought to accommodate US preferences by being willing to consider other possible options. Chinese Ambassador Hu Xiaodi outlined China's overall perspective regarding implications of the weaponization of space in a speech to CD in June 2005, arguing that it would disturb the existing strategic balance, "undermine international security," and damage the current arms control framework for space "thus triggering a new arms race."[22] Finally, Chinese diplomats have noted broader Chinese concerns, arguing that the weaponization of space "risks harming the biosphere of the earth" and that testing of space weapons in low-Earth orbit "will exacerbate the already serious problem of space debris."[23] China's own test of a kinetic-kill interceptor, however, released large amounts of harmful debris, exposing a serious contradiction in current Chinese policy. Overall, China's position seems to indicate a current reluctance to engage in a military space "race" with the United States, but with an implicit threat to meet US plans for space weapons and defenses with a dramatic expansion of its own military space capabilities if need be.

One challenge facing proponents of US cooperation with China in space is that of transparency. China could do more to silence its critics through additional openness regarding its military program, particularly if it wants to help facilitate changes in current US policy. Critics of cooperation make the case that the United States must "assume" Chinese military – and weapons-oriented – intentions and act accordingly. Yet such assumptions can lead to self-fulfilling prophecies, wasteful US spending, and the exacerbation of tensions, if they are not balanced against possibilities that may exist for mutual weapons restraint. Notably, the Chinese have questions about US intentions. Chinese Vice Administrator Luo compared his trips in 1980 (as a young scientist) and 2006 to the United States by observing that America was much more open than China during his first visit and that "Now, it's the other way around."[24]

Given these divisions and US hesitancy to cooperate with China, even in the civilian space sector, it will be a long time before US–Chinese space relations approach the level where the United States and the Soviet Union were during the Cold War. Yet, given current US space hegemony and China's dynamic rise, the future character of US–Chinese space relations will likely to do more to shape the overall nature of international affairs in space than any other single bilateral relationship.

US space policy options: lessons from the Cold War

Policy-makers looking for clues regarding how the United States should face emerging challenges from Russia and China in space may find some useful guidelines regarding the balance of competition and cooperation in history. In the late 1950s and early 1960s, it appeared that space would be an arena bound for direct military confrontation with the Soviet Union. Yet the actual Cold War experience in space after 1963 involved significant mutual restraint, including extensive military uses of space, but also decisions not to deploy space-based weaponry or attack the other side's space assets. Shared fears about the prospects of inadvertent war clearly affected these calculations. The two sides also understood from experience that continued nuclear testing in space and the generation of large amounts of debris in low-Earth orbit would cause problems for other, higher-priority space activities. Mutual respect for each other's satellite reconnaissance and early-warning networks helped ensure that the Cold War remained "cold." These points suggest the need for careful calibration of the risks of deploying twenty-first century space weapons by any of the three powers against possible action–reaction dynamics and implications for existing civilian and military space activities. Still, the indeterminate nature of political relations with Russia and China leave room for considerable speculation regarding future outcomes.

Drawing from the current literature on space policy, we can identify at least four main directions that the United States could take in dealing with Russia and China. Each of them comes with advantages and disadvantages, sometimes with differing implications for the short- versus long-term future. Of course, US space policy will not be made independently of factors in broader US–Russian and US–Chinese relations and will also be affected by how US space initiatives are *interpreted* in Moscow and Beijing. Still, it is worthwhile to lay out the range of possible policy options, if only to provide some general parameters of likely future US debates and considerations regarding these two major space actors.

Reduced cooperation, near-term space weapons deployment

The first possible option for future US strategy toward Russia and China is that proposed by critics of existing policy who argue that current military options are much too constrained and cooperation already too expansive. They believe that the United States has failed to exploit post-Cold War US advantages – including Russian weakness and China's current backwardness – to "seize the high ground" in space while we can. These analysts see current US military space policy as inattentive to eventual Chinese, Russian, and perhaps other foreign threats. As one notes, "we are at risk of relinquishing our military space dominance to competitors."[25] Another analyst from this school argues, using similar reasoning: "We can state with certainty that we will be challenged in space ... simply because it makes military sense to do so."[26] Given these assumptions, this school of thought posits that US cooperation with potential rivals should be strictly limited, that the United States should take the current opportunity to build space defenses, and that it should then seek to build up an insurmountable military advantage. This would include deploying offensive, global-strike weapons, as well as defensive ASAT and missile defense technologies, in space to keep and defend US space dominance. Certain supporters of this perspective also call for a nationalistic commercial space policy, including the dismantling of the Outer Space Treaty in order to free up US companies to claim private property (and related profits) on the Moon.[27]

One weakness in this approach is that history suggests that rivals will inevitably rise against states that attempt to deploy military force in an exclusive manner in any field of human endeavor. Thus, supporters of this strategy must be willing to defend – politically and economically – a long period of struggle, the likelihood of space warfare, and the possible consequences of expanded debris in low-Earth orbit, putting other US assets at risk. Such dynamics could push Russia and China even closer together in coalition against the United States in space, and possibly, depending on perceptions of which side is to blame, European powers as well.

Limited cooperation and preventative space defenses

A second possible strategy would be one based on the main streams of US space policy as practiced by the George W. Bush Administration. This approach would involve some forms of international cooperation (significant cooperation with Russia and perhaps limited space science cooperation with China), in combination with a well-funded research and development strategy aimed at the future testing and deployment of a limited number of space weapons, largely for defensive purposes. This perspective opposes the need to push further for new forms of space arms control, but it does not argue for the abrogation of the Outer Space Treaty. It sees a world where the United States should make the most of its opportunities to exploit military advantages in space. Yet it incorporates an important second track of meaningful civilian and commercial cooperation with selected states. Such a policy leaves open possible non-WMD weapons, as well as the uncertain international military reactions that their deployment might generate.

In considering possible downsides of this option, this approach suffers from a "muddling along" tendency, as it offers only a vague vision of which direction – competition or cooperation – is most desirable in space. On the one hand, it leans toward competition by declining to halt space weaponization or offer new restraints, thus risking a "slippery slope" and copy-cat foreign military programs, thus escalating weapons competition in space. On the other hand, the unexpected flourishing of its cooperative elements (such as with Russia and various US allies) and technical (or cost) problems in the space weapons arena could eventually cause competitive tendencies to decline, thus leading to its possible evolution into the next option.

Moderate cooperation, weapons research only (as a hedge)

A third strategy for responding to Russia and China in the coming decades in space would be one in which the United States and other countries might allow laboratory research on space weapons, but also agree to mutual restraints against their further testing and deployment. Such an approach might allow non-destructive military interference with other states' space assets (such as jamming) in times of war, but put an emphasis on the creation of "rules of the road" governing space behavior and means of reducing mutual space vulnerabilities through "non-offensive" techniques (such as added maneuverability, spares, decoys, and shielding). Such a strategy would likely emphasize commercial and civilian contacts among states and would reach out to other states according to functional rather than political criteria. This cooperation could involve China, if Beijing were willing to provide funding and technologies to contribute to mutual goals of returning humans to the Moon and eventually landing them on Mars. More extensive efforts would be made to prevent arms competition in space and to prevent testing of weapons, particularly in low-Earth orbit. Detailed discussions of what types of harmful behavior to prohibit (such as hostile interference and debris-causing tests) would be needed among major space powers to realize these objectives. Similarly, efforts to develop a common standard for GPS, Galileo, GLONASS, and China's emerging Beidou satellite system would be encouraged.[28]

One weakness of this approach is that the political prerequisites – greater trust between the United States and China – do not yet exist. Achieving it would likely require settlement by Washington and Beijing of a host of issues in the security and economic realms. On the military side, if US intelligence proved poor or export control enforcement weak, a highly nationalist China or Russia might seek to exploit US restraint. Thus, trust would have to be supplemented by greater verification of compliance. On the other hand, excessively aggressive US hedging against possible Chinese or Russian military breakout through US research and possible testing could result in a self-fulfilling military prophecy and renewed space arms competition.

High cooperation, no weapons research, new treaty

Finally, a fourth possible strategy for managing future space relations with Russia and China would seek engagement with both countries and would accept their call for a formal international treaty banning space weapons, including establishment of an international verification system. Such an approach would likely include an active policy of integrating the US civilian space program with those of Russia and China. This vision – some of which can be seen in the concept behind Congressman Dennis Kucinich's proposed Space Preservation Act of 2002 (HR 3616) – would be "transformatory" in its intent, with the goal of truly *internationalizing* space activity and changing space from a realm of competition to one of cooperation. To realize this plan, new forms of intrusive, pre-flight verification might be necessary, requiring considerable organizational efforts to set up a system of cooperative monitoring of space security. While the trust and funding for such a program are not yet available, this concept is not beyond the realm of possibility, if world leaders see a greater global need for enhanced cooperation to solve shared environmental, security, or economic challenges. Of course, achieving such a transition from current, nation-centric space policies would face considerable domestic obstacles in the United States, as well as possibly in China and Russia. Such an outcome would only be feasible after some period of preparation, political rapprochement and steady progress in building new mechanisms for cooperative security. The benefits and cost savings for all countries would be significant, but the United States would again face even greater risks of a possible military breakout by others in space, which would have to be guarded against through extensive monitoring, test limitations, and political-economic means. Much would depend on new forms of military-to-military cooperation on a number of levels. While it is hard to imagine such a shift at present, it is not inconceivable if globalization tendencies continue and expand in the coming years.

Conclusion

While the United States remains the leading actor in space today, it faces challenges to its future supremacy from both Russia and China. A major question is whether future national developments in space will be viewed in zero-sum terms or positive-sum terms. The level of international cooperation will clearly affect these calculations. Due to its extensive experience and reliable launch capabilities, the Russian space program will continue to be a major player in international space policy and a variety of commercial programs. Moscow is also beginning to reconstitute its military functions, although not necessarily in a direction that will be hostile to US interests. Much will depend on broader relations with the United States in such areas as nonproliferation policy, trade, oil/gas development, and human rights. As a space power, Russia has breakout potential against the United States in a variety of military sectors, but such actions by Moscow would cause serious damage to its own commercial space interests and risk stimulating a much more aggressive US policy. China, on the other hand, is a less capable space power currently, although with possibly even greater expansionist potential.

Its direction is somewhat less certain than Moscow's, and rising nationalism and a growing economy in China could stimulate a major space race against the United States, if broader political–military relations between the two sides were to deteriorate.

Fortunately, since the end of the Cold War with Russia and with the general rapprochement in US–Chinese relations since 2001, political prospects for greater cooperation with these two potential US space rivals have been greatly improved. Of the four policy options, the last is arguably the most favorable for all sides, since it reduces the costs of military preparations and increases possible savings from cooperative programs, but it is also the least likely at the present time. Still, if political relations among these three major space powers continue to remain positive and start to improve, it could eventually be realized through the third option, if states can survive the intermediate stage of walking the fine line between a tacit readiness for arms development and the peaceful pursuit of space cooperation. The other two alternatives involve greater or lesser US investments in space weapons, which might defend the United States, but at the risk of stimulating rivals and restricting funds for commercial and scientific space development. Whether weapons scenarios are avoidable, however, remains to be seen. Part of the ultimate outcome, certainly, will also depend on the policies followed in Moscow and, perhaps more importantly, Beijing. As the leader in space today, the United States may still have the ability to *shape* the future outcome toward the option it most desires – that is, if a consensus can be reached in Washington as to which direction best serves long-term US interests in space.

Notes

* The views expressed in this chapter are those of the author alone and do not represent the official policy of the US Navy or the US Department of Defense.
1 See William E. Burrows, *This New Ocean: The Story of the First Space Age* (New York: Random House, 1998).
2 "Russia approves 10-year space plan," *Spacetoday.net*, posted October 26, 2005.
3 "Russia thriving again on the final frontier," *MSNBC.com*, posted September 29, 2005.
4 Roskosmos Director Perminov, as quoted in Aleksey Shcheglov, "Ambitsii Roskosmosa rastut (Roskomos's aims are growing)," Strana.ru website, posted November 11, 2005.
5 Brian Harvey, *Russia in Space: The Failed Frontier* (Chichester, UK: Praxis Publishing, 2001), p. 140.
6 Medvedev, quoted in "Prezident RF stavit zadachu dobit'sya prevoskhodstva Booruzhennikh cil" (Russian Federation President sets goal of obtaining superiority of military forces), *Pravda*, September 26, 2008.
7 This figure is the author's estimate based on the figure of $1 billion for the civilian space budget and an assumption that Russia is now spending proportionally less on military space activity than it did during the Cold War, when the military figure (for 1989, as reported during a period of remarkable transparency) was slightly more than the reported civilian space budget. (On this issue, see Bill Keller, "Soviet Premier Says Cutbacks Could Reach 33% for Military," *New York Times*, June 8, 1989, p. 11.) A supporting calculation can be

made based on the estimated military space share of the total Russian military budget for 2005 of $18 billion (see the Globalsecurity.org website at: www.globalsecurity.org/military/world/russia/mo-budget.htm). Assuming equal proportions to US equivalents (military space budget to total defense budget for the US 2002 request – to control for post-9/11 and Iraq war costs), this yields a ratio of $14 billion/$310 billion (or 5 percent). Multiplying this percentage by the Russian total defense budget yields a military space budget figure of $0.9 billion. For a variety of reasons, the author believes current figures are higher, given the growth in Russian oil wealth and the recent worsening of US–Russian relations.

8 For the text of President Putin's speech to the United Nations on September 25, 2003, see www.un.org/webcast/ga/58/statements/russeng030925.htm.

9 For the text of Ambassador Skotnikov's speech to the Conference on Disarmament on June 9, 2005, see www.cns.miis.edu/research/space/pdf/cdpv984.pdf.

10 Ibid.

11 Ibid., p. 243.

12 Brian Harvey, *China's Space Program: From Conception to Manned Spaceflight* (Chichester, UK: Praxis Publishing, 2004), p. 216.

13 "Chinese experts welcome US offer of warning datum for spacecraft launch," (FBIS document number CPP20051015052021), Xinhua (Beijing), October 15, 2005.

14 On these differences, see Marta Dassu and Roberto Menotti, "How China could divide the West," *Europe's World* (Autumn 2005), available at: www.europesworld.org/EWSettings/Article/tabid/191/ArticleType/articleview/ArticleID/20427/Default.aspx.

15 On these points, see Joan Johnson-Freese, "Scorpions in a bottle: China and the US in space," *The Nonproliferation Review*, Vol. 11, No. 2 (2004): 166–182.

16 William Gouveia, Jr., "An assessment of Anti-satellite capabilities and their strategic implications," *Astropolitics*, Vol. 3, No. 2 (2005): 176.

17 Vago Muradian, "China tried to blind U.S. sats with laser," *Defense News*, September 25, 2006, p. 1.

18 James A. Lewis, "China as a Military Space Competitor," in John M. Logsdon and Audrey M. Schaffer, (eds.), *Perspectives on Space Security* (Washington, DC: Space Policy Institute, George Washington University, December 2005), p. 102.

19 Lt. Col. Scott W. Beidleman, "GPS vs. Galileo: balancing for position in space," *Astropolitics*, Vol. 3, No. 2 (2005): 140.

20 Ibid.

21 See Cheng Jingye, "Treaties as an approach to reducing space vulnerabilities," in James Clay Moltz (ed.), *Future Security in Space: Commercial, Military, and Arms Control Trade-Offs*, Occasional Paper No. 10, CNS, Monterey Institute of International Studies, (July 2002). Available at: www.cns.miis.edu/pubs/opapers/op10/op10.pdf.

22 For the full text of Amb. Hu's remarks on June 9, 2005, see the CNS "Current and future space security" website at: http://cns.miis.edu/research/space/pdf/cdpv988.pdf.

23 Ibid.

24 "China unveils space programme," *Aljazeera.net*, posted on April 4, 2006.

25 Maj. Franz J. Gayl (US Marine Corps, retired), "Time for a military space service," *Proceedings (US Naval Institute)*, Vol. 131, No. 7 (2004): 44.

26 Steven Lambakis, *On the Edge of the Earth: The Future of American Space Power* (Lexington, KY: University Press of Kentucky, 2001), p. 137.

27 On this point, see Everett C. Dolman's *Astropolitik: Classical Geopolitics in the Space Age* (London: Frank Cass, 2002).

28 On this idea, see Lt. Col. Beidleman, "GPS vs. Galileo,": 147.

12 Toward a European military space architecture

Xavier Pasco

Today, space activity is primarily characterized by a dramatic imbalance between spacefaring countries. Considered alone, the United States represents about 80 percent of worldwide public expenditures, while it is responsible for almost 95 percent of military spending on space.[1] This difference between the relative public investments in space of the United States and the rest of the world could bring about other gaps that, if not considered in due time, could generate serious problems concerning the ability of countries to collectively analyze conflicts and conceive military actions in the years to come. Indeed, the latest evolution in space techniques and uses may change the very nature of warfare by introducing new concepts and architectures of systems used by the military forces. But more importantly, it also creates new ways of conceiving and conducting warfighting operations. The real change has to do with the connections between space and global military activity and capabilities. While crises – military or humanitarian – can erupt anytime anywhere, the application of information technology is believed to be a prerequisite for quick, adaptive, and safe responses. Some decision-makers envision an armada of space-, air-, or ground-based sensors as well as advanced command, control, and communication systems to produce the real-time and relevant information needed at any level of the chain of command. In this "information technology" infrastructure, satellite systems obviously play a crucial role.

Faced with this American perspective, European countries wonder about the reality of the foreseen changes. Are we witnessing the birth (in the short-, mid- or long-term?) of substantially new military structures that are characterized as increasingly constituted by and around space applications? Are we becoming vitally dependent on these space assets? If so, do Europeans have to invest heavily in military space?

Choosing space: a strategic decision for Europe

In Europe, decisions concerning relationships with the United States on space issues must be linked with the traditional mistrust about the

US-originated revolution in military affairs (RMA) that has characterized military thinking for more than a decade. Continuing debates about the military relevance of such high-technology oriented thinking, especially for European countries, have not prepared Europe to fully support the development of military space, traditionally viewed as closely associated with RMA. On this side of the Atlantic, the development of space applications has always reflected, above all, a pragmatic approach, conceiving the role of civilian and military space as based on an idea of "space strategic sufficiency."[2] This idea isn't a new one. Many European space leaders assert that the "strategic" role of space has been widely recognized from the start. After all, if we consider the main space accomplishments of Europe since the 1960s and 1970s, it is easy to notice that they all have to do with the idea of preserving autonomy of decision and a high degree of political sovereignty. Whether we consider the Ariane launcher family decided in the early 1970s or the SPOT imagery satellites, to name two of the most dramatic successes of European space,[3] autonomy for European access to space and information comes immediately to mind. For example, Gérard Brachet, a former Head of the French Space Agency, estimated that the main engine for the French space program was the determination of France to "push Europe to acquire autonomy before granting its access to space".[4]

As indicated by many declarations over the years, national motivations have been the main driver of the European effort as far as big programs are concerned. This cannot surprise any serious observer of the European political construction process who knows about the strict delineation between the so-called "three pillars" of the European Union as posited by the Maastricht Treaty signed in 1991. Since then, and in the absence of any genuine higher level constitutional text, the competencies are strictly distributed between the European Commission (the first "pillar") and the European Council (the second "pillar"), while all the issues dealing with the internal European cooperation for police and justice affairs are embedded in the last "pillar." Of course, the dividing line created between the Commission, the only supra-national institution in Europe, and the intergovernmental Council, is of utmost importance for deciding any major collective project in Europe. Basically, any project related to the military is dealt with at the level of the European Council, providing any member state with a possible veto power on any issue regarding national sovereignty.[5] Because of the structure of European decision-making, any "European" military-related space program looks almost impossible as no "communitary" institution (i.e. at the commission level) can interfere with what remains in the hands of 25 states with very diverse interests, resources, and objectives.

The European Commission invests itself in space programs as far as they relate to infrastructures and research financing. Indeed, Galileo, the European navigation satellite program, is being coordinated under the auspices of the Directorate-General "Energy and Transport" from the EC, while the GMES program belongs to the DG Research portfolio. Since 2004 this DG has also been charged with a new domain called "security research" that will structure the next European Union 7-year plan for R&D (called the 7th Framework Programme for R&D in the 2007–2013 Period). Space, in its security dimension, will be part and parcel of this new activity.

For its part, the European Space Agency (ESA) has mainly been devoted to scientific programs for peaceful purposes and has hardly been in a position to be a proactive institution in a field that was traditionally judged to be outside of its mandate. While keeping the institution as a civilian and a scientific body, ESA authorities have made recent efforts to take into account the growing security preoccupations in Europe by choosing to adapt its own structures to be able to address the space security issues more boldly. In this respect, the creation of a joint ESA–Commission body in 2003, the so-called "Space Council," is notable; it is designed to better prepare for future space programs' planning phases, and to better manage the Security flagship program such as Galileo, the European Satellite Navigation program, and GMES (Global Monitoring for Environment and Security).[6]

Despite these institutional reforms, the real issue remains the transformation of a collection of disparate and relatively modest efforts into an organized European endeavor. There is still a long way to go, given the enormous budget disparities that show the persistence of differences between national attitudes toward space.[7] A series of recent reports and actions taken at the Commission level indicate that Europe is gradually becoming aware of the possibilities offered by a more active security space policy. Among other efforts, one can quote the so-called White Paper published in 2003 by the European Commission[8] or even the effort launched in 2004[9] by the Commission to foster preparation to finance a "security and space budget line" in the 7th Framework Programme for R&D in the 2007–2013 Period. In parallel, a recent expert report mandated by the Commission, the Report of the Panel of Experts on Space and Security (or so-called SPASEC Report) published in March 2005 "strongly recommends that the security applications of space be highlighted in the forthcoming European Space Programme" and that "this programme should be fully harmonised with other national and commercial programmes so as to obtain maximum synergy and affordability offering an enhanced capability for all aspects of security."[10]

At this stage, it is important to consider that these different moves demonstrate that the very notion of security (instead of "defense") lies at the heart of any future European strategy. Such a concept, because it deals with broader issues than just the military, proves to be a more efficient federative notion for the 25 member states who may still differ in their respective military policies. By keeping any space decision in the field of security (as demonstrated by Galileo or GMES), the current debates can be held at the so-called "communitary" level, providing then a convenient political avenue to discuss sensitive issues (including dual-use technology issues) without engaging in traditionally more difficult inter-governmental debates on military issues.

Of course, until now, the evolution of military space policy has remained more closely related to national strategic objectives and to the resources consequently devoted to them. France again has invested in space within the same context invoked by General De Gaulle, when he insisted on the need for the country to guarantee its independence vis-à-vis the United States and the USSR.[11] Launchers, remote sensing, and telecommunication satellites were already considered basic components of political sovereignty. At the same time, while it is widely admitted that space has to be part of national military capabilities, it is also recognized as being only one part of any national defense system, without having the central role it has seemed to acquire from the technical point of view these last years in the United States. Surely the discrepancies we see today between the two sides of the Atlantic in military space also reflect differences between global military and national strategies.

A delicate balance

At a time when the European nations have agreed to develop their own collective armed forces, they realize that the United States has engaged in an unprecedented effort to adapt its own military tools to new strategic conditions. Symbolized by the *Bottom-Up Review* launched in 1992 by Les Aspin, then the incoming Secretary of Defense in the Clinton Administration, this enterprise is now facing Europe as member states seek to define their collective military requirements.

First efforts are underway. Some European states have recently felt the need to reflect on the military imperatives created by the new geopolitical context. In particular, they realize how much European countries will have to adapt their intelligence and information resources and systems to more volatile situations and elusive enemies. In accordance with the traditional quest for some degree of military and political autonomy for Europe, a call for a "more intensive" effort to finance space programs at the European level has been launched. In fact, this move took on particular significance in the aftermath of the Kosovo and Afghan military operations, as space assets have almost proved to be prerequisites for both active and fully controlled military participation in allied coalitions. As one French general in charge of military space at the Chief-of-Staff level once put it, "The Kosovo conflict has underscored the considerable gap between the space capabilities of the United States and those of their European allies."

Space has been recognized as a primary area in which development is warranted, especially because of its unique contribution to the development of autonomous intelligence tools for Europe, as called for in the December 1999 Helsinki declaration. In line with this logic that puts space on center stage, early steps have been taken since 1999 through the "BOC" document ("Besoin Opérationnel Commun" or "Common Operational Require-ment") that set common specifications for the future generation of military-observation space systems. Six countries have signed this text already.[12] As demonstrated in a number of European conferences and reports, the question is not whether the European Security and Defence Policy needs some military space applications, but rather how to define in common these "sufficient strategic space capabilities" Europe needs in order to: (1) fulfill its own military requirements; and (2) appear as a partner able to fully share the military burden during coalition warfare. Here also the main difficulty lies obviously in the delicate political and strategic balance European governments must agree on for the definition of a viable ESDP.

The need for a new interaction between military and civilian space

Recent years have shown how difficult it has become to consider military space apart from the general effort toward the building of a global European identity in the field of security in general. It is certainly artificial and unhealthy to separate military space from the mainstream of space technological efforts that tend to fill the traditional gap between military and civilian technologies. For some years, those issues have been raised at the European level, either under the call for more intelligence- and information-gathering programs for military and security purposes, or via the European federative and technologically ambitious programs such as the GMES project or the navigation satellite program, Galileo. This last program exemplifies the delicate balance that exists in Europe between civilian and military uses of high-tech space applications. In 2006, given both the overall civilian nature of this program and the sensitivity of any move potentially related to defense, Galileo was at the heart of a vivid European debate about its possible military use considering its high-performance embedded capabilities. As presented by the European Commission, what is at stake with Galileo is nothing less than the independ-ence of Europe in developing its industry,[13] but also in conducting its own strategic and military actions, given the limits of European missions. In this sense, the program may have been viewed as a way to help the European armed forces deal on their own with military situations that the United States couldn't, or wouldn't want to, be involved in, with the possible con-sequence that the US satellite navigation program GPS wouldn't be avail-able for the Europeans. Still, some member states contend that the specific origins of the program, decided in a "first pillar" framework (i.e. by the European Commission) with the support of the ESA make it ineligible for any European military uses. In any case, Galileo appears today as typically representative of new generation space applications that simultaneously benefit military, technological, and economic dimensions.

Whereas some traditional programs, such as Ariane, have constantly been supported by successive ESA ministerial councils and received support from a "European preference policy" for using European launchers with the objective to parallel de facto US governmental practices,[14] other programs, dubbed "strategic," were given less clear signals. The Galileo program, for example, received a financial go-ahead with a €528 million budget decided in 2001, but was formally stopped shortly after by a no-go decision from the EU (Transportation) Council of Laeken. It was eventually endorsed three months later at the Barcelona European Council and is now fully accepted, as manifested by the first launch of the Galileo Test Bed Satellite in December 2005. The other big European project, a complete architecture devoted to environmental crisis monitoring and security, the GMES program, has also received some continuing financial support, but has remained in its infant stage. Indeed, the GMES project has been slow to reach the planning stage, as it first had to be clearly defined from a scientific perspective. The European Commission that runs the initiative beside the ESA has issued requests for scientific proposals that will have to be answered by willing organizations in the member states and sorted out according to their scientific relevance and rank in European priorities. Only very recently have decisive actions been taken with the setting up in 2005 of a "fast track" policy giving more focused programmatic horizons for the project.[15]

In line with American efforts underway in this area for more than a decade, space can thus be perceived in Europe as a twofold issue: first, it can be considered a keystone of future military architectures around which armed forces capabilities of tomorrow will be structured; second, it can be viewed as a permanently renewed technological activity whose evolution could have a major influence on the way states themselves conceive how they will exert their global power in decades to come.

Asking the question of how to use space as a military tool of tomorrow means addressing issues that are larger than the military dimension itself. European member states will first have to decide if they want and if they can follow the path chosen by the US If so, they will also have to think about how to adapt these orientations to their own requirements, resources, and military culture. In brief, these political decisions will impact their capability to deal collectively with their own security challenges and threats, as well as their interoperative capabilities in allied coalitions with an increasingly technology-oriented American partner.

Europe already possesses a few programs devoted to military uses that can form the basis for evolution in this domain. However, both the limited number of these programs and their relatively moderate interface with real combat show that some decisions remain to be made in Europe about using military space in the conflicts of tomorrow. Given the limited military budgets in Europe, decisions in this direction will have to be carefully weighed against other military options in a global assessment of European objectives and requirements. More specifically, the changes that have occurred in the United States about the use of military space systems[16] will have to be evaluated with precision by the European countries if they want to make the most of their competence and their resources in this domain for their military security.

Short term issues associated with the definition of a "European space military tool": defining a space military tool in the framework of the European defense and security identity

Much has changed since the 1950s and 1960s, when the military space effort was essentially structured around the global nuclear confrontation between the United States and the USSR. While the strategic use of space remains at the heart of military space programs, the threats and military requirements have evolved in parallel, making space a crucial asset for the security of modern nations, i.e. comprising military, economic, and citizen security. In this global framework, reflections on the military side of a future European space architecture can hardly resemble those held in the past. They take place in a political context that has been radically altered by the setting up of the so-called "headline goals" adopted by the European Union in Helsinki in December 1999. These "headline goals" gave birth in 2003 to a 60,000-strong force capable of operating on 60-day notice for at least 12 months. These objectives now constitute the reference points any European military space project will have to reflect. In some respects, the inclusion of the need for strategic intelligence as a priority capacity for the European Reaction Force may be able to revive shrinking, or at best flat, military space budgets. Of course, the delimitation, military or geographic, of the missions attributed to these forces is of significance as soon as the dimension of the military tools is in question.

European capabilities today

Today, Europe as a whole already possesses a variety of space means that can be of significant military value. Essentially, military activities engaged at national levels cover two main areas: Earth observation for strategic purposes, and telecommunications.

Earth observation

Military Earth observation has been initiated in France with the Helios program, started in 1986 on the basis of experience acquired with the first SPOT civilian satellites. Launched in 1995, the first Helios-1A is capable of metric-class observations in the visible spectrum and enjoys some maneuver capacity that allows in theory (i.e. depending on the weather conditions) the satellite to look at the same spot every two days. The participation of Spain and Italy in the program has made it the first military program of its kind to be organized on a multilateral basis in what remains a sensitive domain.[17] Clearly oriented toward the acquisition of data for strategic and political purposes, the Helios satellite has mainly been presented up to now as an independent source of information for the national governments involved in the program.[18] Improvements brought to the imagery-dissemination procedures are already underway to provide better adapted performances to the tactical users' requirements, as they were redefined during the Kosovo conflict in particular.

A second satellite, Helios 1-B, was launched in December 1999. While Helios 1-A's lifespan was not envisioned to exceed five years, its survival has allowed the now two-satellite system to provide a same-site revisit frequency of one day. At last, a new Helios 2 series is now in service with the launch of Helios 2-A, operated in December 2004 (and Helios 2-B envisioned for launch in 2008). The new series should improve the detection and reconnaissance capabilities as well as increase the volume of data transmitted. The new and better performances of the payload (night and day infrared vision, multispectral data collection, geo-referencing for a more precise use of guided weapons systems) along with a more efficient transmission chain (up to three times more images produced for a total acquisition delay reduced by a factor of two)[19] should represent the main improvements made toward a more tactically adapted use of military satellite imagery.

Other European systems are also planned in line with the Helsinki decision to improve the intelligence resources of Europe. Two main projects can be mentioned, one purely military in Germany, the other one, already quoted, dual-purpose and managed by France and Italy. The first one, SAR-Lupe, consists of the development by the German Ministry of Defence of a constellation of five low-orbiting Synthetic Aperture Radar (SAR) mini-satellites allowing an all-weather military-class observation. The deployment phase of this constellation was slated to take place in 2007 for a total cost of €380 billion. This program can be seen as a replacement for the Franco-German cooperation initiative that was linked to the development of Helios-2, with the construction by Germany of the former Horus radar satellite. This cooperative project has been halted by budgetary and political problems, leading to the development of two separate programs in the respective areas of excellence.[20] It must be noted that while such a no-go decision could be made in Germany in 1995, the situation clearly changed after the Kosovo conflict. Two factors seem to have weighed in favor of a resumption of a purely German radar satellite program.

First, as Germany was gradually becoming a true military actor in the coalitions, it would have to develop its own military intelligence capabilities, adapted to its immediate environment, i.e. Central Europe. In this regard, an all-weather permanent capability was a natural choice given the frequent cloud cover of the area, SAR satellites appearing obviously as the best tools for this mission. At the same time, German officials insist that satellites are part of a network of intelligence means and must be integrated in a larger architecture comprising other kinds of sensors. This ambition to possess an efficient global system reinforced the interest for the satellite and secured its budget.[21]

Second, the US-led Kosovo operations, especially at the planning stage, made Germany aware that retaining some degree of autonomous information was important if it was to have this new political and military role. More generally, relying too much on a sole source of information, whatever the source, has inherent limits that can only reduce the autonomy and possibly the efficiency of allied forces on the ground, making it difficult for political authorities to justify the good use of forces at home.

The challenge will be then to address the level of cooperation between the two programs, especially in the context of the future European force. Helios-2 and SAR-Lupe remain nationally managed programs that haven't been commonly organized. At this level, only common use of these systems is envisioned, which means that data exchange procedures will be organized between the two systems in order to enrich the information gathered by each. In this regard, both Helios-2 and SAR-Lupe should be considered "first generation" systems whose basic functioning has not been optimized to define a genuine collective and distributed satellite system on-orbit. Already, some discussions about designing "second generation" systems, i.e. possibly fielded at the horizon of the next decade, are evoked. Still, the data exchange possibilities can be viewed as a first important step in the development of a collective military space system for Europe. Maybe more importantly, this first step will naturally lead to something that would be contrary to a closed "military" system, given the fact that data would be clearly controlled at the ground segment level. It could integrate other non-purely military systems that will be fielded in coming years.

One of these latest projects, Pléiades -Cosmo, very different in nature, allows France and Italy to have access to optical and radar imagery by sharing the data obtained from two nationally separate but coordinated space segments. Concluded in January 2001, the bilateral agreement on this program originated in the will of each country to develop civilian space systems that could be used for military purposes. While France had envisioned preparation of the post-SPOT–Helios technologies for years, with the objective of promoting more affordable technologies providing a sub-metric resolution on small-sized platforms, Italy engaged radar technology with a similar objective, mainly for the monitoring of the Mediterranean basin. The radar observation satellites should be orbited between 2007 and 2009, and the French optical component should follow

between 2008 and 2010 with a global cost of Pléiades-Cosmo evaluated at
€1,070 billion. In this case, the level of cooperation, which is to be
detailed in the definition phase, may be rather ambitious, as the ground
segment will have to be developed in theory to deal with a complex bi-
national/civilian–military requirement.

In any case, for all these reasons, Pléiades-Cosmo may well represent a
real precursor of a collectively designed system on a European scale,
addressing for the first time the complexity of a genuinely strategic
European space system. At the national level, this particular program may
also appear as a test case for the future replacement of national-only
programs such as the Helios series. In this respect, the global cost, i.e.
including the possession and operation cost of the global system, compared with the quality of its results for the participating countries, will be a
measure of its success. Obviously, the increase in the cost of military
systems[22] makes such a success critical in future decisions concerning
the proportion of purely military and dual systems in any prospective
European observation system.

Telecommunications

Telecommunication satellites have been more widely perceived in Europe
as useful tools for military purposes. Great Britain, France, Italy, and Spain
have all developed national capabilities in this field, even if the scale of
the effort is variable. The United Kingdom has developed its own Skynet
capability in conjunction with NATO capacities. In 2002 the United
Kingdom was using a constellation of three dedicated satellites to produce
worldwide coverage for British armed forces. In August 1998, the British
government elected to pursue the effort on a solely national basis with the
development of a new generation of military telecommunication satellites,
Skynet V. This decision, that caused an abrupt halt to the United
Kingdom–French combined effort, revealed how much the UK government was willing to streamline its acquisition processes in accordance with
the then "Smart Procurement Initiative". With this important background
in mind,[23] any cooperative scheme could be seen as an uncertain and
potentially costly endeavor. Moreover, the decisions that were to be made
about the adoption of the future EHF/SHF band systems had a highly
strategic tone and involved actors such as NATO and the United States
who, because of their own status, complicated any cooperative decision
made on these issues between two European Community member states.

Skynet V has finally been developed in the framework of another new
acquisition and development scheme, the Private Finance Initiative (PFI).
According to this initiative, the government has chosen to develop a
telecommunication service fully dedicated to national authorities in cases
of crisis, but possibly given to the managing organization to commercial-
ize the capability in other circumstances.

For its part, France has preferred to use the civilian satellite platforms, Telecom-2, to carry their own military-dedicated transponders. The Syracuse II program is mainly devoted to fulfilling most requirements of the French projection forces, with around 100 ground stations installed between 1992 and 1996. An agreement was signed in 1995 between the United Kingdom and France for an extension of the coverage and a mutual lending of capabilities which will allow one country to use the other's space segment in case of defect of its own. Other agreements of this kind have been signed with NATO and Spain in 2000 and 2001. Given the lifespan of the last Telecom satellite launched in 1996, the technical life of the Syracuse II segment was expected to last until 2006, after which a new fully military satellite, Syracuse III,[24] would continue the service. The exclusive dependency of the military on the civilian system, implying a necessary payment for capabilities even if not really needed, coupled with new requirements for an improved SHF system (higher data rates[25]) and more robust telecommunications have prompted the French military authorities to opt for a fully dedicated system. The full Syracuse III program consists of two satellites, one launched in 2005 to ensure service continuity and the other in 2006 to ensure full coverage. The total cost of the program is estimated at €2.28 billion, almost doubling the amount officially devoted to the British Skynet V program. To some extent, Germany may also take part in the development of the first Syracuse III by leasing some capacities of the first Syracuse II satellite for its own purposes.

Finally, Italy, with the satellite SICRAL-1, and Spain with Hispasat (military transponders) have some limited capacities that will also become the elements of a full coverage system without providing such a possibility by themselves.[26] Confirming this trend, a call for offer during the summer 2002 for NATO Satcom Post-2000, ended up with the selection of a British (Skynet V)–French (Syracuse III)–Italian (SICRAL) joint proposal to provide the first SHF element of this architecture.

Future possible European capabilities

Besides the major areas already dealt with at the level of the ESA and European Union, namely the future observation systems, GMES, and Galileo, three main areas are often cited as potential candidates for future military space developments in Europe:

1   Early warning systems are important given the missile defense debate, not only in the United States but also in Europe. NATO has started to investigate the issue. Considering the political sensitivity of the subject, early warning can be seen today more as a strategic information-collecting activity than merely an element of a comprehensive anti-missile system. Indeed, knowing the threat and being able to obtain

directly from space (or from any other system) the information concerning the launch of an enemy missile will become more and more an indication of political sovereignty. To some extent, given its political significance today, such a system can even be considered an end in itself, notwithstanding the development of any anti-missile system. Only very preliminary work has been engaged in Europe on this subject, while the United States has possessed early warning satellites for more than 40 years. One of the difficulties consists in developing sufficiently sensitive sensors able to detect the launch of short-range ballistic missiles, whose boost phase occurs inside the atmosphere. It requires the mastery of delicate infrared sensing techniques as well as precise missile-tracking techniques. Before deploying any operational system, preliminary experimental work will have to be done in order to gain some experience in the field of detection, especially to acquire signature data and prevent wrong alarms caused by solar reflections in the upper atmosphere or by other phenomena. A complete typical system would consist of the deployment of two or more geostationary satellites above the regions nearing Europe. The cost of such a capability for Europe has been estimated by the French Joint Chiefs of Staff at around €760 million over ten years.[27]

2   Electronic intelligence (ELINT) capability is also frequently evoked as technology in which Europe should invest. As a contribution to the "collective goals" for intelligence underlined in the Helsinki declaration of December 1999, an ELINT space capability could take the form of small satellites orbiting at a low altitude to monitor the electronic activity of the immediate or more remote European environment. Again, only very prospective work has been started in France, with small ELINT-dedicated payloads embarked on the Helios satellites, small experimental DGA[28] programs, and the launch of a small group of four micro-satellites in 2004, called ESSAIM. According to the same calculations made in Paris, such a capability could cost around €1.2 billion over ten years. For the moment, no follow-on would be envisioned to ESSAIM at the French level.

3   If Europe is to make the choice to invest in space for military and security purposes, considering the perspectives of an increased "weaponization" of space and taking into account the fact that space is more and more polluted with debris, the surveillance of space can be considered a crucial element of its future strategy. Here also, only experimental work exists, both in France and in Germany, with no real prospects for future operational developments. It must be noted here, that besides these technological programs, the chances are high that the promotion of more proactive collective security policies would be preferred by Europe. The objective would be then to install a "virtuous" circle among nations, based on some level of increased transparency to keep tension in space as low as possible in the years

Table 12.1 Cost of European military space capability

Program	Programme cost (€ million)	Programme duration (years)	Annual cost (€ million)	French participation (€ million)
Telecom	3,100	15	207	62.1
Observation	2,300	10	230	69
Galileo*	150	8	19	5.7
ELINT	1,220	10	122	36.6
Space surveillance	760	10	76	22.8
Early warning	760	10	76	22.8
Total	8,290		730	219

Source: French Ministry of Defense – Space Bureau.

Note
* The Galileo figures give an estimate of what the cost would be of a jam-resistant signal that would enhance the operational value of Galileo.

to come. Keeping space as a weapon-free zone would certainly make sense for Europe in a context in which spacefaring nations would possibly have much to lose in any future arms race in space.

All these domains present potential for data- or technique-sharing, paving the way toward an eventual European body in charge of implementing space-related intelligence for collective European planning. In this respect Table 12.1, published in October 2001 by the Space Bureau of the French Joint Chiefs of Staff, indicates some rough budgets that would have to be devoted by Europe, with an indication of the French share modeled after the level of the French contribution existing in the framework of the ESA. According to these same military sources, as a whole, such a development scheme would amount to around 32 percent of the annual investment made in the ESA annually.

More recently, the Spasec Group Report to the European Commission delivered its own figures, recommending the following annual investment by categories for a credible military space effort:

- telecommunications: €600 million
- observation: €600 million
- signal intelligence: €200 million
- early warning: €200 million
- space surveillance: €100 million
- generic ground segment: €12 million
- interoperability and standards: €40–50 million
- technology research: €250 million
- new applications: €200 million.

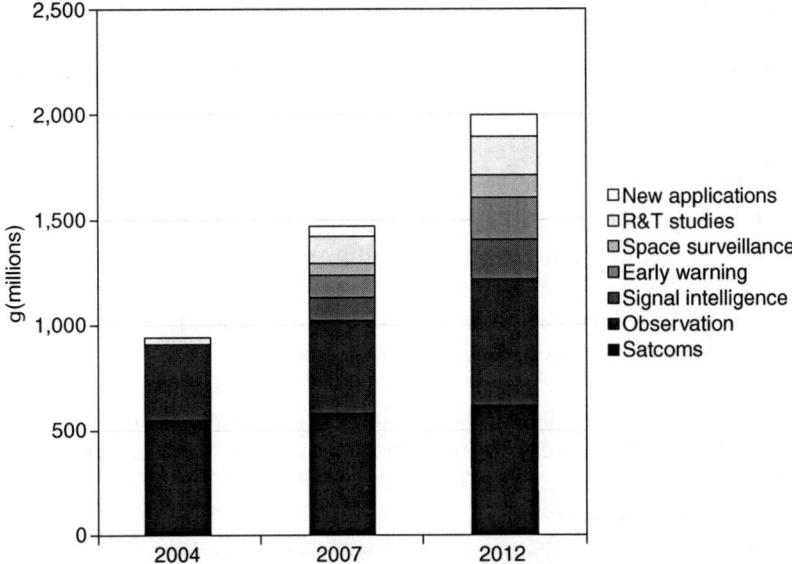

Figure 12.1 Cost estimates of European space systems for ESDP (source: Spasec Report, Appendix C, European Commission, March 2005, p. 52).

Such estimates would make the budget devoted to military space in Europe increase from €1 billion in 2004 to €2 billion from 2012 onwards (see Figure 12.1).[29]

The environment of future European military space programs: political and military challenges

Besides these purely military projects, the recent technological trends, which show an increasing integration capability, make the classical distinction between civilian and military technologies more and more tenuous. In accordance with the general evolution shown by most of the information technologies, space techniques are more and more driven by electronic chips' cost:performance ratio, as well as by the parallel development of system architectures that can mix existing individual systems. Changes in organization could make space systems more responsive to any need that could now arise from a broad spectrum of threats. Indeed, most modern military thinkers consider that armed forces are increasingly dependent on their mastering and using of information flows whatever their source or nature.

Of course, these new requirements coupled with the "dual" evolutions described above put space in the forefront as a potentially fruitful military environment for Europe. The global nature of space applications, their connection with current perceived needs, and the massive usage of generic technologies for civil or military users, are all factors which tend to give European space initiatives strong strategic content that lies beyond the military dimension. Reflections on European space have to be addressed in the wider framework of European integration as they raise a wide array of political issues.

Defining common requirements

The building of genuinely European programs creates new problems relative to a past (and still present) situation where national states control national programs. Thinking about a European military space program in the first place raises the issue of different programmatic and decision levels, mixing national and European jurisdictions.

Issues related to sovereignty are usually dealt with through the framework of multilateral relationships. Military Earth observation has traditionally been the only element of true European military space cooperation. It first took shape through the Western European Union (WEU) Satellite Centre located in Torrejon, Spain. Since its creation in 1991, this center has been in charge of the interpretation of satellite images coming mainly from civilian satellites (SPOT and Landsat) as well as from Helios (in a more limited number of cases) for military purposes. Experience has shown that sharing intelligence data remains a delicate activity conducted in a very constrained environment that limits its usefulness. Moreover, the heterogeneity of the member states in terms of national resources and strategies has complicated the task of the WEU Satellite Centre. The latest creation of a national satellite-image interpretation center by Belgium shows how much the relationship with the WEU Satellite Centre remains unclear for a number of European governments. However, the interest in Europe's development of a common activity in the field of military Earth observation cannot be disregarded. This was confirmed in 1997 when it was decided to reinforce the mission of the WEU Satellite Centre. Since 2001, it has been integrated as a military agency in the European Union, showing how much its missions are now fully connected with the European security policy, even if the new organization has yet to become a genuine common security organization.[30]

Paralleling this institutional approach, a model has recently been created to exploit the originally French military observation Helios-2 on a multilateral basis. To summarize what it is about, it must be recalled that some multilateral cooperation had existed on the first Helios-1A/1B projects with the participation of Italy and Spain (at respective levels of 14 percent and 7 percent) in the development of the program. These national participations were transformed into timeshares in the use of the system on a proportional basis. As far as Helios-2 is concerned, national partners have been more reluctant to invest in the development itself, with a mere 6 percent of the development cost finally shared between Belgium and Spain (respectively

2.5 percent and 3.5 percent). On its side, Germany decided not to embark on the project given its own investment into the SAR-Lupe satellite.

In this context, a document referred to as the Common Operational Requirements for Global European Earth Observation System by Satellites[31] has been elaborated since 1999 in common by the French and German governments. The Italian and Spanish governments quickly agreed on the approach and signed the document; Belgium eventually followed. Despite the classified nature of the document, it is well known that the aim was to define the expected performances of military or dual Earth-observation systems in the visible, infrared, and radar domains in the short- and long-term. This document, which "could be enlarged to other European countries"[32] according to the French brigadier general then in charge of space at the French Joint Chiefs Level, tends to be interpreted as a first step of a revamped strategy for cooperation, with the objective of making future multinational agreements more durable.

In some ways, such a document isn't without political significance. It recognizes that intra-European agreements in military space can barely avoid discussing higher levels of military cooperation. It opens the door to further common technical or financial engagements that can in theory become the logical result of converging collective needs. In this respect, the procedure differs radically from purely financial agreements that used to be the foundation of multilateral cooperation. Of course, a great deal of effort is needed in order to turn this attempt to organize a political and military demand into a European reality.

Still, the BOC model may appear as a relevant "bottom-up" mechanism to foster a truly European integration process. The creation in 2002 of a European military imagery group called Strategic Imint Action Group, with military representatives from Belgium, France, Germany, United Kingdom, and Spain, must be noted as a further step in this direction. It must also be remarked that this approach deliberately chooses to bet on the development of military and dual-use systems such as the Franco-Italian Pléiades-Cosmo to build an enlarged security-oriented space architecture. This can be legitimately recognized as the result of the product-oriented approach promoted in the BOC model. Obviously then, the specifications and the sharing of respective ground segments by military and civilian users remain key issues to help realize future European security satellite architecture.

Defining common procedures and doctrines

Another difficulty lies in the necessity of mixing different military systems and doctrines into a unique defense system. The degree of integration remains an important variable that usually decides the fate of cooperative programs. Telecommunication satellites symbolize the perfect current example of the importance attached to integration, which finds expression in this case in the level of interoperability between national systems. As already mentioned, some cooperative projects have been envisioned between European nations during the 1990s to share the development of common satcom systems. In fact, several cooperative architectures have been suggested, from a US–European option (dubbed Inmilsatcom) to an all-European option (Eumilsatcom) to a reduced version, Trimilsatcom that was co-planned by France, Germany, and the United Kingdom. One reason for that last arrangement was the converging replacement schedule for both the British and the French space segments, Skynet and Syracuse. These cooperative projects were finally abandoned as the United Kingdom was facing increasing financial constraints and implementing new procurement strategies (such as the Smart Procurement Initiative and the Private Finance Initiative). NATO meanwhile had to define a new space segment for its own telecommunications, NATO Satcom Post-2000. This latter program specifies interoperability conditions that will govern future use of allied information systems and will drive the choices for common technical standards guiding the future Alliance. For example, common architecture allowing the use of SHF and EHF innovative bands should accommodate increasing needs for bandwidth at the combined level for allied coalitions.

For the United States, employing such advanced techniques in modern combat conditions has concrete advantages. Communications in this wavelength can use portable antennas in a very precise manner, which make them very well suited to special-operation forces. In the same vein, this makes communications very difficult to jam and intercept while allowing a large amount of data flow, thanks to the broadband used by these systems. More generally, the United States also sees the wide use of these systems as a way to keep an edge in very advanced telecommunications techniques which would make them the real master of any future allied nation's Satcom or broader information architecture.

Apart from its purely industrial or technical aspect, this perspective does not go unquestioned in Europe as to its impact on individual military doctrines and procedures. As mentioned above, the massive use of information technology on the battlefield goes hand-in-hand with an adapted military tool that would bet on the use of aerial superiority – or that would lead to the creation of a "new soldier" connected in real-time to all the information available, making him theoretically better informed and more efficient. This "sensor-to-shooter" approach of future military operations has a direct impact on the way armed forces will deal with future situations, especially when they have to act in coalitions. Already called into question in the United States by some high-ranking military officials, the RMA remains controversial, not only because of its debated military efficiency in some situations, but because of the consequences it would imply for national doctrines and warfighting techniques.

More specifically, many military staff fear that too much use of integrated information technologies (IT) during combat operations would inevitably lead to a "flattening" of the chain of command. Beyond resistance to this sort of change exhibited by any social organization, such positions express more deeply the reluctance to see human factors disappear gradually from the conduct of military operations that often require subtle decisions and actions. As a consequence, different assessments exist today about the relevance and the desirable level of use of IT in modern combat. These doctrinal reflections contribute largely to the cautious approach adopted toward any telecommunications integration process that would be based on a unique technical scheme.

Over the years, there has been considerable discussion concerning the development of a common US–European IT infrastructure. The justification for such a plan is that Europe would therefore not be obliged to deal with enormous data flows that would hardly be useful for her forces, considering the state of her programs and doctrines. These discussions have been at the heart of debates about future NATO Satcom architectures and have already produced rough consensus on a structured view that relies upon the complementarity of systems such as Skynet, Syracuse III, SICRAL and the US EHF satellite. For years to come, one of the main challenges will be that of building a European military space architecture that can be self-sustainable, i.e. that can fulfill Europe's own requirements and can adapt to her own common doctrine, without precluding necessary interoperability with US systems already fielded or under development.

Conclusion

For Europe, it is now politically vital to define a future military space program. But in this respect, it is no less important to first conceive what kind of global security is needed. Space strategy must correspond to political as well as military interests and maintain relevance in a changing geopolitical framework dominated by a fully integrated US space vision. Some level of creativity will be required to allow Europe to find her own way with new and original initiatives that will answer her future security needs and help her define her own vision distinct from that of the United States.

Two main guidelines may be suggested in order to make military space applications relevant tools for the European defense effort as defined in the Helsinki declaration: first, utilization must be well conceived in accordance with the missions Europe is assigning to itself and with the resources it possesses. Military space is generally twofold. It comprises, on the one hand, applications directly relevant to the soldier on the ground or to the conduct of the war (i.e. intelligence systems, high precision navigation systems, or telecommunications) that could be called "force application space systems," and, on the other hand, space systems that aim at defending the orbital segment and that do not directly control force application on the ground but rather relate to an indirect "space control" mission (comprising space surveillance and anti-satellite systems). Realistically, this "do it all" approach possibly elected by the United States cannot be a template for Europe. As mentioned earlier, given the "Petersburg tasks," military missions the European forces are likely to fulfill collectively and on an autonomous basis, and given the more limited role assigned to technology in the European armed forces, investing now in "space control" systems is obviously irrelevant. On the contrary, developing systems that would give Europe more efficiency on the ground seems like a prerequisite for any collective military endeavor.

As mentioned earlier, intelligence space systems rank first on the list along with telecommunications. The existence of the BOC, which has enlarged awareness of the need for federating military and civilian space systems in global security monitoring, as well as the nascent telecommunication satellite infrastructure involving NATO, give a good sense of the intensity of efforts in this direction. Current improvements brought to the systems and to the common procedures can even be considered a first for the European military effort. These must be pursued and certainly expanded to early warning missile infrared detection satellites, which must be considered part and parcel of the intelligence systems. Such systems provide strategic and politically sensitive information on neighboring regions and, through the monitoring of test campaigns for example, could collect first-hand information on the ballistic threat before it becomes a reality. Considering the current debates about anti-missile systems, such space capabilities would give Europe – at a minimum – a better position as an intelligence provider on this delicate issue.

A second principle of conduct for Europe should be to promote innovative ways of considering space applications for security. First, Europe must be at the forefront of the effort to integrate dedicated military systems and other relevant space systems in a global space architecture. Satellites dedicated to monitoring natural evolutions could also be helpful when needed to deal with more short-term threats, whether they come from deadly terrorist actions or from insufficiently watched, aging industrial or armament installations in the eastern countries, for example. Here civilian security and defense agencies can find common interest in accessing space systems that are more polyvalent and supported by more responsive and more flexible architecture.

Space systems will also have to be considered in connection with other kinds of tools, including other kinds of sensors and intelligence collection systems that make a global "system of system" approach the recipe for success. In other words, military space thinking doesn't have to be kept a prisoner of the still-existing separation between space and non-space systems but must rather become a transparent element of an all-inclusive European security architecture that will be both civilian–military and space and non-space based. Only such orientations can make space appear as an affordable tool for European countries legitimately worried by financial and resource constraints.

At the present time, the debate on a future European military space program is in search of some kind of legitimacy. An effort in this sense could build on the gathering of still-dispersed national interests as well as on the demonstration of its cost:benefit ratio for national purposes. For this reason, European states may find it in their own interest to think about military space through a more global security-oriented approach, even if this remains to be more clearly defined. In parallel, defining an institutional arrangement to deal with the new security dimension of space remains a central issue for Europe as a whole. Several ideas have already been aired, ranging from the adaptation of the ESA founding convention to make it a central agency for security space if needed,[33] to the creation of a common imagery and mapping agency, a possible "European NIMA," as supported by Belgium that, for some others, should ultimately become a European military space agency. Whatever the solution, making the most of the new technologies in innovative ways and devising integrated organizations to deal with our future global security needs at the European level are two prerequisites for the future of European military space. These are crucial challenges that, if met, will enhance our future cooperative security and at the same time will help Europe reassert itself as a determined and capable partner beside the United States.

Notes

1 European military space investments amount to less than 4 percent of the total.
2 For an attempt to define the notion of "space strategic sufficiency", see Xavier Pasco, "The Transformation of Space: From Peripheral Asset to Core Capability," *RUSI Journal*, 144 (1999): 43–46.
3 In recent years Ariane has gained around 60 percent of the international commercial launches, while Spotimage remains today the world leading satellite imagery firm.
4 Gérard Brachet, "Une Politique Spatiale Nationale à Dimension Européenne," *Défense Nationale*, (October 2001): 42.
5 In this respect, any military use of European space assets are dealt with at the level of the Council by Mr. Javier Solana, the European Union High Representative for the Common Foreign and Security Policy. The European arrangements in the field of security and defense policy are also complemented with agreements signed with NATO at this same level, with, as a most recent event, mutual support decided in 2002 between the EU and NATO (the "Berlin Plus" agreement).
6 See below.
7 For significant and relatively recent concatenated figures, see the document of the French National Assembly: Bernard Grasset, *PLFI 2002, Avis 3323, Tome III Espace, Communications et Renseignement*, (October 2001), available at: www.assemblee-nationale.fr/budget/plf2001/a2627–03.asp.
8 http://europe.eu.int/comm/space/whitepaper/pdf/whitepaper_en.pdf.
9 This "preparatory action for security research" has already financed one pilot project in the field of space, ASTRO+, devoted to a better understanding of the possible use of combined Earth observation, navigation, and telecommunication space techniques to support remote humanitarian relief operations. A combined experimental platform was produced and demonstrated in early February 2006 in Poland.

10 European Commission, *Report of the Panel of Experts on Space and Security*, (Brussels, March 2005), p. 41. Available at: http:/europa.eu.int/comm./space/news/article_2262.pdf.

11 It must be noted that this political position led to the creation of the French space agency, CNES (Centre National D'études Spatiales) in December 1961, i.e. a few years after the creation of the space administration (NASA) in the United States (1958), and a few months after the first manned space experiments realized by the USSR.

12 See below.

13 "Galileo will enable Europe to acquire the technological independence that it wants in this area, as it did with other initiatives such as Ariane or Airbus," in *Galileo, The European Project on Radio Navigation by Satellite* (Directorate General Energy and Transport, European Commission, March 26, 2002).

14 Which have been reaffirmed through a Pentagon decision to help both Boeing Delta IV launcher and Lockheed Atlas V by awarding them sufficient public cash flow through already-planned military launch orders. See Marcia Smith, "Space Launch Vehicles: Government Activities, Commercial Competition, and Satellite Exports," Issue Brief for Congress IB93062 (Washington, DC: Congressional Research Service, April 17, 2003), p. 8. Available at: www.fas.org/asmp/resources/govern/crs-IB93062.pdf.

15 See http://europa.eu.int/rapid/pressReleasesAction.do?reference—EMO/05/420&type=HTML&aged+0&language+EN&guiLangague=en.

16 Reflected by the increasing use of satellite-guided munitions in the latest conflicts or the crucial role fulfilled by wideband telecommunication satellites in the transmission flows of data coming from diverse systems and interconnected widespread forces.

17 Each programming request is done at the national level, then transmitted to the Main Helios Centre in France and the data collected are redistributed directly to the national reception centers, located in Colmars (France), Lecce (Italy), and Mas Palomas (Spain). As noted in a French parliamentary report, it seems that the practice has allowed more flexible rules in terms of time and image sharing (35 percent of the imagery). See Jean-Michel Boucheron, *Le Renseignement par L'image*, Rapport n°3219, Assemblée Nationale, July 4, 2001, available at: www.assemblee-nat.fr/rap-info/i3219.asp.

18 For example, the French position opposing the US strategy in 1996 toward Iraq was reported to be decided after Helios intelligence had been obtained by France about the reality of the moves of the Iraqi armed forces. Helios imagery has also been mentioned in some occasionally differing situation assessments between the US and French authorities during the Kosovo campaign. See Jean-Michel Boucheron, Rapport n°3219.

19 In line with these specifications, a Helios-2 satellite is envisioned to take up to 200 images per day.

20 However, the development of a civilian one-meter class radar satellite, *TerraSAR-X*, was decided upon in April 2002 by the German Aerospace Center (DLR) teamed with Astrium, the European aerospace firm. Prepared for 2006, this commercial satellite was to reinforce European capabilities in radar imagery.

21 Speech by Brig. Gen. Kurt Herrmann, Strategic Reconnaissance Command, German Ministry of Defence, "European Satellites for Security," June 18–19, 2002.

22 As an example, the total cost for the Helios-1 series amounted to €1.5 billion while the Helios-2 total cost is expected to exceed €2 billion. However, at the same time, operation costs have constantly diminished from €56 million per year for Helios-1A to less than €40 million in year 2000 for the two satellite systems Helios-1A and 1B. The objective was to reduce it to the level of €30 million for Helios-2, while dramatically improving performances and capabilities.

23 In 1997, any armament decision made in the United Kingdom was filtered through two main political exercises, the "Comprehensive Spending Review" (CSR), which aimed at sorting investments, obviously conducted in connection with the "Strategic Defence Review" (SDR), another major reflection then underway.

24 The first satellite was launched October 14, 2005, and the second launched less than two years later, August 12, 2006.

25 Syracuse III satellites have also been equipped with some experimental EHF payloads authorizing higher data rates, as well as transmissions better protected against jamming and better suited to the needs of small and remote military units.

26 Spain in particular may play an increasing role in the future European/NATO Satcom architecture with the beginning in 2004 of a dual-service capability managed by a Spanish firm, Hisdesat. A contract for a new broadband telecommunication satellite was signed for this purpose at the end of 2001.

27 As published in Brig. Gen. Daniel Gavoty, "L'Espace Militaire, Un Projet Fédérateur pour l'Union Européenne," *Défense Nationale*, (October 2001): 79–96. A summary of these prospective costs for a European military space capability is presented above. New figures produced since more than double the indicated price tag.

28 Délégation Générale à l'Armement (French Procurement Agency).

29 SPASEC Report, p. 40.

30 Sarah Mattocks, Administrator in the Directorate of Defense Aspects at the European Council, conceded that fully devoting the Torrejon Centre to military uses remained a subject of controversy between the governments. Conference on the "European Satellites for Security," June 18–19, 2002.

31 Dubbed BOC after its French acronym (Besoin Opérationnel Commun).

32 See Daniel Gavoty, "L'Espace Militaire, Un Projet Fédérateur pour l'Union Européenne." According to declarations made in Brussels in June 2002, nations adhering to the BOC could either do so with full "partner" status through the exchange of information or through the acquisition of programming rights, or with "third party" status, which means that they would just request imagery.

33 According to Jean-Pol Poncelet, the current ESA Director for Strategy, and a former Belgian Minister of Defence who decided the creation of the recently opened Belgian satellite center, ESA is "prepare[d] to act in the security field" thanks to the flexibility of its convention and of its so-called "optional programs" that would allow more security-oriented programs. For ESA-specific missions being addressed outside of the European Union treaty, an institutional "bridge" would have to be built between the EU and the ESA.

13 The impact of foreign space developments on US defense policy

David A. Turner and James Vedda

As the previous chapters have made clear, space capabilities are proliferating. Today's environment related to space activities – in the civil and commercial sectors as well as in national security – is far different from what it was during the Cold War. This chapter looks at how the US is responding to this changing environment, and discusses factors to be considered in the coming years.

Trends in the post-Cold War era

Since *Sputnik 1* inaugurated the space age, the exploration, development, and utilization of space have generally been elite activities. Participants traditionally have been a subset of the world's industrialized nations that have made space investments for security, economic, scientific, and political reasons. Since the end of the Cold War, the reasons for going into space have remained essentially the same, but the motivation to do so has spread across the globe.

There are several reasons for the recent expansion of the spacefaring club. On the technical side, hardware and know-how that used to be solely in the hands of the innovators and early adopters have become widely available through military coalitions, scientific collaboration, and commerce. The technologies are more affordable to more nations, who have been observing the benefits that have accrued to the early spacefarers from their evolving space capabilities. This realization feeds the aforementioned motivations. Nations realize that if they are to be players in the modern, information-driven world – economically successful, politically relevant, respected in the scientific community, etc. – they must plug into the global information network (in part, using satellite communications), gather information about their environment and their potential adversaries (via satellite remote sensing and weather monitoring), and take advantage of technologies to improve the efficiency of commerce and travel (through satellite navigation). To improve national prestige, autonomy, and domestic industrial capabilities, many nations choose to cultivate these space applications indigenously, even when sufficient capacity is already available on the world market.

The expansion of the club of spacefaring nations brings both benefits and drawbacks to the United States. On the plus side, US participants in all space sectors now have more potential partners who can function on a more equal level for space science missions, military coalition operations, or commercial development of new space-related products and services. The increasing sophistication of other nations means there are also more potential trading partners for US exports of space and other high-tech products. Politically and diplomatically, US foreign policies and military actions can be justified to the global community through the greater transparency enabled by widely available satellite monitoring and communications.

Of course, greater transparency is a concern as well, since it is not a window shade that can be raised or lowered at will, and individual nations are finding it increasingly difficult to control access to potentially sensitive information on security and commerce. Increasing numbers of national and commercial remote sensing systems give the world transparency whether the US likes it or not. Characteristics associated with the spy satellite systems of the superpowers only a few years ago – a variety of sensor types, short revisit times, and especially, high spatial resolution – are now available to nations, news media, and other non-state actors around the world, including those with unfriendly or hostile intentions. While this situation presents challenges for US national security, it has the potential to yield a net benefit if the legitimacy and significance of US activities, such as military coalition and peacekeeping actions, can be demonstrated to the world by multiple independent information sources.

Another downside of a large and growing spacefaring club is the likelihood that the behavior of the participants will become less predictable, often in dangerous ways. In addition to behaviors deliberately intended to be contrary to the interests of the US or its allies in other regions of the world, poor practices by inexperienced or careless spacefaring entities, and inadequate supervision by national or international authorities, could be detrimental to space access and operations due to circumstances such as launch accidents or increased space debris. Unpredictable behavior also has the potential to cause the destabilization of regional security due to reactions to space activities perceived as hostile acts.

As discussed in Chapters 4 and 5, an assortment of bilateral and multilateral treaties and agreements related to space activities has emerged since the 1960s to prevent international misunderstandings and preserve the ability of nations (including less-developed nations) to access and use space. Some have questioned whether these documents continue to be relevant in the post-Cold War environment.[1] But the issues addressed by the treaties are still with us today. The difference is that instead of just two spacefaring nations, there are many, and more will follow.

An example of a traditional space issue that needs to be expanded and made relevant to current circumstances involves the responsibilities of

314 D.A. Turner and J. Vedda

space launch operators to third parties. During the Cold War, the US and the Soviets sought to prevent misunderstandings that could lead to major conflict, such as a space launch being misinterpreted as a missile attack. The two nations agreed to exchange information on the timing and trajectory of space launches[2] and managed to avoid the accidental initiation of aggression over such an incident.

As missile and space launch capabilities proliferate, pre-launch notification becomes important to many nations beyond just the US and Russia. The August 1998 flight of a North Korean Taepo Dong missile over Japan (claimed to be a space launch attempt) was a clear demonstration of the panicked reactions that can ensue after an unannounced launch. It not only provoked heated protests from the Japanese government, but also prompted them to initiate the Information Gathering Satellite program to deploy four satellites featuring high-resolution optical and radar sensors.[3] The Taepo Dong launch is unlikely to be the last such incident, and the consequences could be much more dire depending on the circumstances and location (for example, an Iranian launch over Israel).

The Pre- and Post-Launch Notification Agreement (PLNA)[4] is the current arrangement between the US and Russia under which the two countries provide each other with pre- and post-launch notifications for ballistic missiles that have a planned flight range or a planned altitude greater than 500 km and, with rare exceptions, pre- and post-launch notifications for space launch vehicles. A communications mechanism, the Joint Data Exchange Center (JDEC) in Moscow, was established specifically for this agreement. The PLNA was intended to be open to participation from other countries, although no other parties have joined to date. This agreement, or something like it, may be an effective mechanism for bringing launching states together to avoid dangerous misunderstandings related to space launches.

Earlier multinational agreements continue to have applicability to nations' responsibilities to each other in space launch and operations. The Outer Space Treaty[5] and the Liability Convention[6] address liability for damages caused by a nation's launch operations and by satellites reentering Earth's atmosphere or encountering other spacecraft in orbit. Most of the signatories had no space capabilities when they became parties to the treaties, but many have since developed them. More experienced space-capable nations have set high standards for safety and reliability that new spacefarers will be expected to meet if they are to honor the treaty provisions they agreed to decades ago. Launch accidents and spacecraft failures historically have occurred more frequently among new, unproven systems, so catastrophic accidents become more likely with the addition of new players in the space arena.

Even if new spacefarers behave as conscientiously as existing ones, there are still concerns related to the additional traffic that will increase the likelihood of frequency interference and conjunction of orbital paths, and exacerbate the problem of space debris. There are no international treaties directly addressing these problems, but in the case of space debris there are "best practices" that have been proposed internationally as operational guidelines.[7] These guidelines cover practices such as the depletion of energy sources (e.g. propellants and batteries) to prevent accidental explosion of derelict spacecraft and upper stages, and various methods for post-mission disposal of spacecraft. Although the US government created these guidelines and encourages their voluntary adoption internationally, the US currently opposes a multilateral treaty on space debris due to the restrictions it may place on hardware design, and security concerns presented by sharing of information on spacecraft, launchers, orbital element sets, and other space-tracking details.

Growing interdependence of national interests

As the previous discussion indicates, a substantial number of the world's nations want to guarantee their ability to access and operate in space to serve their national interests. For security applications, the desired capabilities are similar across nations: payload launch, satellite operations, communications, navigation, intelligence, early warning, and space situational awareness. The difficulty that individual nations and groups of allies must wrestle with is how to achieve the right level of autonomy while still maintaining interoperability.

Individual nations and the multinational security organization, NATO, exhibit different rates of technological development relevant to space systems. To maintain interoperability, the US must consider how to align its upgrades with those of its allies. However, since the US is the largest investor in new national security space technologies, it may be reluctant to slow its own pace of deployment to match its partners.[8] Alternatives include transferring technology to allies to increase their pace, or designing for backward compatibility so allies are not left behind. Each alternative has its own set of trade-offs and costs.

The nations of Europe so far have been the most successful grouping of countries attempting to move toward integrated space systems. This is likely due to their long experience with NATO for military planning, the European Union for commerce and other governance issues, and the European Space Agency (ESA) for research and development. Europe has independent capabilities in launch vehicles, communications satellites, and spacecraft for remote sensing and weather monitoring, and is developing a constellation of navigation satellites. Yet even the Europeans have a long way to go.

Typically, pan-European civilian projects take the technical risks to develop space infrastructure. Military programs are left to individual nations, where space projects must compete with other defense priorities for limited funds. As in the United States, military planners in European countries are concerned about whether their investment in a space system will yield more benefits to another service or agency, or even another country, than to their own nation or military organization. Current and projected remote sensing systems, for example, are developed by individual countries that are willing to make arrangements for the sharing of data, but not the sharing of the systems themselves.[9] This culture compounds the difficulty of working with European countries, individually or as a group, on interoperable military space systems.

In the past, the US has been the hub of international civil and commercial space activities due to its large government spending and its dominance in critical technologies. This is changing as other countries develop their own indigenous capabilities and decide to wean themselves away from dependence on the United States. Some recent examples of new cooperative arrangements between foreign countries pursuing civil and commercial space objectives include:

- *Europe and Russia working together on launch vehicles.* For several years, Germany and Russia have offered launch services for small payloads aboard Rockot, which is based on the Russian SS-19 ballistic missile.[10] Starting in 2010, the Russian Soyuz vehicle will begin launching from the European spaceport in French Guiana in an effort financed by the ESA.[11] Additionally, the ESA has been talking to Russia about cooperative development of future launch vehicles, although no final decision has been made as of this writing.[12]
- *Galileo program participants outside Europe* include China, India, Israel, Ukraine, Morocco, and South Korea, and there have been discussions with nine additional countries.[13]
- *Russia and Indonesia* are studying development of a joint multimedia satellite system.[14]
- *China's outreach to developing countries* includes ground tracking station agreements and training in the construction of satellite systems. For example, since the late 1990s, China has been working with Brazil on the China–Brazil Remote Sensing (CBRS) satellite program, which has launched two satellites and plans at least two more. On the commercial side, China has sold its first two commercial communications satellites to Venezuela and Nigeria.[15]
- *India* has scientific and commercial ambitions in space that will involve partnerships with many countries. A recent example is the agreement between EADS Astrium of France and ANTRIX, the commercial arm of the Indian Space Research Organization (ISRO) to provide communications satellites for the international market. Their first contract together was signed in February 2006 to build the W2M satellite for Eutelsat Communications, the world's third-largest commercial satellite services company.[16]

As new cooperative arrangements develop between foreign countries pursuing civil and commercial space objectives, the likelihood of military space cooperatives increases.

Foreign space partners frequently voice concerns about the character and implementation of their partnerships with the United States. There have been many resounding successes in international space collaborations, particularly in science missions such as the Hubble Space Telescope and the Cassini-Huygens probe to Saturn. However, potential partners for new projects have long memories of a wide variety of cooperative efforts where they have had unsatisfying experiences, such as the International Space Station and the Joint Strike Fighter. They have felt like junior partners, kept out of the critical development path and left out of important decisions. As their space capabilities grow, so too will the resentment from nations who see the United States as a domineering or unreliable partner. The United States must work to counter this perception, through actions as well as words, since it is likely to drive away potential partners – possibly in a direction that prompts concern for US national interests.

Another key concern of foreign allies is the current US export control regime. A noteworthy reaction is the European Component Initiative, started in 2004, which seeks to eliminate reliance on US components for satellites and other space systems that can be delayed or denied to European projects by the US government. Other indicators of foreign efforts to distance themselves from US technology include Alcatel Space of France weaning itself away from US components and the belief by European scientists that successful collaborations with the United States such as the Cassini–Huygens mission are no longer possible under the current export regime.[17]

Another foreign perception that presents a possible obstacle to cooperation deals with US military ambitions in space. From the foreign perspective, the United States is seen as the dominant space power by far, yet still seems to be unsatisfied with its leading position. Foreign powers, who perceive that even greater superiority is sought through space weapons, speculate on how the US will use such weapons and worry that conflict in orbit will ruin the operating environment for everyone else.[18] This perception can be found among allies as well as potential adversaries. Allies observing recent US actions, such as initial opposition to the Galileo system and unfavorable treatment under the tightening of export controls since 1999, often come away with the message that the United States is not supportive of their advancement in space, and may be trying to hinder it – a message that is not necessarily what the United States is trying to convey.

Proactive versus reactive approaches to US defense space policy

There is a tendency to think of proactive approaches as "good" and reactive approaches as "bad" or at least "sub-optimal." But history shows that this view is not necessarily accurate for national security space. Major actions can set precedents, as the United States and the former Soviet Union often discovered during the Cold War. Proactive and reactive approaches both have had successes and failures.

US space planning in the Cold War era was dominated by reaction to the Soviet threat. Fears of nuclear weapons in orbit and vast numbers of powerful missiles aimed at the United States prompted the development of space-based eyes and ears to penetrate the closed Soviet society and its large landmass. These efforts began even before the launch of *Sputnik 1* in October 1957, but didn't win the strong support of mainstream military planners and the public until after that milestone event. At the same time, the United States realized that in addition to looking down from above, the military and intelligence communities needed to know what was going on overhead, so a comprehensive system for space surveillance took shape. These reactive programs proved their worth beyond our ability to measure, playing a major role in keeping the peace throughout the remainder of the Cold War.[19]

In another example, the United States began work on anti-satellite (ASAT) weapons in 1959. This can be viewed as a proactive move, since the Soviets had no ASAT weapons of their own and had not deployed any threatening satellites by that time. After several iterations of ASAT concepts over the next three decades, including one system using nuclear-tipped interceptors that was declared operational for a brief period, the United States abandoned ASATs, as did the Soviets, after they had failed to prove their military utility and affordability.[20] Thus, proactive efforts to develop ASATs have yet to result in weapon systems demonstrating lasting strategic or tactical importance.

To a large degree, civil space planning during the Cold War was also a reaction to the Soviet Union. The US human spaceflight program, and Project Apollo in particular, responded to the fear of Soviet dominance of Earth orbit and eventually the Moon. With the July 1969 mission of *Apollo 11*, the United States spectacularly achieved its goal of demonstrating to the world the superiority of the American economic and political system, at least with regard to its scientific and technological prowess. In the early 1980s, another program, originally dubbed Space Station Freedom and known today as the International Space Station, was initiated partially as a reaction to Soviet space achievements.[21] So far, however, it has proven far less successful than Project Apollo in reaching its goals: its completion has been delayed a decade and a half beyond the original plan, and all US research projects, other than human physiology in space, have been dropped (including materials processing, plant and animal studies, and Earth and space sciences).

Traditionally, the US response to evolving foreign space capabilities has been a combination of cooperation and competition in space research and applications, assisted by export control, diplomacy, and arms control.

The balance between these techniques has shifted with changes in leadership and the policy environment. In recent years export control has been emphasized while diplomacy and arms control have declined, and the appropriate balance between cooperation and competition has been elusive as developments in the civil, commercial, and national security sectors – including foreign concerns about the direction of US space efforts – affect behaviors across all sectors.

In 2001, the Pentagon under Secretary of Defense Donald Rumsfeld made a deliberate shift from threat-based to capabilities-based defense planning.[22] The intention was to focus more on how an adversary might fight rather than on whom the adversary might be or where a war might occur, recognizing the need to plan for circumstances other than large conventional wars. This can be viewed as a proactive move, adjusting to the post-Cold War environment. Alternatively, this shift can be seen as reactive to the asymmetric threats posed by terrorists and other non-state actors that became particularly evident in the 1990s. A more cynical interpretation would view this approach as a rationale for building warfighting capabilities that cannot be justified by any existing or anticipated threat. This latter interpretation seems to be embraced (at least in rhetoric) by some foreign powers, often in venues such as the United Nations and the Conference on Disarmament, and by many international analysts.[23]

For national security space, capabilities-based planning means a focus on concepts such as "space superiority" and "space dominance."[24,25] To foreign entities, this is sometimes interpreted as submission to US authority and the possibility of denial of access to or use of space.[26] This view of US national security space efforts, combined with the difficulties presented by stricter US export controls, has contributed to an acceleration of foreign efforts to achieve autonomy in space manufacturing and operations. The Galileo navigation satellite system, numerous projects in militarily useful satellite imaging, and the European Component Initiative have already been mentioned, and there is also keen foreign interest in space situational awareness – a prerequisite for protection of space assets.

Space protection

Continuation of the Pentagon's capabilities-based (rather than threat-based) planning, coupled with the proliferation of militarily significant space applications, would appear to mandate protection of satellite systems from possible unspecified future threats to guard against a "space Pearl Harbor."[27] But protection of space assets supporting vital US interests from all possible attacks and hazards is a complex and expensive task.

Access to services from national security space systems, once the province of senior decision-makers, has gradually filtered down to the lower echelons, with soldiers in the field gaining direct access to communications, weather, and intelligence services. Combined with information-intensive innovations such as unmanned aerial vehicles, the result has been dramatic expansion of the US military's need for both fixed and mobile satellite services since the 1991 Gulf War. Increasing bandwidth requirements have outpaced deployment of dedicated military systems, pushing military forces toward greater reliance on commercial service providers. Estimates of commercial satellite use in the Afghanistan and Iraq campaigns range as high as more than 80 percent of the total bandwidth used by deployed forces.[28]

As a result, the space protection challenge involves more than just US military space assets. Military planners must decide whether to extend protection to the commercial satellites that routinely carry large volumes of US national security traffic. But what type and level of protection should be sought, and how will it be paid for? Should commercial satellites receive the same level of protection as military systems? Should US commercial satellite operators be required in their licenses to maintain certain physical and electronic safeguards? What about operators that are partially or wholly owned by foreign entities? Should satellites that are critical to allied governments fit under a US protection umbrella in some way, even if they don't carry any US traffic? These questions are being discussed, but have not been resolved.

To further complicate matters, telecommunications systems have been identified in US homeland security policy as critical infrastructure worthy of protection,[29] a status that also has been extended to the Global Positioning System (GPS) navigation satellites.[30] Although these two classes of satellites have been designated as critical infrastructure, they have not received attention in homeland security protection programs. The Department of Homeland Security currently lacks the expertise and resources to address protection of space systems, so other approaches to the problem will need to be considered.

The nature of intentional attacks on commercial systems, and the probability of their occurrence, are poorly understood, complicating owners' and operators' decisions on how, and how much, to protect their systems against these threats. Industry makes a distinction between natural hazards and threats from hostile actors, with natural hazards being a much greater concern thus far. ASAT attacks against the commercial space segment are perceived as a very low probability, yet the cost of protection is high. Attacks against the ground and link segments are considered much more likely, and companies have taken affordable steps to mitigate this concern.[31] However, the lack of high-consequence attacks to date and the cost of preparing for them have provided little incentive for industry investment in more robust security measures to satisfy government customers, who constitute a minority of their market.

Some in the space community have speculated that adequate protection of space assets could be addressed through international cooperative measures such as a formal code of conduct or informal "rules of the road" that would safeguard the "space lanes."[32] This leads to the question of whether such arrangements would require a ban on space weapons, and whether the US would agree to this restriction.

Weapons in space

US policy relevant to space weapons has been very consistent during the past two decades. It recognizes the right of self-defense, but does not advocate deployment of any particular space weapons on any specific timetable. However, it leaves the door open for research and development and, if directed by the president, deployment and use of space weapons. As expressed in the National Space Policy of 1996:

> Consistent with treaty obligations, the United States will develop, operate, and maintain space control capabilities to ensure freedom of action in space and, if directed, deny such freedom of action to adversaries.[33]

The reference to space control clearly indicates that this statement applies to an array of activities that do not involve weapons (surveillance, prevention, and protection) as well as those that do (negation).

Attempts to assess the existence, character, and seriousness of potential threats to US and allied space systems are limited by the classified nature of most national security space activities. Detailed, authoritative information on US vulnerabilities and potential adversaries' plans to exploit those vulnerabilities understandably is not available in the open literature. Quite simply, there is not enough unclassified data to make an informed judgment on whether passive defenses (such as hardening and redundancy) are adequate to counter the threat, or whether defensive weapon systems are needed. However, numerous unclassified publications issued in recent years by US military organizations provide general descriptions of plans and visions for combat operations in, to, through, and from space.[34] Emphasis on concepts such as "space dominance," "space superiority," "space power," and "space control" has left both domestic and international observers with the impression that the United States has an established policy, an implementation plan, and a coherent set of programs aimed at the deployment and use of space weapons. This is reinforced each year when the DoD budget request is released containing a variety of programs directly or indirectly linked to the development of space weapon systems.[35]

Many analysts and critics perceive that US pursuit of space weapons would initiate a space arms race. As they see it, potential adversaries will respond in kind. Due to its high reliance on space systems, the US is seen as a net loser in this race, so from this perspective US interests are better served by not allowing weapons in space. Additionally, US space weapons could have a negative impact on relations with allies and potential foes alike due to the perception that the US is on a quest for military dominance.

Other observers[36] look at space weapons from the practical viewpoint of space system operators. They believe that increased military activity in space could jeopardize the unprecedented level of peaceful space cooperation and commercial space development we now enjoy and that non-military space activities might be disrupted by the hazards and constraints imposed by expanded military space operations. For example, weapons tests – particularly those involving physical destruction of the target – would threaten nearby satellites and increase orbital debris, and military platforms would require more access to scarce orbital slots and electromagnetic spectrum.

On the other hand, some commercial space advocates see military space activities, particularly space situational awareness and satellite protection measures, as a positive development that would protect their investments. This view envisions future commercial space platforms of various types protected in the same manner as ships at sea are protected today.

Space weapons must be assessed as part of a larger picture that includes all US capabilities for global power projection. By some estimates, US nuclear forces have enhanced their capabilities since the end of the Cold War (including increased accuracy for Minuteman and Trident II D-5 missiles, integration of nuclear-tipped cruise missiles on B-52 bombers, and development of new B-2 bomber avionics that enhance stealth), while Russia's forces have deteriorated and China's have struggled to grow from their comparatively small size.[37] In this context, space weapons (in conjunction with missile defenses) may be perceived to be more valuable as an adjunct to a US first-strike capability than as defensive systems.

Nations and organizations around the world run the gamut of philosophies on space weapons. In influential international forums and much of the foreign press, most views express at least some concern, and at most, vehement opposition to perceived US plans for weapons in space. Each year, the UN General Assembly approves, by an overwhelming majority, a resolution on prevention of an arms race in outer space, which urges the development of bilateral and multilateral agreements for this purpose.[38] Meanwhile, the members of the UN Conference on Disarmament, led by China and Russia, have been pressing since 1999 for discussions leading to negotiations for a multilateral treaty banning space weapons. The United States has agreed to discussions, but not negotiations. The US position has been that there is no arms race in space, nor is one on the horizon – therefore, it is premature to begin negotiations on a treaty. But a position that is too rigid in international forums may prove untenable if foreign perceptions of the future threat environment continue to be at odds with US beliefs.

Given the high level of controversy at home and abroad, the United States must carefully consider the benefits and costs of space weapons deployment as a response to anticipated threats against vital space systems.

The polarization of the debate, and the lack of complete information in the open literature, has left many important questions unanswered, and possibly unanswerable. Would space weapons increase US and allied security and ensure space dominance when needed, or would they invite competition and international scorn sufficient to cancel out any military advantage? Exactly what military requirements is the nation seeking to fulfill, and is there a more cost-effective way of fulfilling them without using space weapons? Is it less controversial, but still effective, to deploy systems on the ground that target space assets rather than basing weapons in orbit? Can agreement be reached on a clear definition of what constitutes an offensive versus defensive space weapon, and would it make any difference to those who seek to prevent their deployment?

For as long as the United States remains at the forefront of military space capabilities, it holds the best bargaining position in any international forum that would consider measures to preserve the peaceful uses of space. As the dominant player, the United States could suggest a strategy that would allow a compromise between the extreme positions that so far have dominated the space weapons debate. This could involve confidence-building measures such as the "rules of the road" mentioned earlier. As another example, the United States could pledge that it would not be the first to deploy space weapons, but still press ahead with research that would allow it to be a "fast follower" if other nations proceeded with deployment.[39] Under this scenario, the United States' existing space infrastructure, expertise, and potential for rapid mobilization are seen as a strong deterrent to potential space rivals. Additionally, the onus would be on those rivals to convince the world that their efforts to be the first weaponizers of space were appropriate and non-threatening to others' access and use of space. As so often happens in the business world, those who are first to market must do the hard work of educating the public and setting the stage, while a later entrant to the market can devote all resources to becoming dominant. Of course, the feasibility of this strategy must be judged in the context of the prevailing circumstances, which may involve open conflict with a peer rival or evidence of a significant imminent threat.

The road ahead

The United States has a range of possible options to pursue in its ongoing interactions with existing and emerging foreign space powers. One end of the spectrum could be characterized as the embrace of full and equal partnership where all parties abide by "rules of the road" and renounce the deployment and use of weapons in space.[40] On the other hand, the United States could claim "space hegemony," seizing control of space as soon as possible, while it still can, and then declaring itself the world's space policeman for the good of all, as defined by US authorities.[41] The most feasible and desirable option likely lies somewhere between these extremes.

Any course that is chosen must consider the cultural differences between US and foreign space and defense communities. Other countries distinguish between civil, commercial, and national security space activities just as the United States does, but often do not see the government role in the same way. When foreign governments want a particular capability, or wish to build an export market for their nation's industry, direct investment (in US parlance, a subsidy) is a normal way of doing business, and often results in space systems that are designed from the start to serve both government and commercial needs. In contrast, the United States considers such arrangements to be exceptions to the rule, requiring thorough case-by-case consideration. If a US system is applicable to both government and commercial needs, government requirements, as a result of their special status, usually drive the design and may conflict with requirements for a viable business case.

The US is slowly becoming acclimatized to the post-Cold War era of "globalization." There is no universally accepted definition of what constitutes globalization, which many observers see as primarily an economic phenomenon. But it is more than that – it involves the diffusion (some would say "democratization") of technology, information, economic power, and international influence. At the same time, "world politics is changing in a way that makes it impossible for the strongest world power since Rome to achieve some of its most crucial international goals alone."[42] In such a world, international coalitions must be mobilized to address shared threats and challenges. "Superpower" status, as it was conceptualized in the Cold War, will eventually lose its meaning or be substantially redefined.

The United States came of age as a world power in World War II and then spent the next 45 years shaping itself in the Cold War paradigm. At the same time, the United States was helping to strengthen its allies so they could resist the Eastern Bloc. This pervasive influence drove the building and evolution of institutions across the public and private sectors, and these organizations are not able to quickly adjust their doctrine, thinking, and behavior to the twenty-first century globalized world. The process may take decades, and space and defense policy are caught up in this new evolution.

Space technology, particularly in communications, has been one of the enablers of the globalization era. Now that we are in the midst of this transitional period, we must ask: What effect will globalization have on space? Do doctrinal concepts like "space dominance" and "space superiority" and associated efforts to sustain US hegemony in space still make sense in a world where space capabilities are more widespread, and space-capable entities around the globe are highly interdependent? The United States currently far surpasses other nations in its national security space capabilities, and may even widen the gap in the near-to-medium term. But in the long-term, will globalization lead to proliferation of space capabilities and a relative decrease in US power and influence? Theoretically, the diffusion of power and influence will not affect the United States alone – other nations will be prevented from claiming a dominant position similar to that which the United States holds today.

Nonetheless, there will be no shortage of nations seeking a dominant position, or at least substantially increased influence on the world scene. In particular, we are still in the early stages of the rise of Asia, which could continue for decades. The rising powers tend to be nationalistic, seek redress of past grievances, and want to claim what they see as their rightful place in global politics and economics. This will dramatically change the substance and the context of international challenges.[43]

Space projects involve very long timelines, forcing us to adjust our thinking to timescales measured in decades. If we're heading into a world where power and influence will become more and more dispersed, we need to consider questions that, indeed, appear at various junctures in this volume on space and defense:

- In general, what geopolitical, economic, and legal changes affecting space activities are likely in the next ten, 20, or 30 years? More specifically, how will increased global involvement in space activities affect cooperation, competition, and the international legal regime for space? How will the space-related goals and capabilities of other nations, multinational organizations, and non-state actors change in the decades ahead?
- Where do the US civil, commercial, and national security space communities want to be by mid-century? Are they on track to get there, or are fundamental changes needed?
- Will long-term US requirements be better satisfied by incremental advances in current technology or by revolutionary developments in operational systems?
- Will growing concerns about the US space industrial base drive changes in the way we view and implement public/private relationships?
- Will the US need to increase its reliance on foreign technologies and products?

In one long-term scenario, economic power (closely aligned with technological prowess) could be the defining element of national security interests rather than military power. Economically powerful nations would position themselves and be viewed as world leaders not only because they attained wealth, but because of the way they used it: to improve the lives of their own people, to advance science and technology, to help other less fortunate nations improve their standard of living, and to protect the Earth's environment.

If economic power dominates, and commercial space activities grow and proliferate, the debate over weapons in space as we know it today could become moot. Disagreements between nations might be sparked most often by economic competition, not military or ideological rivalry. To protect business investments and maintain the flow of space commerce, there might be wide agreement that military forces must be deployed in space. But those forces would probably be more akin to police and the Coast Guard than traditional military services. The dark side of economic competition in space could involve industrial espionage, theft and destruction of property, smuggling, and piracy.

In a contrasting scenario, the globalization movement could end relatively soon, perhaps suddenly. This is precisely what happened with the onset of World War I, at which time the United States had just achieved the status of the world's largest economy. There were many similarities to the current situation: unprecedented international movement of capital, raw materials, and people; revolutionary technological innovation, including the telephone, radio, and internal combustion engine; and an ongoing struggle for balance between protectionism and free trade. There were also flaws in the international order that seem quite familiar today: imperial overstretch, great power rivalry, an unstable alliance system, rogue regimes sponsoring terror, and the rise of a revolutionary terrorist organization (the Bolsheviks) hostile to capitalism.[44]

Is there a disruptive event on the horizon that could take us off the globalization path? What would it be, and when will it come (if ever)? So far, events in Iraq and Afghanistan and the war on terror have not turned the world away from globalization, although they have revealed strong anti-globalization forces. But we could still face large-scale terrorism, a crisis with China over Taiwan, or revolutionary regime change in a country that had been considered a key ally. Whether or not such an event ended globalization, the national security space community would need to determine how to adapt to new challenges. Would this be the tipping point that justifies the move beyond current space applications to the use of force in and from space?

Whatever scenario comes to pass, the United States must protect its national interests, and its space capabilities are critical tools for doing so. US policy will be better served if it recognizes that this is increasingly true for other nations as well.

Notes

1 For example, see Michael J. Listner, "It's Time to Rethink International Space Law," *The Space Review*, May 1, 2005. Available at: www.thespacereview.com/article/381/1.

2 Agreement on Measures to Reduce the Risk of Outbreak of Nuclear War between the United States of America and the Union of Soviet Socialist Republics, September 30, 1971.

3 Paul Kallender-Umezu, "Japan Discloses Plans to Launch Three National Security Satellites," *Space News*, August 30, 2004.

4 Memorandum of Understanding between the United States of America and the Russian Federation on Notifications of Missile Launches, December 16, 2000. The agreement is to remain in force for ten years after the signature date and may be extended for successive five-year periods.

5 Treaty on Principles Governing the Activities of States in the Exploration and Use of Outer Space, including the Moon and Other Celestial Bodies, January 27, 1967. Ratified by 98 countries and signed by 27 as of 2005.

6 Convention on International Liability for Damage Caused by Space Objects, March 29, 1972. Ratified by 82 countries and signed by 25 as of 2005.

7 US Government Orbital Debris Mitigation Standard Practices, December 2000.

8 Robert Dickman, "Space Superiority," in John Logsdon and Audrey Schaffer (eds.), *Perspectives on Space Security* (Space Policy Institute, George Washington University, 2005).

9 Peter B. deSelding, "One European Space System Unlikely for Foreseeable Future," *Space News*, October 18, 2004.

10 See www.eurockot.com.

11 Peter B. deSelding, "Liability at Guiana to be Shared," *Space News*, December 19, 2005.

12 Peter B. de Selding, "EADS Space Exec Questions Benefit of Euro-Russian Launch Alliance," *Space News*, January 23, 2006.

13 Heinz Hilbrecht, "Galileo – European Commission," presentation to the 2006 Munich Satellite Navigation Summit, February 21, 2006. Available at: www.munich-satellite-navigation-summit.org/PresOpening.htm.

14 "Russian, Indonesian Firms Study Joint Satellite System," *Space News*, December 19, 2005.

15 Tariq Malik, "China Advances Spaceflight, Commercial Market in 2005," *Space News*, December 19, 2005.

16 "EADS Astrium–ISRO alliance sealed first contract with Eutelsat for W2M satellite," EADS Space press release, February 20, 2006.

17 "Is Anyone Paying Attention?" *Space News*, February 21, 2005.

18 For example: Novosti Russian News and Information Agency, "Weapons in Space: How Can Russia Respond to U.S. threat?" June 1, 2006; Hui Zhang, "Action/Reaction: U.S. Space Weaponization and China," *Arms Control Today*, December 2005; Robert Evans, "U.S. Bid to Dominate Invites Disaster – Gorbachev," Reuters Limited, May 30, 2005; Peter B. deSelding, "European Officials Skeptical of U.S. Space Motivations," *Space News*, May 25, 2004; Alain Dupas, "Dual Space Dominance: The Hidden Agenda Behind George W. Bush New Space Vision," unpublished paper, first quarter 2004.

19 For an informative open-source narrative on this period, see William Burrows, *Deep Black* (New York: Random House, 1986).

20 Paul Stares, *The Militarization of Space* (Ithaca, NY: Cornell University Press, 1985).

21 National Aeronautics and Space Administration, "Final Report: Study Group on Space Station," September 13, 1982.

22 US Department of Defense, "Quadrennial Defense Review Report," September 30, 2001.

23 Some examples include: Foreign Affairs Canada paper, "The Non-Weaponization of Outer Space," March 31, 2002, available at: www.dfait-maeci.gc.ca/arms/outer7-en.asp; China State Council Information Office, "China's Endeavors for Arms Control, Disarmament and Non-Proliferation," September 1, 2005; Novosti Russian News and Information Agency, "Weaponization of Space Will Have Unpredictable Consequences," April 5, 2006; Thomas Graham, "Space Weapons and the Accidental Risk of Nuclear War," *Arms Control Today*, December 2005.

24 US Air Force Space Command, "Strategic Master Plan FY06 and Beyond," October 2003.

25 US Air Force, "Transformation Flight Plan," November 2003.
26 Michel Bourbonnière, "Space Access and Use Denial: The Law of Neutrality and European Space Security," Royal Military College of Canada (no date, but no earlier than December 2003).
27 Report of the Commission to Assess United States National Security Space Management and Organization, January 11, 2001.
28 Amy Butler, "Heavy DoD Reliance on Commercial SATCOM Prompts Questions of Protection," *Defense Daily International*, April 16, 2004.
29 National Strategy for the Physical Protection of Critical Infrastructures and Key Assets, February 2003; Homeland Security Presidential Directive (HSPD) 7, "Critical Infrastructure Identification, Prioritization, and Protection," December 17, 2003.
30 National Security Presidential Directive (NSPD) 39, "U.S. Space-Based Positioning, Navigation, and Timing Policy," December 15, 2004.
31 National Defense Industrial Association (NDIA), "CINC Space Study: Protection of Commercial Space Assets," December 1998.
32 For example, see the statement by UK Ambassador John Freeman at the Conference on Disarmament plenary meeting on Prevention of an Arms Race in Outer Space, June 30, 2005.
33 Presidential Decision Directive 49, "National Space Policy," September 19, 1996.
34 For example, see US Space Command, "Vision for 2020," February 1997; US Space Command, "Long Range Plan," March 1998; US Army, "Space Master Plan," March 2000; US Air Force Space Command, "Strategic Master Plan FY06 and Beyond," October 2003.
35 Center for Defense Information, "Space Weapons Spending in the FY 2007 Defense Budget," March 6, 2006.
36 For example: Theresa Hitchens, "U.S. Weaponization of Space: Implications for International Security," Center for Defense Information, September 29, 2003, available at: www.cdi.org/friendlyversion/printversion.cfm?documentID=1745; statement of Russian Ambassador Valery Loshchinin at the Conference on Disarmament, CD/PV.1001, February 2, 2006, available at: http://daccessdds.un.org/doc/UNDOC/GEN/G06/608/92/PDF/G0660892.pdf?OpenElement.
37 Keir A. Lieber and Daryl G. Press, "The Rise of U.S. Nuclear Primacy," *Foreign Affairs*, March/April (2006): 42–54.
38 In 2005, the UN General Assembly vote was 180 in favor of the resolution and two against (the United States and Israel). This was the first time in this annual exercise that the United States (or anyone else) voted against the resolution. In years past, the United States and a small group of allies abstained in both the First Committee and General Assembly votes.
39 Wesley Hallman, "A Fast-Following Space Control Strategy," *Astropolitics* (Spring 2005): 35–42.
40 For something close to this position, see: Henry L. Stimson Center, "Model Code of Conduct for the Prevention of Incidents and Dangerous Military Practices in Outer Space," May 2004; Henry L. Stimson Center, "Space Security or Space Weapons? A Guide to the Issues," February 2005.
41 Everett C. Dolman, *Astropolitik: Classical Geopolitics in the Space Age* (London: Frank Cass Publishers, 2001).
42 Joseph S. Nye, "U.S. Power and Strategy after Iraq," *Foreign Affairs* (July/August 2003): 60–73.
43 James F. Hoge, "A Global Power Shift in the Making: Is the United States Ready?" *Foreign Affairs* (July/August 2004): 2–7.
44 Niall Ferguson, "Sinking Globalization," *Foreign Affairs* (March/April 2005): 64–77.

Conclusion

Ambassador Roger G. Harrison

This first textbook on space and defense has sought to delimit the academic parameters of the subject and to find within those parameters some intellectual coherence. In the former purpose, the effort is notably successful. There is certainly no other more comprehensive treatment available of all aspects of those subjects which might be grouped under the rubric "space policy." But those who came to this text hoping to discover a core of rationality underlying the various initiatives taken by the governmental and commercial sectors in space will doubtless be disappointed.

The fault is not with the authors, who have faithfully represented the universe of strategic, economic, commercial, and intelligence policies and programs which constitute America's efforts in space. But the very precision and comprehensiveness of their analysis has served to highlight the lack of coherence underlying American space policy. The picture that emerges from a number of authors is of a wilderness of governmental bureaucracies, each pursuing its own interests, sometimes in coordination but more often in competition for shrinking budgets, operating under congressional mandates which are themselves often contradictory, under administrations which have not shown the sort of persistent interest in space which might have elevated key policy issues from out of the bureaucratic trenches.

Of course, the problem may be in the nature of the situation we face. With the expansion of space-related activities, it is possible that no single rationale could cover a range of activities, some inspired by "pure" science, some by the profit motive, some – notably the manned space program – by national self-image, some by the hardheaded demands of national security, some by entrepreneurial enthusiasm, and some by awkward combinations of these factors. Reading Damon Coletta's excellent introduction, it occurs that strategic thinkers of the past could advance convincing, single-factor theories only by ignoring such lurking complexities; thus, the Italian fascist theorist Douhet focused on the decisive use of air power to defeat the enemy's "will" by terrorizing civilian populations, and Mahan argued that control of sea lines of communication was the key to world power. History would prove neither correct in full; both visions functioned better as politically mobilizing metaphor than policy prescriptions. Their legacy survives in the school of thought which sees space as a strategic "high ground," which, if seized, provides its possessor with unchallengeable military supremacy. But, as this textbook illustrates, many other, sometimes contradictory, interests are at play in space policy.

The ambiguities of any discussion of "grand strategy," and particularly any regarding space, are explored by Jim Wirtz (Chapter 1). He concludes that what renders our space strategy something less than grand is any clear concept of how future space activity will contribute to scientific and economic gains – or even to fulfillment of human destiny. That space does – or, at least, soon will – contribute mightily and in entirely new ways in all of these areas is asserted loudly and often; Dr. Wirtz finds hard evidence lacking.

If evidence is lacking, interests are not – as Professors Sadeh and Vallance point out in Chapter 6. They describe the interest groups which affect space policy, some of which fit a standard decision-making model, but some of which do not. The process itself, they argue, is anything but rational, and the competing interests often contradictory. Few other areas of government activity have to justify themselves, for example, by their contribution to both national defense and human destiny. Moreover, the interplay of economic, scientific, security, commercial, and spiritual interests can lead to programs, like the space shuttle, that serve the interests of no group very well, but by their existence and cost prevent development of alternative systems.

Other dilemmas are described in succeeding chapters. Several chapters, for example, point to the vacillation of control over commercial space exports between the Department of Commerce and State Department – the first (when it has the bureaucratic upper hand) focusing on the promotion of commercial activity and trade, the second, particularly after the Cox Report of 1999, treating both space hardware and intellectual capital as potential weapons of war, to be guarded from allies and potential adversaries alike. This conflict mirrors the schizophrenia of Congress, which has sought over the past decade to both encourage commercial space producers to exploit foreign markets, and clamped tight restrictions on the information flows which alone would make such market expansion possible. Many in industry argue that the International Traffic in Arms Regulations (ITAR) – now administered by the DoS – results in a shrinkage of the United States' share of the international market for satellites and launch services, encourages competitors like the French to develop – and consumers to buy – new satellites free of "American content," discourages American companies from pursuing foreign buyers and dampens the intellectual exchange with foreign space industries that might have expanded our own capabilities. A situation has thus emerged – these observers contend – that reflects neither the intent of Congress, nor of the administration, nor the DoS, nor serves national security interest.

Meanwhile, we are conflicted about what additional role, if any, international regulation should play in our future space policy, as comprehensively described by authors Waldrop and Kasku-Jackson in Chapter 4. The persistent dream of our species is that the rule of law might somehow be applied so as to restrict the actions of others while leaving our own freedom of action intact – our hope, in other words, that given the purity of our motives, the spirit of justice might lift her blindfold occasionally to wink at our transgressions. But systems of law have general applicability, and this confronts us with a dilemma – one that is at the heart of any discussion of a grand strategy for space. We wish to exercise complete freedom of action within a space environment which is ordered and predictable. But creating an ordered and predictable environment requires restraint on the part of all spacefarers according to rules – however established – that all recognize as legitimate.

Some argue that our interests are best served by a laissez-faire approach to space competition. But as space becomes more crowded, with commercial operators (many multinational) jostling for position with each other and with national space programs, an unregulated environment threatens chaos: conjunctions in orbit, threats to the lives of astronauts, overcrowding of orbits and electromagnetic spectra, a decrease in space situational awareness, an increase in suspect payloads, and an expansion of space debris. In short, we cannot have it both ways: hoping for an ordered environment while ourselves standing outside it. At some point we will have to decide whether creation of some additional regulations, beyond the Outer Space and Moon treaties, serves our interests better than its lack. Indeed, it may be that from such a decision will flow the intellectual coherence that our space policy presently lacks. Talk of "dominance" will perhaps delay our serious consideration of this reality; it will not alter it.

The task of creating regulation – and, indeed, of developing some comprehensive intellectual foundation for space policy – is complicated by the presence of a new breed of space entrepreneurs. As Bruce Linster points out in Chapter 3, the space commercial sector has attained central importance. The policy of the United States government is to impose as few regulations as possible on these new operators, and in many cases to subsidize their operations in some fashion. Encouragement to private operators is, in part, a result of frustration at the paralyzing effects on our space activities of gargantuan infrastructure linked to impenetrable bureaucracy, and the conclusion of some in Congress and within NASA that the primary hope for innovation and energy lies in the private sector. Private companies have the potential to be more nimble in this regard than the large space launch providers; but will they be resilient enough in the face of potential accidents and periodic reductions in demand for space launch? Perhaps cheaper launch will bring forth large increases in demand, as currently uneconomic projects become economic given cheaper cost to orbit. The suspicion lingers, however, on two issues: whether the entrepreneurial companies will survive once they are forced to demonstrate a genuine return on capital (and give up the open and disguised government subsidies many now enjoy); and whether commercial space may have transitioned from an entrepreneurial to a 'mature' industrial sector, liable to experience stable but only moderate rates of growth.

Indeed, the specter which haunts all sectors of the space community, both commercial and military, is that of comparative advantage. Space is a harsh environment; operating there is expensive. It may be that we have already discovered the sorts of space activities which have a comparative advantage over similar activities within the atmosphere. Intelligence collection is one of these; mass communication is another; geographic surveys and various kinds of basic scientific exploration and research are still others. These areas have been clear for decades, but suggestions for new and productive uses have been less successful. Zero gravity manufacturing and experimentation was once advanced as an economic incentive for the International Space Station; but it has not been as productive as its backers hoped, and such experiments on the space station have been abandoned. Solar power generation in space is possible in theory, but would depend for viability on a huge increase in the cost per BTU of energy from terrestrial sources – including solar and nuclear energy. Mining in space would have to overcome enormous technological and financial obstacles in order to produce what are, in effect, raw materials. Space tourism will be hostage to a single accident; even if safety were not an issue, creating a new form of tourism for the rich will hardly sustain substantial private investment in space exploration, or satisfy those who see space as fulfilling human destiny. Nor does lowering the cost of putting payloads in orbit seem an unambiguous gain. From the national security perspective, the cost overruns and time delays in all next-generation satellite systems outlined in some detail by Pete Hays (Chapter 7) mean that the cost of orbiting the satellites has become an insignificant factor. A $14 billion program will hardly be affected by a $100 million reduction in cost for launch. Even for commercial systems, cost of launch is only one – and a relatively small – component of development, manufacturing, operation, and maintenance costs. Cheaper launch vehicles might indeed put some now marginal or uneconomical payloads in orbit. They also provide incentive to explore the idea of constellations of satellites for various purposes, and would make systems of operational responsive space (the replacement of space assets destroyed by accident or hostile attack) more feasible. But a massive increase in launch activity would also have diseconomies – for example, worsening the problems associated with space situation awareness, overcrowding in orbit, space debris, and substantial environmental pollution.

Even the military uses of space are faced with issues of comparative advantage. If the issue is putting warheads on target in a timely and cost effective way, there are now other competitors emerging, particularly unmanned terrestrial systems. The latter have a considerable cost advantage, since the technology is known, and the development costs already largely incurred. Although UAVs will never have the worldwide coverage that potential orbiting systems theoretically provide, they have the crucial advantage of being real rather than theoretical. Three or more generations of such systems will appear before an orbiting system could be designed, tested, and deployed. Moreover, "worldwide coverage" is more a slogan than a necessity; most of the Earth's surface is not of security interest, and we will soon have the ability to swarm UAVs in areas where our interest is intense. Finally, the conclusion which might be drawn from Pete Hays' analysis is that the immediate military future of space – out perhaps a decade and more – is already determined. The investment needed to complete satellite programs already in the pipeline has become so overwhelming – and the development stretched to such an extent – that simply completing what we have already decided to do will crowd out any other possible programs for some years. In these circumstances, it is no wonder that even the Bush Administration, more committed than its predecessors to military dominance in space, has never requested from Congress anything beyond small research and development funding for anything that might be described as a space weapon.

As for manned exploration of the near solar system, both Moon base and Mars missions are now national policy, and NASA is directly increasing resources to those programs. Whether the cost of these efforts can be sustained over the long run without massive increases in NASA's budget remains in doubt. International cooperation in exploring the Moon and Mars would be one solution; it might also be a way of encouraging a more multilateral and cooperative approach to space. But the experience of the International Space Station is cautionary in this regard; and some would argue that robust competition in space – especially with the Chinese – is the only way to create a political atmosphere that will sustain long-term, massive public investment in space exploration.

In the face of these difficulties, what does the future hold? More precisely, what, if anything, is liable to bring coherence to our efforts in space? One answer is leadership from a new administration that has a clear idea of what it wants to accomplish in space, is willing to fund programs adequately, and imposes discipline on the bureaucracy to achieve its goals. Even if a new administration has other priorities, however, it is likely that some difficult but long-avoided decisions will be forced by more practical considerations: the necessity to either accept or reject the notion of more extensive international regulation of space activities, the pressures to do triage among space programs because of budget shortfalls, and the possibility of the unexpectedly quick emergence of some new threat or competition, either from Europe (where our policies are leading to rapid development of rival capabilities) or more likely from China, whose capabilities and intentions in space are still a mystery.

The China ASAT test of January 2007 came too late in the day to be intensively considered by our authors; however, it reminds us of the other factor which may concentrate our minds in developing a grand strategy. Although we know our own policy process is often confused, sometimes mistaken, frequently uncoordinated, and even self-contradictory, we tend to impute to our competitors a ruthless and rational single-mindedness. This is especially true if, as in the case of China, little hard information about the rationale for their actions is available. Thus, the wave of interpretation which followed the Chinese test has sought to place it in some coherent and understandable (albeit perhaps ominous) pattern. For all we know, however, it may be the illogical act that on the surface it appears to be. After all, why would a power which is moving to expand its presence in orbit choose to exhibit a rudimentary technology abandoned 20 years ago by the competitors in the Cold War? The space debris created endangers all orbiting devices equally. Perhaps it was, as some have said, a wake-up call to the United States that China is a force to be reckoned with in space. Perhaps it was intended as a concrete refutation of the Bush Administration's space policy document, with its claim to a right to "deny access" to space to potential adversaries – a reminder that a right is a hollow thing in the absence of a capability. Perhaps it represented the triumph of one bureaucratic player in China over other more cautious players. In reality, we can't know the purpose of the Chinese test until they explain it, which so far they have refused to do.

Of course, in this and other areas, we may simply be dealing with the "Rumsfeld conundrum," i.e. the undeniable fact that we don't know what we don't know. The technological progress we have imagined in the past has many times been illusory; but what we failed to imagine has been life altering. One thinks of the Internet in this connection. Perhaps the American space effort will revive, this time funded by private sources; perhaps an undreamt of technological innovation will give us capabilities in space we cannot now foresee; perhaps the Chinese will do these things, and challenge our national self-image.

In short, as of this writing, the area of space policy is in greater flux than it has been at any time since Ronald Reagan gave focus to our efforts with his Strategic Defense Initiative. There is no immediately obvious intellectual construct to give similar energy to our present endeavors. There are also new players, and new potential uses of space, to complicate the picture. Technologically, we seem to have reached a plateau; it will be some time before new propulsion technologies become economically feasible, and new national security satellite technology has shown by huge cost overruns that payloads are pushing up against, and possibly beyond, the limits of presently available technology. Some promising technologies can be imagined; none promises to basically alter our situation in space either militarily or commercially in the near- or medium-term. Sunk costs mean that our course in regard to deploying new systems for the immediate future is already determined.

All this will, of course, be clearer when the readers of this textbook emerge into positions of authority within the space community. Will they be able to formulate the comprehensive intellectual foundation for space which has so clearly eluded the present generation? Or will they, perhaps, conclude that we were naïve to make the attempt, and that while we struggled to make sense of events, a wholly new era of space was emerging from technological innovation and international cooperation – or competition – that we could neither foresee nor control. Time, as they say, will tell.

Appendix
Landmarks in US space policy

Preston Arnold

Date Event

January 1865	Jules Verne published his novel, entitled *From the Earth to the Moon*.
January 1901	H.G. Wells published his book, *The First Men in the Moon*, in which a substance with anti-gravity properties launched men to the Moon.
November 6, 1918	Robert Goddard fired several rocket devices for representatives of US Signal Corps, Air Corps, Army ordinance and others at Aberdeen proving grounds.
April 1924	Soviets establish the Central Committee for the Study of Rocket Propulsion.
March 16, 1926	Goddard tested the world's first successful liquid-fueled rocket.
January 1927	Rocket enthusiasts in Germany formed the Society for Space Travel (VfR). Hermann Oberth was among the first several members to join.
April 1930	The American Interplanetary Society was founded in New York for the purpose of promoting interest in space travel. Later renamed the American Rocket Society.
January 1932	Von Braun demonstrated a liquid-fueled rocket to the German Army.
January 1933	Soviets launched a rocket fueled by solid and liquid fuels to a height of 400 m.
December 1934	Von Braun and associates launched two A-2 rockets, both to heights of 1.5 miles.
January 1935	Russians fired a liquid-powered rocket that achieved a height of over 8 miles.
March 1935	A rocket of Robert Goddard's exceeded the speed of sound.
September 1944	The first of over 1,000 operational V-2 rockets was launched against London from Germany.

January 24, 1945	Germany successfully launched the A-9, a winged prototype of the first intercontinental ballistic missile, which was designed to reach North America.
May 5, 1945	Von Braun was captured by the United States and relocated to the White Sands Proving Grounds.
May 9, 1945	At the time of the German collapse, over 20,000 V-1s and V-2s had been fired.
August 10, 1945	Goddard died due to cancer at the University of Maryland Hospital in Baltimore.
December 1945	Fifty-five German scientists arrived at Fort Bliss and White Sands Proving Grounds.
January 1946	The US outer space research program was started with captured V-2 rockets.
March 15, 1946	The first American-built V-2 rocket static-fired at the White Sands Proving Grounds.
March 22, 1946	The first American-built rocket to leave the Earth's atmosphere launched.
April 1946	The first actual launch of an American-built V-2 took place at White Sands.
October 24, 1946	A V-2 with a motion picture camera was launched. It recorded images from 65 miles above the Earth, covering 40,000 square miles.
December 1946	German rocket engineers arrived in Russia to begin work with Soviet rocket research groups. Sergei Korolev built rockets using technology from the V-2.
January 1947	Russians began launch tests of V-2 rockets, at Kapustin Yar. Meanwhile, telemetry successfully used for first time in a V-2, launched from White Sands.
May 13, 1948	The first two-stage rocket launched in Western Hemisphere took off from White Sands.
June 11, 1948	The first of the "Albert" missions was launched, containing live animals, named for the monkey that rode, and suffocated, in the first rocket.
May 11, 1949	President Truman signed the bill for 5,000-mile test range off of Cape Canaveral, FL.
July 24, 1950	The first rocket was launched from Cape Canaveral, number eight of the two-stage rockets.
June 5 1953	A missile was fired from an underground launch facility in White Sands.
July 29, 1955	President Eisenhower announced plans to launch unmanned satellites as participation in the International Geophysical Year. Russians soon made a similar announcement.

November 8, 1955	The Secretary of Defense approved Jupiter/Thor Intermediate Range Ballistic Missile (IRBM) programs.
October 4, 1957	The Soviet Union launched the first satellite, 184-lb *Sputnik 1*, into space.
November 3, 1957	The Soviet spacecraft *Sputnik 2* was launched with a dog named Laika on board. The first animal in space, Laika did not survive the voyage.
January 31, 1958	The United States launches its first satellite: *Explorer 1*.
October 1, 1958	US Congress Space Act established NASA (National Aeronautical Space Agency) to assume responsibilities of National Advisory Committee on Aeronautics.
December 13, 1958	The Committee on the Peaceful Uses of Outer Space was first established.
January 2, 1959	Soviet satellite *Luna 1* launched in an attempt to hit the Moon. It was unsuccessful but became the first man-made object to achieve an orbit around the Sun.
September 12, 1959	Soviet satellite *Luna 2* launched, becoming first man-made object to hit the Moon.
October 4, 1959	Soviet satellite *Luna 3* launched, orbiting the Moon and photographing 70 percent of its far side.
April 1, 1960	*Tiros 1*, the first successful weather satellite, launched by the United States.
August 19, 1960	Two dogs in *Sputnik 5* became first animals to be successfully returned to Earth.
April 12, 1961	Soviet cosmonaut Yuri Gagarin became first human in space aboard *Vostok 1*.
May 5, 1961	Astronaut Alan Shepard became the first American in space.
May 25, 1961	President Kennedy challenged the country to put a man on Moon by end of decade.
August 1961	Gherman Titov spends 25 hours in space on the second Vostok manned flight.
January 1962	Transatlantic television programs relayed by the 171-lb *Telstar* satellite.
January 1962	Communications Satellite Act passed.
February 20, 1962	Astronaut John Glenn became the first American in orbit.
June 16, 1962	Valentina Nikolayeva Tereshkova became the first woman in space.
October 10, 1963	Limited Test Ban Treaty entered into force.
December 13, 1963	Declaration of Legal Principles Governing the Activities of States in the Exploration and Use of Outer Space adopted as UN General Assembly resolution.

June 14, 1964	ESRO established in Europe with ten member nations.
April 6, 1965	NASA launched first commercial communications satellite, 87-lb *Intelsat 1* (Early Bird).
December 15, 1965	*Gemini VI* launched and successfully space-docked with *Gemini VII*.
March 18, 1965	Tethered to his spacecraft, Soviet Alexi Leonov became first man to walk in space.
June 3, 1965	Astronaut Ed White became the first American to walk in space.
July 14, 1965	The spacecraft, *Mariner 4*, transmitted the first pictures of Mars.
February 3, 1966	The Russian spacecraft *Luna 9* became the first spacecraft to land on the Moon.
June 2, 1966	*Surveyor 1* became the first American spacecraft to land on the Moon.
January 27, 1967	Three US astronauts killed in accidental fire in command module on the launch pad.
April 24, 1967	Soviet cosmonaut killed when parachute on his *Soyuz 1* spacecraft failed to deploy.
May 5, 1967	First ever all-British satellite successfully launched into orbit from the United States.
October 10, 1967	Treaty on Principles Governing the Activities of States in the Exploration and Use of Outer Space, including the Moon and Other Celestial Bodies (Outer Space Treaty) entered into force. US acceded on the same date.
October 18, 1967	Descent capsule from Soviet probe *Venera 4* collected data about Venus' atmosphere.
September 15, 1968	Soviet *Zond 5* launched, becoming the first spacecraft to orbit the Moon and return to Earth.
October 11, 1968	*Apollo 7*, the first manned Apollo mission, orbited the Earth once.
December 3, 1968	Agreement on the Rescue of Astronauts, the Return of Astronauts and the Return of Objects Launched into Outer Space (Rescue Agreement) entered into force. The United States acceded on the same date.
December 21, 1968	*Apollo 8* launched; her crewmembers became the first men to orbit the Moon.
January 16, 1969	The first Russian manned docking maneuver (*Soyuz 4* and *Soyuz 5*).
July 20, 1969	Neil Armstrong and "Buzz" Aldrin became the first men on the Moon on *Apollo 11*.
April 11, 1970	*Apollo 13* was launched. Fifty-six hours into the flight, an explosion in the fuel cells resulted in the loss of the main power supply.

September 12, 1970	Soviet *Luna 16* launched; became first automatic spacecraft to return Moon soil sample.
December 15, 1970	Soviet *Venera 7* became the first probe to land on Venus.
January 1971	Soviets launched *Mars 2* and *Mars 3* for first impact on Mars (the planet) by *Mars 2* and first soft landing on December 2, 1971 by *Mars 3*.
April 19, 1971	Soviet space station Salyut 1 was launched, remaining on-orbit until May 1973.
June 30, 1971	Three Soviet Cosmonauts found dead in Soyuz space capsule on return to Earth following 22-day visit to first space lab, Soviet 41,580-lb. Salyut 1.
July 30, 1971	Moon rover driven on the Moon for first time by *Apollo 15* astronauts Scott and Irwin.
November 13, 1971	The *Mariner 9* probe became the first craft to orbit another world – Mars.
May 26, 1972	ABM Treaty signed by President Nixon and General Secretary Brezhnev.
September 1, 1972	Convention on International Liability for Damage Caused by Space Objects (Liability Convention) entered into force. The United States acceded on October 9, 1973.
December 11, 1972	Eugene Cernan and Harrison "Jack" Schmitt became last men [so far] to walk on the Moon.
May 14, 1973	United States launched its first space station, Skylab. The first crew visited it 11 days later.
July 17, 1975	American *Apollo 18* and Soviet *Soyuz 19* dock in the Apollo–Soyuz Test Project.
July 20, 1976	Pictures of the Martian surface taken by *Viking 1*, the first US attempt to soft land a spacecraft on another planet.
September 15, 1976	Convention on Registration of Objects Launched into Outer Space (Registration Convention) entered into force. The United States acceded on the same date.
March 5, 1979	*Voyager 1* began transmitting images of Jupiter and her moons. *Voyager 2* joins in from July 9, 1979.
July 11, 1979	Skylab I plunged to Earth, scattering debris across Indian Ocean and Western Australia.
September 1, 1979	The US probe, *Pioneer 11*, reached Saturn and began transmitting images.
December 24, 1979	The ESA launches its first satellite, *Ariane 1*, from French Guiana.
November 13, 1980	*Voyager 1* reached Saturn and began transmitting images.

December 6, 1980	The first of the powerful *Intelsat V* communications satellites launched.
April 12, 1981	*Columbia* became the first space shuttle to be launched.
August 26, 1981	*Voyager 2* reached Saturn and began transmitting images.
March 1, 1982	Soviet *Venera 13* landed on Venus, providing the first Venusian soil analysis.
May 13, 1982	Soviet cosmonauts launched to become the first team to inhabit space station *Salyut 7*.
November 11, 1982	Shuttle *Columbia* began its first operational mission, deploying two satellites.
December 10, 1982	Principles Governing the Use by States of Artificial Earth Satellites for International Direct Television Broadcasting adopted as UN General Assembly resolution.
April 4, 1983	Second space shuttle, *Challenger*, launched.
June 13, 1983	*Pioneer 10* became first probe in interstellar space, beyond outermost planet, Neptune.
June 18, 1983	Sally Ride became the first American woman in space on *Challenger*'s second mission.
February 3, 1984	Astronaut Bruce McCandless made first untethered space walk.
July 11, 1984	Agreement Governing the Activities of States on the Moon and Other Celestial Bodies (Moon Treaty) entered into force. The United States is not yet a party to this treaty.
August 30, 1984	The third space shuttle, *Discovery*, launched.
January 1985	The first probe reaches a comet (USA).
October 3, 1985	*Atlantis*, the fourth space shuttle, was launched.
January 24, 1986	*Voyager 2* began transmitting images from Uranus.
January 28, 1986	The space shuttle *Challenger* exploded seconds after lift-off.
February 20, 1986	The core section of the Space Station Mir was launched.
February 20, 1986	Soviets launched space station Mir; essentially continuously occupied until November 2000.
December 3, 1986	Principles Relating to Remote Sensing of the Earth from Outer Space adopted as UN General Assembly resolution.
August 25, 1989	*Voyager 2* began transmitting images from Neptune.
April 24, 1990	The space shuttle *Discovery* deployed the Hubble Space Telescope.
August 10, 1990	Magellan spacecraft began mapping the surface of Venus using radar equipment.

May 7, 1992	The space shuttle *Endeavor* was launched for her maiden voyage.
December 14, 1992	Principles Relevant to the Use of Nuclear Power Source in Outer Space adopted as UN General Assembly resolution.
December 2, 1993	Shuttle *Endeavor* made the first servicing mission of the Hubble Space Telescope.
February 3, 1994	Sergei Krikalev became the first Russian cosmonaut to fly on a space shuttle.
June 29, 1995	US shuttle *Atlantis* docked with Russian space station Mir, delivering relief crew.
December 7, 1995	The *Galileo* probe began transmitting data on Jupiter.
May 18, 1996	X PRIZE competition announced: $10 million for the first successful launch and recovery of a non-governmental spacecraft.
December 13, 1996	Declaration on International Cooperation in the Exploration and Use of Outer Space for the Benefit and in the Interest of All States, Taking into Particular Account the Needs of Developing Countries adopted as UN General Assembly resolution.
July 4, 1997	The *Mars Pathfinder* arrived on Mars and later began transmitting images.
October 29, 1998	John Glenn became the oldest man in space.
November 20, 1998	Russian-built Zarya Control Module, the first component of the new International Space Station, is launched on a Russian Proton rocket.
December 4, 1998	Shuttle *Endeavor* lifted off carrying the Unity module for International Space Station.
February 14 2000	US Near Earth Asteroid Rendezvous spacecraft began sending images of asteroid, Eros.
April 28, 2001	American Dennis Tito became first space tourist, paying Russian space program $20 million.
February 1, 2003	Space shuttle *Columbia* exploded on reentry into the Earth's atmosphere.
October 15, 2003	China sent its first manned spacecraft into orbit.
January 3, 2004	NASA rover Spirit landed on Mars, soon joined by Opportunity, seeking signs of water.
June 21, 2004	*Spaceship One* became first non-government spacecraft to be flown into space.
July 1, 2004	*Cassini* probe arrived at Saturn to begin photographing the planet and its moons.
October 7, 2004	American Robert Bigelow offered $50 million to first private spacecraft to achieve orbit.
January 14, 2005	*Huygens* probe landed on Saturn's largest moon, Titan, sending images back to Earth.

July 4, 2005	*Deep Impact* space probe impacted a comet known as Tempel 1.
July 26, 2005	First shuttle launch since *Columbia* disaster; fleet re-grounded for fuel tank redesign.
January 15, 2006	NASA's *Stardust* capsule returned to Earth after collecting samples from comet Wild 2.

Index